NATURAL HISTORY
UNIVERSAL LIBRARY

西方博物学大系

主编：江晓原

AMERICAN ORNITHOLOGY

美国鸟类

[美] 亚历山大·威尔逊 著

华东师范大学出版社

图书在版编目(CIP)数据

美国鸟类 = American ornithology : 英文 / (美) 亚历山大 · 威尔逊著. — 上海 : 华东师范大学出版社, 2018
(寰宇文献)
ISBN 978-7-5675-7719-0

Ⅰ. ①美… Ⅱ. ①亚… Ⅲ. ①鸟类-美国-英文
Ⅳ. ①Q959.7

中国版本图书馆CIP数据核字(2018)第095406号

美国鸟类
American ornithology
(美) 亚历山大 · 威尔逊著

特约策划　黄曙辉　徐　辰
责任编辑　庞　坚
特约编辑　许　倩
装帧设计　刘怡霖

出版发行　华东师范大学出版社
社　　址　上海市中山北路3663号　邮编 200062
网　　址　www.ecnupress.com.cn
电　　话　021-60821666　行政传真　021-62572105
客服电话　021-62865537
门市（邮购）电话　021-62869887
地　　址　上海市中山北路3663号华东师范大学校内先锋路口
网　　店　http://hdsdcbs.tmall.com/

印 刷 者　虎彩印艺股份有限公司
开　　本　16开
印　　张　111.75
版　　次　2018年6月第1版
印　　次　2018年6月第1次
书　　号　ISBN 978-7-5675-7719-0
定　　价　2200.00元 (精装全三册)

出 版 人　王　焰

《西方博物学大系》总序

江晓原

《西方博物学大系》收录博物学著作超过一百种，时间跨度为 15 世纪至 1919 年，作者分布于 16 个国家，写作语种有英语、法语、拉丁语、德语、弗莱芒语等，涉及对象包括植物、昆虫、软体动物、两栖动物、爬行动物、哺乳动物、鸟类和人类等，西方博物学史上的经典著作大备于此编。

中西方“博物”传统及观念之异同

今天中文里的“博物学“一词，学者们认为对应的英语词汇是Natural History，考其本义，在中国传统文化中并无现成对应词汇。在中国传统文化中原有“博物”一词，与“自然史”当然并不精确相同，甚至还有着相当大的区别，但是在“搜集自然界的物品”这种最原始的意义上，两者确实也大有相通之处，故以“博物学”对译 Natural History 一词，大体仍属可取，而且已被广泛接受。

已故科学史前辈刘祖慰教授尝言：古代中国人处理知识，如开中药铺，有数十上百小抽屉，将百药分门别类放入其中，即心安矣。刘教授言此，其辞若有憾焉——认为中国人不致力于寻求世界“所以然之理”，故不如西方之分析传统优越。然而古代中国人这种处理知识的风格，正与西方的博物学相通。

与此相对，西方的分析传统致力于探求各种现象和物体之间的相互关系，试图以此解释宇宙运行的原因。自古希腊开始，西方哲人即孜孜不倦建构各种几何模型，欲用以说明宇宙如何运行，其中最典型的代表，即为托勒密（Ptolemy）的宇宙体系。

比较两者，差别即在于：古代中国人主要关心外部世界“如何”运行，而以希腊为源头的西方知识传统（西方并非没有别的知识传统，只是未能光大而已）更关心世界“为何”如此运行。在线性发展无限进步的科学主义观念体系中，我们习惯于认为“为何”是在解决了“如何”之后的更高境界，故西方的分析传统比中国的传统更高明。

然而考之古代实际情形，如此简单的优劣结论未必能够成立。例如以天文学言之，古代东西方世界天文学的终极问题是共同的：给定任意地点和时刻，计算出太阳、月亮和五大行星（七政）的位置。古代中国人虽不致力于建立几何模型去解释七政“为何”如此运行，但他们用抽象的周期叠加（古代巴比伦也使用类似方法），同样能在足够高的精度上计算并预报任意给定地点和时刻的七政位置。而通过持续观察天象变化以统计、收集各种天象周期，同样可视之为富有博物学色彩的活动。

还有一点需要注意：虽然我们已经接受了用“博物学”来对译 Natural History，但中国的博物传统，确实和西方的博物学有一个重大差别——即中国的博物传统是可以容纳怪力乱神的，而西方的博物学基本上没有怪力乱神的位置。

古代中国人的博物传统不限于“多识于鸟兽草木之名”。体现此种传统的典型著作，首推晋代张华《博物志》一书。书名“博物”，其义尽显。此书从内容到分类，无不充分体现它作为中国博物传统的代表资格。

《博物志》中内容，大致可分为五类：一、山川地理知识；二、奇禽异兽描述；三、古代神话材料；四、历史人物传说；五、神仙方伎故事。这五大类，完全符合中国文化中的博物传统，深合中国古代博物传统之旨。第一类，其中涉及宇宙学说，甚至还有“地动”思想，故为科学史家所重视。第二类，其中甚至出现了中国古代长期流传的“守宫砂”传说的早期文献：相传守宫砂点在处女胳膊上，永不褪色，只有性交之后才会自动消失。第三类，古代神话传说，其中甚至包括可猜想为现代“连体人”的记载。第四类，各种著名历史人物，比如三位著名刺客的传说，此三名刺客及所刺对象，历史上皆实有其人。第五类，包括各种古代方术传说，比如中国古代房中养生学说，房中术史上的传说人物之一“青牛道士封君达”等等。前两类与西方的博物学较为接近，但每一类都会带怪力乱神色彩。

“所有的科学不是物理学就是集邮”

在许多人心目中，画画花草图案，做做昆虫标本，拍拍植物照片，这类博物学活动，和精密的数理科学，比如天文学、物理学等等，那是无法同日而语的。博物学显得那么的初级、简单，甚至幼稚。这种观念，实际上是将“数理程度”作为唯一的标尺，用来衡量一切知识。但凡能够使用数学工具来描述的，或能够进行物理实验的，那就是“硬”科学。使用的数学工具越高深越复杂，似乎就越“硬”；物理实验设备越庞大，花费的金钱越多，似乎就越“高端”、越“先进”……

这样的观念，当然带着浓厚的“物理学沙文主义”色彩，在很多情况下是不正确的。而实际上，即使我们暂且同意上述“物理学沙文主义”的观念，博物学的“科学地位”也仍然可以保住。作为一个学天体物理专业出身，因而经常徜徉在“物理学沙文主义”幻影之下的人，我很乐意指出这样一个事实：现代天文学家们的研究工作中，仍然有绘制星图，编制星表，以及为此进行的巡天观测等等活动，这些活动和博物学家“寻花问柳”，绘制植物或昆虫图谱，本质上是完全一致的。

这里我们不妨重温物理学家卢瑟福（Ernest Rutherford）的金句：“所有的科学不是物理学就是集邮（All science is either physics or stamp collecting）。”卢瑟福的这个金句堪称“物理学沙文主义”的极致，连天文学也没被他放在眼里。不过，按照中国传统的“博物”理念，集邮毫无疑问应该是博物学的一部分——尽管古代并没有邮票。卢瑟福的金句也可以从另一个角度来解读：既然在卢瑟福眼里天文学和博物学都只是“集邮”，那岂不就可以将博物学和天文学相提并论了？

如果我们摆脱了科学主义的语境，则西方模式的优越性将进一步被消解。例如，按照霍金（Stephen Hawking）在《大设计》（*The Grand Design*）中的意见，他所认同的是一种“依赖模型的实在论（model-dependent realism）”，即“不存在与图像或理论无关的实在性概念（There is no picture- or theory-independent concept of reality）”。在这样的认识中，我们以前所坚信的外部世界的客观性，已经不复存在。既然几何模型只不过是对外部世界图像的人为建构，则古代中国人干脆放弃这种建构直奔应用（毕竟在实际应用中我们只需要知道七政“如何”运行），又有何不可？

传说中的“神农尝百草”故事，也可以在类似意义下得到新的解读：“尝百草”当然是富有博物学色彩的活动，神农通过这一活动，得知哪些草能够治病，哪些不能，然而在这个传说中，神农显然没有致力于解释“为何”某些草能够治病而另一些则不能，更不会去建立“模型”以说明之。

“帝国科学”的原罪

今日学者有倡言“博物学复兴”者，用意可有多种，诸如缓解压力、亲近自然、保护环境、绿色生活、可持续发展、科学主义解毒剂等等，皆属美善。编印《西方博物学大系》也是意欲为“博物学复兴”添一助力。

然而，对于这些博物学著作，有一点似乎从未见学者指出过，而鄙意以为，当我们披阅把玩欣赏这些著作时，意识到这一点是必须的。

这百余种著作的时间跨度为15世纪至1919年，注意这个时间跨度，正是西方列强“帝国科学”大行其道的时代。遥想当年，帝国的科学家们乘上帝国的军舰——达尔文在皇家海军“小猎犬号”上就是这样的场景之一，前往那些已经成为帝国的殖民地或还未成为殖民地的“未开化”的遥远地方，通常都是踌躇满志、充满优越感的。

作为一个典型的例子，英国学者法拉在（Patricia Fara）《性、植物学与帝国：林奈与班克斯》（*Sex, Botany and Empire, The Story of Carl Linnaeus and Joseph Banks*）一书中讲述了英国植物学家班克斯（Joseph Banks）的故事。1768年8月15日，班克斯告别未婚妻，登上了澳大利亚军舰“奋进号”。此次“奋进号”的远航是受英国海军部和皇家学会资助，目的是前往南太平洋的塔希提岛（Tahiti，法属海外自治领，另一个常见的译名是“大溪地”）观测一次比较罕见的金星凌日。舰长库克（James Cook）是西方殖民史上最著名的舰长之一，多次远航探险，开拓海外殖民地。他还被认为是澳大利亚和夏威夷群岛的“发现”者，如今以他命名的群岛、海峡、山峰等不胜枚举。

当“奋进号”停靠塔希提岛时，班克斯一下就被当地美丽的土著女性迷昏了，他在她们的温柔乡里纵情狂欢，连库克舰长都看不下去了，“道德愤怒情绪偷偷溜进了他的日志当中，他发现自己根本不可能不去批评所见到的滥交行为”，而班克斯纵欲到了“连嫖妓都毫无激情”的地步——这是别人讽刺班克斯的说法，因为对于那时常年航行于茫茫大海上的男性来说，上岸嫖妓通常是一项能够唤起“激情”的活动。

而在“帝国科学”的宏大叙事中，科学家的私德是无关紧要的，人们关注的是科学家做出的科学发现。所以，尽管一面是班克斯在塔希提岛纵欲滥交，一面是他留在故乡的未婚妻正泪眼婆娑地“为远去的心上人绣织背心”，这样典型的“渣男”行径要是放在今天，非被互联网上的口水淹死不可，但是“班克斯很快从他们的分离之苦中走了出来，在外近三年，他活得倒十分滋润”。

法拉不无讽刺地指出了“帝国科学”的实质：“班克斯接管了当地的女性和植物，而库克则保护了大英帝国在太平洋上的殖民地。”甚至对班克斯的植物学本身也调侃了一番：“即使是植物学方面的科学术语也充满了性指涉。……这个体系主要依靠花朵之中雌雄生殖器官的数量来进行分类。”据说“要保护年轻妇女不受植物学教育的浸染，他们严令禁止各种各样的植物采集探险活动。”这简直就是将植物学看成一种“涉黄”的淫秽色情活动了。

在意识形态强烈影响着我们学术话语的时代，上面的故事通常是这样被描述的：库克舰长的“奋进号”军舰对殖民地和尚未成为殖民地的那些地方的所谓“访问”，其实是殖民者耀武扬威的侵略，搭载着达尔文的“小猎犬号”军舰也是同样行径；班克斯和当地女性的纵欲狂欢，当然是殖民者对土著妇女令人发指的蹂躏；即使是他采集当地植物标本的“科学考察”，也可以视为殖民者“窃取当地经济情报”的罪恶行为。

后来改革开放，上面那种意识形态话语被抛弃了，但似乎又走向了另一个极端，完全忘记或有意回避殖民者和帝国主义这个层面，只歌颂这些军舰上的科学家的伟大发现和成就，例如达尔文随着“小猎犬号”的航行，早已成为一曲祥和优美的科学颂歌。

其实达尔文也未能免俗，他在远航中也乐意与土著女性打打交道，当然他没有像班克斯那样滥情纵欲。在达尔文为“小猎犬号”远航写的《环球游记》中，我们读到：“回程途中我们遇到一群黑人姑娘在聚会，……我们笑着看了很久，还给了她们一些钱，这着实令她们欣喜一番，拿着钱尖声大笑起来，很远还能听到那愉悦的笑声。”

有趣的是，在班克斯在塔希提岛纵欲六十多年后，达尔文随着“小猎犬号”也来到了塔希提岛，岛上的土著女性同样引起了达尔文的注意，在《环球游记》中他写道：“我对这里妇女的外貌感到有些失望，然而她们却很爱美，把一朵白花或者红花戴在脑后的髮髻上……”接着他以居高临下的笔调描述了当地女性的几种发饰。

用今天的眼光来看，这些在别的民族土地上采集植物动物标本、测量地质水文数据等等的“科学考察”行为，有没有合法性问题？有没有侵犯主权的问题？这些行为得到当地人的同意了吗？当地人知道这些行为的性质和意义吗？他们有知情权吗？……这些问题，在今天的国际交往中，确实都是存在的。

也许有人会为这些帝国科学家辩解说：那时当地土著尚在未开化或半开化状态中，他们哪有“国家主权”的意识啊？他们也没有制止帝国科学家的考察活动啊？但是，这样的辩解是无法成立的。

姑不论当地土著当时究竟有没有试图制止帝国科学家的“科学考察”行为，现在早已不得而知，只要殖民者没有记录下来，我们通常就无法知道。况且殖民者有军舰有枪炮，土著就是想制止也无能为力。正如法拉所描述的：“在几个塔希提人被杀之后，一套行之有效的易货贸易体制建立了起来。”

即使土著因为无知而没有制止帝国科学家的“科学考察”行为，这事也很像一个成年人闯进别人的家，难道因为那家只有不懂事的小孩子，闯入者就可以随便打探那家的隐私、拿走那家的东西、甚至将那家的房屋土地据为己有吗？事实上，很多情况下殖民者就是这样干的。所以，所谓的“帝国科学”，其实是有着原罪的。

如果沿用上述比喻，现在的局面是，家家户户都不会只有不懂事的孩子了，所以任何外来者要想进行“科学探索”，他也得和这家主人达成共识，得到这家主人的允许才能够进行。即使这种共识的达成依赖于利益的交换，至少也不能单方面强加于人。

博物学在今日中国

博物学在今日中国之复兴，北京大学刘华杰教授提倡之功殊不可没。自刘教授大力提倡之后，各界人士纷纷跟进，仿佛昔日蔡锷在云南起兵反袁之“滇黔首义，薄海同钦，一檄遥传，景从恐后”光景，这当然是和博物学本身特点密切相关的。

无论在西方还是在中国，无论在过去还是在当下，为何博物学在它繁荣时尚的阶段，就会应者云集？深究起来，恐怕和博物学本身的特点有关。博物学没有复杂的理论结构，它的专业训练也相对容易，至少没有天文学、物理学那样的数理“门槛”，所以和一些数理学科相比，博物学可以有更多的自学成才者。这次编印的《西方博物学大系》，卷帙浩繁，蔚为大观，同样说明了这一点。

最后，还有一点明显的差别必须在此处强调指出：用刘华杰教授喜欢的术语来说，《西方博物学大系》所收入的百余种著作，绝大部分属于“一阶”性质的工作，即直接对博物学作出了贡献的著作。事实上，这也是它们被收入《西方博物学大系》的主要理由之一。而在中国国内目前已经相当热的博物学时尚潮流中，绝大部分已经出版的书籍，不是属于“二阶”性质（比如介绍西方的博物学成就），就是文学性的吟风咏月野草闲花。

要寻找中国当代学者在博物学方面的“一阶”著作，如果有之，以笔者之孤陋寡闻，唯有刘华杰教授的《檀岛花事——夏威夷植物日记》三卷，可以当之。这是刘教授在夏威夷群岛实地考察当地植物的成果，不仅属于直接对博物学作出贡献之作，而且至少在形式上将昔日“帝国科学”的逻辑反其道而用之，岂不快哉！

2018 年 6 月 5 日
于上海交通大学
科学史与科学文化研究院

亚历山大·威尔逊
（1766-1813）

被誉为“美国鸟类学之父”的亚历山大·威尔逊（Alexander Wilson）生于苏格兰佩斯利一个毛纺商家庭。他年幼时曾想继承副业，进纺织工坊当学徒，出道后成为毛纺匠人。受罗伯特·伯恩斯影响开始创作诗歌，虽然也出版过若干诗集，却一直没能出名。威尔逊生性风流，言语尖刻，在国内闹出一连串名誉诉讼和道德事件。硝烟散尽后，他决定远走高飞移民美国。

1794年，威尔逊抵达费城，当过临时工，干过印刷和毛纺。好不容易在迈尔斯通谋得一个稳定的教师职务，又因勾搭有夫之妇而声名扫地，不得不卷铺盖走人。在此期间，他开始学画，起初专注于林木和走兽，后结识鸟类学家乔治·奥德，决定巡游美国全境，撰写、绘制一本美国鸟类图鉴。在美国博物学大家威廉·巴特拉姆的支持下，他完成了书籍的征订。

1807年万事俱备，威尔逊着手实施出版计划，并于1808年推出了《美国鸟类》的第一卷，图文均由他执笔。至1813年，本书已出版七卷，威尔逊却在这一年因过度劳累而病死，乔治·奥德在其身后刊行最后两卷，并在卷末为威尔逊树传。本书涉及鸟类近270种，其中二十余种为首度描绘记载。除精美准确的鸟类画外，威尔逊优美的文体诗意盎然，将北美林海中各色鸟类栖息、争斗、繁衍的生态娓娓道来，美不胜收，是美国鸟类学的先驱之作。

本次影印的底本，为《美国鸟类》1870年版（三卷本）。

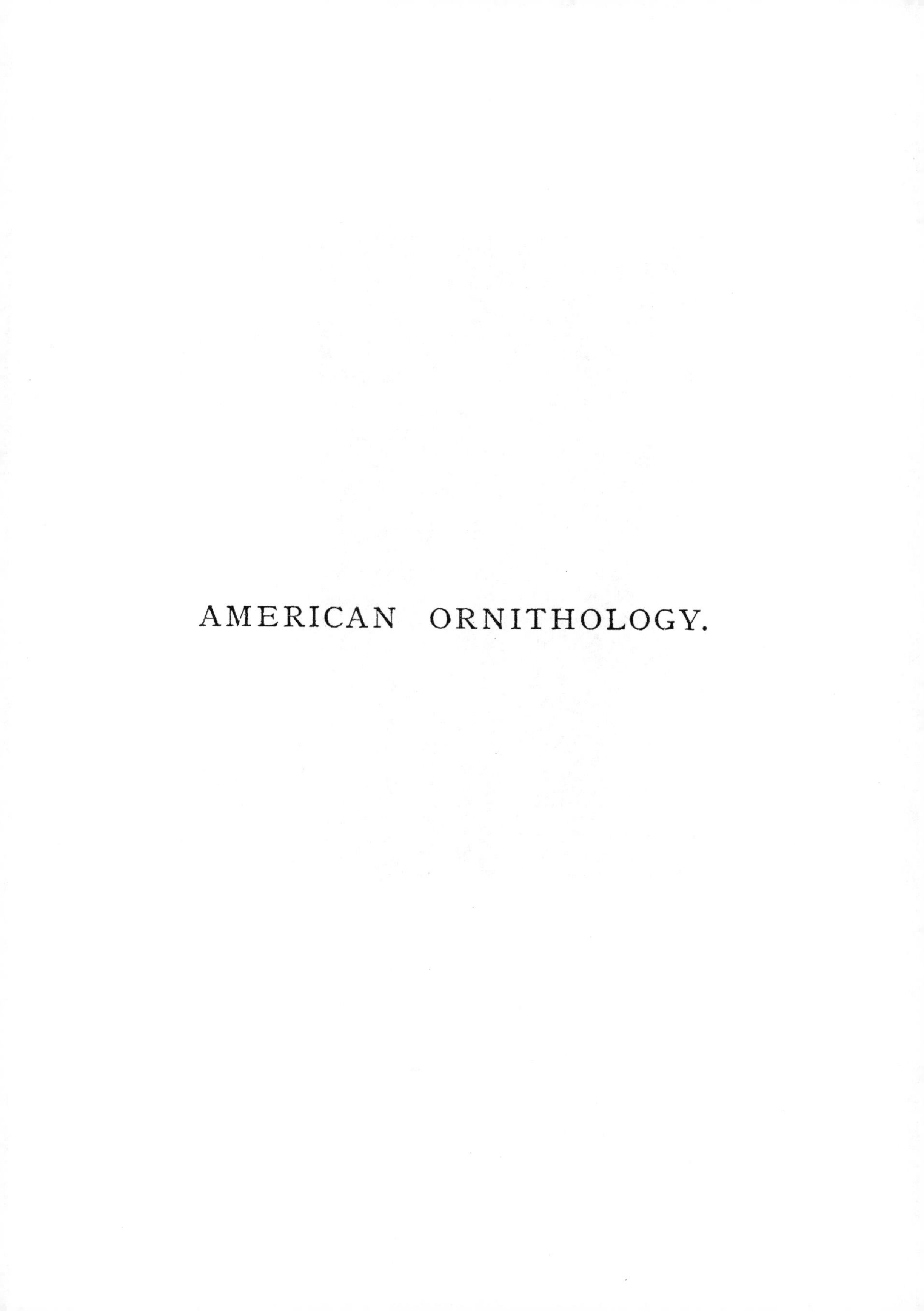

AMERICAN ORNITHOLOGY.

ALEXANDER WILSON.

Engraved for Sir William Jardine's edition
of Wilson's American Ornithology. 1832.

AMERICAN ORNITHOLOGY;

OR,

THE NATURAL HISTORY

OF THE

BIRDS OF THE UNITED STATES.

BY

ALEXANDER WILSON

AND

PRINCE CHARLES LUCIEN BONAPARTE.

The Illustrative Notes and Life of Wilson

BY SIR WILLIAM JARDINE, BART., F.R.S.E., F.L.S.

IN THREE VOLUMES.—VOL. I.

CASSELL PETTER & GALPIN:

LONDON, PARIS & NEW YORK.

CONTENTS OF THE FIRST VOLUME.

The names printed in italics are species not contained in the original, which have been introduced into the notes.

LIFE

OF

ALEXANDER WILSON.

In looking at our present knowledge of the natural history of any vast country, we generally lose sight of a very important circumstance,—the value which its early naturalists, and their sources of information, should hold, in the opinions and deductions that we form regarding it. Mines, as it were, of the relics of animal creation are daily discovered, containing forms we have never seen or imagined—of whose shape and figure even the slightest tradition does not exist; and we possess later records of animals and birds, whose truth we cannot substantiate, or of whose present existence we can find no trace. Independent, however, of the great changes which have taken place upon the surface of the earth, embracing, in their convulsions, all creation, whether animate or inanimate, there is one powerful existing cause, which tends sometimes to render of no avail, and at other times to vary, those laws, which would regularly influence the distribution of animal life in a natural or wild state,—civilisation, and, consequent upon it, the extirpation of some, and the introduction and naturalisation of other, species, to countries and climates originally not their own. Of these, the former is most to be dreaded. Introduction will destroy the exclusive locality; but it may benefit the nations or individuals who are at the trouble and expense of it; and the animals introduced being generally conducive to some want or luxury, we are less liable to lose the tradition of their first uses. When destroyed, the species is either entirely lost, or its locality becomes more limited. This takes place, by the cultivation of the country destroying the natural productions, and introducing others not adapted for the sustenance of the native animals; or by the great and indiscriminate destruction of the different species, either for food, or articles of commerce; or it arises from the jealousy and hatred of the native tribes towards their more civilised aggressors. Thus, late travellers in the interior of North America, frequently complain of the scarcity of various sorts of game, in parts where they previously abounded. Major Long particularly notices this, in his expedition to the source of the St Peter's River. Speaking of the native Indians there, he says,—"They hunt without reserve, and destroy the game more rapidly than it can be reproduced. They appear, since their intercourse with the white men, to have lost the sagacious foresight which previously distinguished them. It was usual for them formerly, to avoid killing the deer during the rutting season. The does that were with young were, in like manner, always spared, except in cases of urgency; and the young fawns were not wantonly destroyed: but, at present, the Indian seems to consider himself as a stranger in the land which his fathers held as their own. He sees his property daily exposed to the encroachments of white men; and therefore hunts down indiscriminately every animal that he meets with, being doubtful whether he will be permitted to reap, the ensuing year, the fruits of his foresight during the present; and fearing lest he may not be able to hunt undisturbed upon his property for another season." The department of nature about which we are now more immediately interested, points out similar instances, occurring almost to our personal observation. In our own little islands, many of those birds formerly esteemed common, are now hardly to be met with. The bustard is almost extinct from our plains; and the noble capercalzie of the Scottish forests, has disappeared for nearly a century. The ostrich and large bustards of Africa, the rhæa and emu of their respective countries, are driven by the settlers and colonists, to seek for new and undisturbed abodes and feeding grounds; and in "Canada, and the now densely peopled parts of the United States," says the continuator of Wilson, "wild turkeys, like the Indian and buffalo, have been compelled to yield to the destructive ingenuity of the white settlers." *

In those rapid changes, then, appearing to our view slow from their constancy, we should be in danger of losing all trace of some species, or of possessing a tradition or description, valuable only according to the station which the author of it at the time held in the science he professed. In this respect, the ornithology of North America has been most fortunate. Her naturalists have wrought from observations, the fruits of their own laborious researches, and have not trusted to the hearsay evidence of their predecessors; and the species of that continent are at this time better known than those of any part of the world, northern Europe excepted. Passing from the more primitive ornithologists, from Edwards and Catesby, embracing in their histories the birds of the islands belonging to the southern continent, and also from the gentlemen engaged in the fur establishments, who furnished our earliest information regarding the more arctic inhabitants, we find, in the United States, the venerable Bartram and the elder Peale, warm admirers of nature in all her forms, paving the way for the favourable reception of the arduous, and then novel, labours of our countryman, ALEXANDER WILSON. He was the first who truly studied the birds of North America in their natural abodes, and from real observation; and his work will remain an ever-to-be-admired testimony of enthusiasm and perseverance—one certainly unrivalled in descriptions; and if some plates and illustrations may vie with it in finer workmanship or pictorial splendour, few, indeed, can rival it in fidelity and truth of delineation. Since his untimely decease, the labours of his admirable continuator and commentator, the Prince of Musignano, of Mr Ord, and the younger Peale, and the extensive journeys of Messrs Say and Long to the interior and the Rocky Mountains, have done much to fill up what was wanting to this deparment of the fauna of North America; while the materials collected, during the different arctic expeditions, undertaken by this country, and particularly the last overland, under the charge of Captain Sir J. Franklin, have brought down our knowledge to the present date, and are beautifully illustrated in the important second volume of the "Northern Zoology," by Dr Richardson, and Mr Swainson.

* Bonap. Continuation, Part I. p. 80.

To these may be added, though last, certainly not least, the magnificent work of Mr Audubon, now in course of publication, while its enterprising author is traversing the wilds of America in search of objects for its completion.

The naturalists who have contributed to our knowledge of the ornithology of North America, bearing this high rank, on account of the merit of their respective works, let us endeavour to trace the life of him who set the example, and, who, by perseverance amidst many difficulties, nearly brought to a conclusion one of the most extensive of these works.

The population of the manufacturing districts differs most essentially in its constitution and character, from the other parts of the community of Great Britain. Composed of men primarily devoted to the acquisition of independence, activity and decision form a strong feature in their dispositions, and stamps them with vivid impressions of the worldly changes which may ultimately conduce to the loss or prosperity of their engagements. In Paisley, the largest manufacturing town in Scotland, the middle and operative classes, in whose sphere the individual about whom we are now interested chiefly moved, are respectable and industrious—or idle, of restless dispositions, and preferring dissipation and revelry, according to the habits they had formed on their early entrance into the world. The greater part of the employment in this important town is given out in piece, which permits the labourer to enjoy a greater proportion of leisure than he could do by the more usual method of working a fixed time. A little exertion and assiduity will allow him some hours of relaxation, and the manner in which this is spent, often bears a high influence on the future prospects of the individual. Numerous clubs have at various times been instituted, to which most of the operatives resort, and spend their leisure time, according to their inclinations. In some, intemperance prevails, with the high and wrangling discussion of the affairs and conduct of their different masters, and the politics of the day. What are called liberal sentiments are promulgated; and with the assistance of the more radical newspapers, the foundation is laid for that jealousy of the welfare of their superiors, and discontentment with their own lot, which so often causes the distress of friends, and the destruction of their neighbours' properties, and which, if it does not always bring themselves to ruin, can never promote either their happiness or welfare. In others, though, to a certain extent, the same courses are run, they are generally conducted with moderation—intemperance is avoided—politics and literature are freely entered upon and keenly contested, and the argument is conducted with a ready conviction to the truth, and more for the sake of information, than the obstinate maintenance of any untenable opinion. Libraries of considerable extent belong to some of these clubs ; the taste for reading and study is gradually increasing, and many of the more sedate members avail themselves of their privilege, to advance their knowledge of some favourite subject, and occasionally launch into debates with ardour and penetration, and talent, not often found so varied among this class of society. Other sources of information and improvement also occupy the unemployed hours of the more respectable operatives. The different branches of mechanics are eagerly pursued, and often practically applied, with such success, as to raise the individual from dependency to the higher ranks of society. Natural history is also studied, and the more interesting works on the subject sought after and eagerly perused ; and botany, or rather the culture of flowers, forms one of the most favourite and universal recreations. Almost every one possesses his flower garden, and, as his taste directs, enters warmly into the culture of what are called florist's plants,—polyanthuses, ranunculi, anemonies, hyacinths, pinks, and carnations ; and as much pleasure is afforded, and emulation excited, in watching the success of a favourite bulb or seedling, as could arise from an indulgence in more common, but less innocent amusements. The naming of their flowers, too, is a matter of great importance, and serves to draw their attention to the history of the individuals whom they wish to commemorate. Political characters, and men renowned for great talents or learning, every townsman of any celebrity, or friend with some endearing qualifications, has a favourite tulip, or pink, or carnation dedicated to his praise, and the memory of those who have long departed are yet called to remembrance by the same fleeting emblems.

Such is a rude sketch of the nature of the society in which Wilson spent his infancy and early youth. The son of honest and industrious parents, whose circumstances were never such as to enable him to procure those branches of education at length so eminently displayed, he was indebted almost to his own exertions, and the generous assistance of some friends, who were capable and willing to direct his mind in the occasional sallies of his younger days, and allowed him access to their larger libraries, for the rudiments, at least, of all his after acquirements.

The father of our author was a respectable gauze weaver in Paisley, where he spent the greater part of his early life; but having married, he removed to Auchinbathie Tower, near the village of Lochwinnoch, thinking that more extensive and varied employment would improve the condition of an increasing family. He now rented a piece of ground, which he cultivated himself, keeping, at the same time, employment for several looms, and commencing a sort of trade in distilling and smuggling. He would thus seem to have been of a somewhat speculative disposition, but, in other respects was well informed, and bore the character of a shrewd, upright, and independent man. His eldest son, Alexander, now better known as the "American Ornithologist," was born at Paisley on the 6th of July 1766, previous to the removal of his father from that town. His childhood was most likely passed as that of many others; nor can it be supposed that the boyish pranks of one born in comparative obscurity, should have been treasured up and converted into the dawning promises of future greatness, or his young mind charged

> ——"With meanings that he never had,
> Or, having, kept conceal'd."

In the earliest notice of his youth, we find him near Lochwinnoch in the capacity of a herd (to Mr Stevenson of the Treepwood), a circumstance accidentally recorded, from his having already attempted to celebrate the beauties of Castle Semple in a song,* extending to six verses, curious as the first specimen, and remarkable for its truisms, the characteristic of his later poems.

The ready quickness, and mild disposition of the boy now induced his father to hope that some profession higher than that of

* "Castle Semple stands sae sweet,
The parks around are bonnie, O;
The ewes and lambs ye'll hear them bleat,
And the herd's name is Johnnie, O," &c.
—*Paisley Magazine for November* 1828, p. 583.

an operative would be more suitable to the character of his son; and the laudable pride shown by a great part of the Scottish peasantry, that one of their offspring, at least, should embrace a learned profession, medicine or the church, confirmed his parents in deciding upon the latter as his future avocation, and he was placed under the charge of Mr Barlas, then a student of divinity, to whom, I believe, many of the youth of Paisley have been since indebted for scientific acquirements. We are not informed how long Wilson enjoyed the tuition of this divine; it could not, however, be for any great period, as at the age of ten he suffered the loss of a kind and affectionate mother; and to this melancholy bereavement may, perhaps, be traced the whole bent and inclinations of his varied life. His father, feeling the care of conducting his household and young family, and, at the same time, of attending to his different occupations, too great a burden without assistance, again married. His family still increasing, the funds sufficient to defray the expenses of an education suited for a learned profession were found too limited, and Wilson, upon the recommendation of his relatives, but much against his own wishes, was, at the age of thirteen, bound an apprentice to Mr William Duncan, a respectable operative weaver in Paisley.

While he remained with Mr Duncan, every attention was paid to his business and tasks, apparently from an honourable motive, though an opportunity was never let slip in which he could gratify his taste for reading, or indulge his romantic fancy in wandering about the beautiful vicinity of his native town.

His perhaps otherwise dormant mind had been roused during his short acquaintance with Mr Barlas, and the stealthy snatches which he now obtained of his favourite authors served only to inflame his desire for information. At the conclusion of his apprenticeship he wrote the following quaint lines upon his indenture, showing distaste for his business, and the reluctance with which he filled up his time:—

"Be't kent to a' the warld in rhyme,
That wi' right mickle wark an' toil,
For three lang years I've ser't my time,
Whiles feasted wi' the hazel oil."—August 1782.*

* His indenture for three years, dated 31st July 1779, is now in possession of Mr Clark, Seedhill Mills, Paisley.

He now laboured at the employment of a journeyman only when necessity urged. Such books as the kindness of his friends supplied him with were kept about his loom, and much time was occupied in perusing them, and in attempts to turn his ideas into verse. His enviable faculty of seizing upon the strong and bearing points of any subject or incident had become apparent, and the sallies of boyish wit and ridicule among his companions, gained for him a superiority far beyond what was due to his years.*

Having spent some time in this manner at Paisley, he became a journeyman gauze weaver to his father, who resided sometimes at Lochwinnoch, sometimes at Auchinbathie Tower; and though he now wrought more diligently, and bore the character of being "the most sober and *tramping* journeyman that had ever entered the village," the thought that he had been disappointed in his prospects of a higher profession—his utter distaste for the trade that had been chosen for him—and the higher feelings which his slight literary education had awakened—bore the mastery over his anxiety to perform his allotted tasks, and he was sometimes seduced from them by the pleasure he experienced in rambling among the woods of Castle Semple, or by the banks of the river Calder—one of the most beautiful and romantic mountain streams I have ever seen. These solitary walks confirmed the pensive and diffident turn of his mind, but fitted it to enjoy the deeper solitudes he was afterwards destined to traverse. It was here that he brooded over what he then considered his ill-fated lot, or formed and reformed schemes for his future advancement; where he saw nature as she was in her mild and soothing aspects and more placid skies, her

* While Wilson wrought at Lochwinnoch, he was much importuned by one of his shopmates to write him an epitaph. This individual had excelled in little except *daundering* upon Sundays about the hedgerows and whin bushes in search of birds' nests. Wilson for a long time resisted the entreaties of his companion, for this best reason, that there was nothing in his character that could entitle him to a couplet; but being hard pressed, he burst forth with the following extemporaneous *hit*, which at once silenced the inquirer, and set his shopmates into a roar of laughter at his expense:—

"Below this stane John Allan rests,
An honest soul, though plain,
He sought hail Sabbath days for nests,
But always sought in vain."

green woodlands and brawling brooks; and he afterwards loved to contrast them with the glowing lights and majestic rivers of another hemisphere, with forests where

> ———"The century-living crow,
> Whose birth was in their tops, grew old, and died
> Among their branches."———

These scenes, and incidents connected with them, are often portrayed in his poems with considerable beauty and simplicity, and always show that they were the free effusions of what he at the moment felt.*

About this period his father removed permanently to Auchinbathie Tower, that he might more easily superintend the different occupations in which he was engaged, and Wilson again betook himself to Paisley, where he wrought in a *two-loom* room, having Mr Brodie, afterwards schoolmaster at Quarleton, for his companion; and his diffidence was so great, that nearly three weeks elapsed before a regular conversation and acquaintance took place, which, however, ripened into an ardent friendship that neither distance nor his various pursuits could obliterate. His love for study became now confirmed. Sallust, Virgil, and other Latin authors, and many of the more esteemed English poets, supplied from various sources, were the companions of his loom. Much of his time was occupied in their perusal, and they were the cause of many a broken thread. Brodie, from his better education, being able to appreciate the merits of these authors, soon found the way to the heart of our young enthusiast, and yet expresses his lively pleasure and recollection of the many days which they spent together. He describes him as of a very thoughtful turn of mind, constantly thinking aloud, and giving vent to poetical effusions, which his keen imaginative mind applied to the leading incidents of the day, or to the beauties of his last country ramble. He would often indulge in abstraction, or reverie, and delighted in *dreaming;* and such was his pleasure in following his fancies while asleep, that he would frequently go to bed during the day, or at an early hour in

* The banks of the Calder, near the Loups, furnished the incidents for the Disconsolate Wren, a tale told with great feeling and simplicity, evincing accurate observation of the nature and manners of the birds introduced in it.

the evening, with the hope of following up the impression left on his mind by a dream which he imperfectly recollected in the morning. His solitary musing walks were still continued, and often extended to Auchinbathie Tower, in visits to his father and family, to Lochwinnoch, and his favourite Calder. As the game laws were not so strict, or the individual preservation so much attended to, as they are at present, a gun was his frequent companion, and his deeds in this way were often extolled by his fellow-workers. To these poaching expeditions may, perhaps, be assigned his first lessons in discriminating the various game that occurred, and which showed him at once, when in a new country, the difference of its birds from those to which he had been accustomed.

Wilson now left Paisley to visit his brother-in-law, William Duncan, at Queensferry on the banks of the Forth, where he remained for a few months assisting his relative at his employment, and afterwards accompanied him on a mercantile travelling excursion over the eastern districts of Scotland. This was the greatest distance he had yet been from his birth-place, and the new scenes and variety of incidents which he encountered, induced him to think that the occupation of a travelling merchant would be far preferable to the sedentary, and, to him, irksome employment of a weaver. He resolved, therefore, to attempt "the establishment of his good fortune in the world,"* and, being able, by the kindness of his friends, to provide the requisites for a small pack, and having "fitted up," as he tells us,† "a proper budget, consisting of silks, muslins, prints, &c., &c., for the accommodation of those good people who may prove his customers," he commenced a new and more varied life, with a light heart, and sanguine expectations of success.

> "'Ralph the pedlar'————
> ————bore a curious pack,
> With trinkets fill'd, and had a ready knack
> At coining rhyme."

In this itinerant life he for some time persevered, alive both to the beauties of the country he travelled through, and the repulses he often met with when displaying his wares. His attention was attracted by everything of worth, and he would often leave his tract

* Journal, Poems, 1st Edit. † Ibid.

to visit some place of antiquity, or the former residences of his favourite authors and poets. He visited, also, every churchyard which lay in his way, transcribing those epitaphs which struck his fancy, and had thus collected above three hundred, many of them highly curious; but, with his desultory writings, these have been long since lost. From several of the poems written about this time, during the unoccupied hours of this journey, in which many of the incidents that befel him are described, we learn that he began to feel the life of a pedlar was not all ease and comfort, and that many petty annoyances, besides cold, fatigue, and hunger, awaited him. In an "Epistle from Falkland to Mr A—— C——,"* he designates himself a

> ———"Lonely pedlar,
> Beneath a load of silk and sorrows bent;" †

and, in another, compares his former more comfortable bed with his ensconcement in a barn, where

> "The dark damp walls—the roof, scarce cover'd o'er—
> The wind, wild whistling through the cold barn door,"‡

were too real to allow room for playful fancies or delicious reveries.

Somewhat disgusted, he returned again to Paisley, and commenced the publication of his poems, which had now accumulated to a considerable stock, and which he fondly thought would bring him both fame and fortune. Desiring that some one, better qualified than himself, should correct any inaccuracies in the MSS., he fortunately applied to Mr Crichton of the Town's Hospital, a man of great worth, who became a faithful friend in adversity, and still is his enthusiastic admirer. Wilson introduced himself, "and with a great deal of modesty, expressed his wish for a little conversation. He told me his name, and informed me that he had a volume of poems in manuscript, which he intended for the press, and requested that

* Alexander Clark.

† "His lonely way a meagre pedlar took,
Deep were his frequent sighs, careless his pace,
And oft the tear stole down his cheerless face,
Beneath a load of silk and sorrows bent."
—1st Edit. p. 96.

‡ Morning, 1st Edit. p. 1.

I would look over them at my leisure. He put the small volume, which was neatly written, into my hand, and left it with me, saying that he would soon call again to hear my opinion." * Mr Crichton having given his sanction to the volume, the very sanguine disposition of our now about-to-be young author, in addition to his being also a poet, bore him through all preliminary difficulties. He contracted with his printer in Paisley, circulated his prospectus, and having, he tells us,† "committed the contents of my pack to a hand-bill,‡ in a style somewhat remote from any I have yet seen," he a second time sallied forth, "to make one bold push for the united interest of pack and poems."

He had now, by the solicitation of his friends, been induced to keep a journal. We learn from it, that, after travelling to Edinburgh,

* Biographical Sketches of the late Alexander Wilson, in a Series of Letters by Senex (Mr Crichton, Town's Hospital, Paisley), p. 13.

† Journal, Poems, 2nd Edit.

‡ Advertisement Extraordinary.

Fair ladies, I pray, for one moment to stay,
Until with submission I tell you,
What muslins so curious, for uses so various,
A poet has here brought to sell you.

Here's handkerchiefs charming; book-muslins like ermine,
Brocaded, striped, corded, and check'd;
Sweet Venus, they say, on Cupid's birthday,
In British-made muslins was deck'd.

If these can't content ye, here's muslins in plenty,
From one shilling up to a dozen,
That Juno might wear, and more beauteous appear,
When she means the old Thunderer to cozen.

Here are fine jaconets, of numberless sets,
With spotted and sprigged festoons;
And lovely tambours, with elegant flowers,
For bonnets, cloaks, aprons, or gowns.

Now, ye Fair, if ye choose any piece to peruse,
With pleasure I'll instantly show it;
If the Pedlar should fail to be favour'd with sale,
Then I hope you'll encourage the Poet.

—Journal, Poems, 2nd Edit.

he proceeded along the coast to Dunbar, and, crossing to Burnt-island, coasted that side of the Firth to Kinghorn, where his memoranda stops. This was his first attempt at prose writing which appeared publicly, and is remarkable for the clear observation upon human nature, of incident, and the appearance of the surrounding country. The ill success of this journey disgusted him not only with the pack, but showed him, that hawking poems was not a more profitable trade; and, annoyed at the failure of his plans, he returned to his native town nearly penniless, and much depressed in spirits, thinking that a pedlar and poet stood lower in the scale of rank than he was previously inclined to place them. "A packman," he writes, in a letter from Edinburgh, to his friend Mr Brodie, "is a character which none esteem, and almost every one despises. The idea which people of all ranks entertain of them is, that they are mean-spirited, loquacious liars, cunning and illiterate, watching every opportunity, and using every low and mean art within their power to cheat. When any one applies to a genteel person, pretending to be a poet, he is treated with ridicule and contempt; and even though he should produce a specimen, it is either thrown back again, without being thought worthy of perusal, or else read with prejudice." Ill success rendered him severe on these often eccentric, very generally harmless, professions.

The sale of his poems being insufficient to procure for him the ordinary necessaries of life, he was obliged to resume occasionally the labours of his loom at Lochwinnoch, at which, by his expertness and diligence when willing, he could always raise a temporary supply. He possessed, however, no "scheming foresight," and sometimes allowed himself to be so *hard run* as to be unable to procure paper and other writing materials, and the same cause also sometimes produced a scantiness in his wardrobe. An anecdote of his resources, upon an occasion which would have given great annoyance to many, was thus related to me by one of his best friends. Wilson was fond of music and dancing, and, in the latter branch, bore the character of a neat and light performer. In those days, the fashionable ball-dress among persons in his sphere of life was knee-breeches, white stockings, and black gaiters, or, as they are called, *kutikens.* Being one evening invited to a ball given by some young companions, he found himself reduced to a single pair of white thread stockings, rather the worse for wear, and not improved in purity of colour. Knowing that he was looked up to as a pattern for neatness, and unwilling to be disappointed, he chalked the upper parts of his stockings, and finished the deception by painting upon the lower part a pair of black gaiters. He spent the evening to his satisfaction, and returned to his home undiscovered.

A continuance of this unsettled life threw him into ill health, and a state of great mental despondency. Of the depressing effects of the latter he was fully aware. In a letter to Mr Crichton, he says, "Among the many and dismal ingredients that embitter the cup of life, none affect the feelings or distress the spirit so deeply as despondence." In another, written nearly a year after, he appears to have been really ill,* and still more diseased in mind; yet amidst the distresses which he thought were crowding round him, and the sorrows which came,

> "Not single spies,
> But in battalions!"

he never loses sight of the consolations he expected to receive from his religion, and endeavoured, as he elsewhere expresses himself,

> "To lift his thoughts from things below,
> And lead them to divine." †

The extracts from this letter will show that the principles of Divine Revelation had been well and early implanted in his mind, and his future writings bear witness that they were never uprooted. "Driven by poverty and disease to the solitudes of retirement,‡ at the same period when the flush of youth, the thirst of fame, and the expected applause of the world, welcomed me to the field,"—"I feel my body decay daily, my spirits and strength continually decrease, and something within tells me that dissolution—dreadful dissolution, is not far distant. No heart can conceive the terrors of those who tremble under the apprehension of death. This increases their love of life, and every new advance of the King of

* His complaint was an inflammatory cold, which threatened to fix upon his lungs.

† Hymn VI. Poems, p. 110.

‡ He was now at Auchinbathie Tower.

Terrors overwhelms them with despair. How hard—how difficult—how happy to prepare for eternity! and yet, how dreadful to live or to die unprepared! Oh! that I were enabled to make it my study to interest myself in His favour, who has the keys of hell and of death. Then all the vanities of life would appear what they really are, and the shades of death would brighten up a glorious path to everlasting mansions of felicity!"—"These are the sincere effusions of my soul, and I hope that, through the divine aid, they shall be my future delight, whether health shall again return, or death has left the commissioned dart." For his recovery from this low state, he was as much indebted to the kind and salutary counsels of Mr Crichton, as to the prescriptions of his physicians; but, as that gentleman remarks, he was "soon up and soon down, and the air of the country, and temporary removal from Paisley, the scene of his distress," in some measure recovered his bodily strength and wonted spirits.

There is another circumstance which may have weighed on his mind. Although it has been said, by most of his biographers, "that female attachments he had none," were there no other proof, it would be almost impossible to conceive a young man of ardent temper and keen perceptions, totally insensible to the charms of female beauty. In his younger days, I have good authority for saying that he had several liaisons, and for some time had been attached to the sister of Mrs Witherspoon, a pretty and respectable girl, to whom he made frequent allusion in his poems, though two only of those published contain any reference to her; and there can be little doubt that Martha Maclean bore an influence in his fits of despondency. In the New World he formed new attachments, and, had he lived, was to have married Miss Miller, daughter of a considerable proprietor in the vicinity of Winterton, and whom he appointed his executrix.

His spirits being roused by the counsels and exertions of his friends, he again commenced travelling, still carrying with him the pack and poems; and, as another resource, endeavoured to procure some employment by writing for the periodicals of the day. He also projected a work, to be edited and conducted by himself, and to be called the *Paisley Repositary.* Of this, a prospectus was printed and circulated; but the advice of Mr Crichton and of Mr

Brodie saved him from farther embarrassment as a publisher. He contributed several pieces, in prose and poetry, to the *Glasgow Magazine*, and wrote "The Solitary Philosopher," * as a specimen, for the *Bee*, under charge of Dr Anderson, hoping, by its merits, to gain farther employment. The character of the singular being who formed the subject of this memoir,—a "botanist, philosopher, naturalist, and physician," is thus sketched; and I have transcribed a part, as I consider the intimacy of a young man with secluded characters, possessing such eccentricities as the philosopher and Tippenny Robin, † must have cast their influence over his after feelings, and laid the first shade in his love for seclusion.

"On the side of a large mountain, in a little hut of his own rearing, which has known no other possessor these fifty years, lives this strange and very singular person. Though his general usefulness, and communicative disposition, require him often to associate with the surrounding rustics, yet, having never had an inclination to travel farther than to the neighbouring village, and being totally unacquainted with the world, his manners, conversation, and dress, are strikingly noticeable. A little plot of ground that extends round his cottage, is the narrow sphere to which he confines him-

* Published first in the *Bee*, and afterwards in a "Collection of Ancient and Modern Characters," printed at Paisley in 1805, p. 250.

† This very eccentric character, whom Wilson had discovered during his rambles, and frequently visited, was an Irishman, named Robert Carswell, and received his nickname from the circumstance of his never accepting more than twopence for a day's work, except during harvest, when he allowed it to be doubled. He lived in a small thatched house, at the Kaim, on the Calder; but was very anxious to possess another dwelling, objecting to that in which he lived, on account of a loft, which he said prevented his prayers from reaching heaven. The inside was very dirty, filled with peats and potatoes, and was never allowed to be swept, unless by himself. He had hoarded up some money, which was kept in paper parcels, of a few shillings each, generally scattered about the floor, and which, at his death, he bequeathed to the parish poor. His dress was a plain plaiding doublet, the waist girt with a rope of straw or tow, in the one side of which was always hung the key of his door, and in the other stuck a *bountree sheath*, for holding his knitting wires. Notwithstanding these habits, he had received a better education, could read and write, and possessed a considerable number of books: he could also fence. He was a Cameronian; and every Friday left his house early for some wild elevated ground, carrying with him a creelful of books, and remained abroad the whole day.

self; and in this wild retreat, he appears to a stranger as one of the early inhabitants of earth, ere polished by frequent intercourse, or united in society.

"In this vale, or glen, innumerable rare and valuable herbs are discovered, and, in the harvest months, it is his continual resort. He explores it with the most unwearied attention,—climbs every cliff, even the most threatening, and from the perplexing profusion of plants, collects those herbs, of whose qualities and value he is well acquainted. For this purpose, he has a large basket with a variety of divisions in which he deposits every particular species by itself. With this he is often seen labouring home to his hut, where they are suspended in large and numerous parcels from the roof, while the sage himself sits smiling amidst his simple stores.

"About six months ago, I went to pay him a visit along with an intimate friend, no less remarkable for a natural curiosity. On arriving at his little hut, we found, to our no small disappointment, that he was from home. As my friend, however, had never been in that part of the country before, I conducted him to the glen, to take a view of some of the beautifully romantic scenes, and wild prospects, that this place affords. We had not proceeded far along the bottom of the vale, when, hearing a rustling among the branches above our head, I discovered our hoary botanist with his basket, passing along the brow of a rock that hung almost over the centre of the stream. Having pointed him out to my companion, we were at a loss for some time how to bring about a conversation with him. Having, however, a flute in my pocket, of which music he is exceedingly fond, I began a few airs, which, by the sweetness of the echoes, was heightened into the most enchanting melody. In a few minutes this had its desired effect, and our little old man stood beside us with his basket in his hand. On stopping at his approach, he desired us to proceed, complimented us on the sweetness of our music, expressed the surprise he was in on hearing it, and, leaning his basket on an old trunk, listened with all the enthusiasm of rapture. He then, at our request, presented us with a sight of the herbs he had been collecting, entertained us with a narrative of the discoveries he had made in his frequent searches through the vale, 'which,' said he, 'contains treasures that few know the value of.'"

At this time Wilson wrote the well-known ballad of *Watty and*

Meg. It was nearly contemporary with the *Tam o' Shanter* of Burns; and its great success, with the thought that, for a time, the productions of the Paisley poet could be taken for those of the Ayrshire bard, raised his spirits to their usual pitch. There is a difference of opinion regarding the residence of the heroes of this piece. One of my correspondents is of opinion, that the couple resided at Lochwinnoch, and that the male actor was a drunken coachman, "one of Smithie's drunken core," in the service of Mr M'Dowell of Garthland, and thinks this confirmed by his real name, Peter Thomson,* being mentioned in the poem. By the attention of Mr Lang, I was introduced to the Paisley Watty and Meg, residing near the Seedhills, and one might easily conceive they had performed the parts so graphically described by our author, whose persons and domicile would work well under the pencil of Hervey. It was now, also, that he wrote the review of *Tam o' Shanter* for the *Bee,* but which Dr Anderson refused to publish. On receiving this refusal, he sent the paper, in the height of his indignation, to Burns himself, who answered his communication with it, assuring him, that he had received innumerable criticisms, and had never answered any of them, but as Wilson's was of a superior order, he would reply to it, and proceeded to justify his poem. This was Wilson's first introduction to a short acquaintance with Burns; previously, he had gone to Ayr on purpose to visit him, but found him from home, and had only the satisfaction to converse for a considerable time with his sister, who must have made a favourable impression upon his mind, as he was heard to remark, on his return, that "Burns must be a very superior man if like his sister." The poets met, some years after, at Burns' farm, spent a pleasant evening, and made an exchange of the poems which occasioned their introduction. This was their first and last meeting.

It was natural to suppose that a disposition, bearing a stamp so superior and different from his fellow-operatives in a similar station, would form associations with somewhat kindred minds. Mr Crichton, Dr Barlas, and Mr Brodie, he looked up to with a certain

* "Dyster Jock was sitting crackin'
Wi' *Pate Tamson* o' the Hill,
'Come awa,' quo' Johnny, 'Watty!
Haith, we'se hae anither gill.'"

awe, and held them more as preceptors, or friends, to whose steady guidance he could trust, than as participators in all his whims and youthful frolics; and we find him in frequent society with Gavin Turnbull, E. Picken, and James Kennedy, who formed part of his companions in his song of "The Group." It was by their means that he was introduced to a debating society held in the Edinburgh Pantheon, where the merits of various questions given out for discussion were contested in speeches by individuals taking different sides, and decided by the votes of the audience, which consisted of both ladies and gentlemen. Wilson made all his addresses in poetry, generally in the form of a tale applicable to the subject, and wrote for this society several pieces, which, considering the time he devoted to them, may rank among the best of his juvenile performances.

The first address he delivered there was unpremeditated; he had gone without any intention of mingling in the debate; but, after others had spoken upon the subject, "Whether affection or interest was the greatest inducement to matrimony,"—his feelings had been warmed, and a pause ensuing, during which the audience seemed to expect some other orator, he availed himself of the opportunity, and delivered an address which astonished the audience as much as it surprised himself. The ice was now broken unawares; he began to throw aside his natural diffidence, and appeared a frequent disputant. He took part in the questions, "Whether is diffidence or the allurements of pleasure, the greatest bar to the progress of knowledge?" "Whether suffering humanity received most assistance from the male or female sex?" and "Whether is disappointment in love, or the loss of fortune, hardest to bear?" But, of all his poems written for this society, the best perhaps is the "Laurel Disputed," or a comparison of the merits of Allan Ramsay and Robert Fergusson. I received the following history of this poem:—Wilson one day called upon Mr Brodie, at Quarleton, to ask his advice regarding a letter he had received from James Kennedy, at Edinburgh, intimating that a prize was to be awarded at the Pantheon for the best essay upon "Whether have the exertions of Allan Ramsay or Robert Fergusson done most honour to Scottish poetry." Kennedy intended to compete, and strongly urged Wilson to do the same, and offered to present his essay, as those in poetry might be delivered by deputy. Mr Brodie, however, advised Wilson to go himself, and deliver his essay in person, if he intended to compete; but the poet declared his total inability to perform the journey for want of funds; and, besides, said, "It is long since I read Ramsay, and, as to Fergusson, I never saw a copy of his poems in my life; I am at a loss, therefore, which side of the argument to espouse." He was answered, that "Ramsay would most probably be the favourite. His Gentle Shepherd had made a deep impression on the public, and would always continue popular; he, besides, enjoyed superior advantages over Fergusson,—Thomson, the author of the Seasons, having seen and criticised the Pastoral previous to its publication. The poems of Fergusson possess great merit; but the difficulties the young poet had to contend with, and his untimely and melancholy death, left it difficult to speculate upon what he would have attained had he lived to the years of his predecessor." "For these reasons," replied Wilson, "will I espouse his cause; I shall go to Edinburgh myself;" but, stopping suddenly, he exclaimed, "Where am I to get the siller, or see a copy of his poems?" Mr Brodie kindly furnished him with his own copy of Fergusson, and advised him to finish his web at Lochwinnoch, which, if he could get done in time, would yield him the means to undertake the journey. With a light and joyous heart he set out; and, in one week, returned from Lochwinnoch, visiting his father at Auchinbathie by the way; having in that short period, woven forty ells of silk gauze (of itself a good week's work), read and studied Fergusson, and had composed, written, and learned to recite his poetical address of the "Laurel Disputed."

Our youthful poet walked to Edinburgh, and was one of the seven candidates on this, to him, eventful evening. To use his own expression, "We were ranged on a front seat in the Pantheon, like so many pamphlets on a shelf." Drawing lots to regulate the precedence of the orators, Wilson found his turn about the middle; and when the first had concluded his address, he thought that he could do better than it. The audience, who were also the judges, consisted of nearly five hundred persons. Six of the competitors took the side of Ramsay, Wilson stood singly and alone for Fergusson.

The medal was adjudged to a Mr Cumming; Wilson was declared second in merit, and Picken third; the essays of the others were not taken into consideration. Wilson was both elevated and provoked at this award; for he considered that he had the true majority. Cumming was only first by seventeen votes; and his friends shrewdly suspected these were obtained by presenting forty tickets to the *ladies*, which, although only sixpence each, was a mode of canvassing beyond the humbler means of our author, who, even had he possessed them, would have spurned the idea of such corruption.*

We have now reached the only period in the history of our author's career, which is tarnished by the performance of actions discreditable to him, and totally at variance with his real disposition, and the whole tenor of his former and after life. Entirely discouraged with the ill success of all his undertakings, and his habits of application being little improved during the unsettled life he had for some time led, he found still greater difficulty in applying to any sedate occupation. Though he perceived the necessity of this course, and had expressed his wishes to various friends for some regular employment, more suited to his taste than that which he had hitherto professed, his unsteadiness would not permit him to follow up those better resolves. He was recommended to qualify himself for a clerk to some mercantile establishment. A letter from Mr Gavin, Commercial Academy, Paisley, to whom he applied for instruction in some of the branches necessary for his occupation, will best show the state of his mind at this time :—"Wilson came to my school, and requested that he might be taught some branches of arithmetic, in which he was deficient, in order, he said, to qualify him to become a clerk to some merchant or manufacturer. I cheerfully undertook to teach him all I knew, and he sat down at a desk, apparently in good earnest. Before to-morrow, another thought had struck him,—he never returned to school." † He tried it again, while on a journey, at Callander, but with no better success. He records the circumstance himself in a letter to Mr Brodie :—"Having agreed with the town schoolmaster," he says, "I accordingly went, purchased a slate, and set seriously about it. Two days elapsed; I persevered; but the third day gave it up." Thus, a mind, too roving for a respectable and stable business, led him into the society

* The poem delivered by Picken, and the "Laurel Disputed," were published together in a shilling pamphlet.

† Letters of Senex, p. 42.

of those companions, whose temporary applause made him forget the precepts of his father, afterwards so strongly impressed upon his mind."*

The west country was now in such a state as to require the more immediate attention of government. The general depression of trade, occasioned by the wars incident to the French Revolution, threw into comparative idleness many of the young operatives, who began so openly to promulgate revolutionary principles, and sentiments of discontent against the steadier manufacturers, who would not yield to all their unreasonable demands, that many of them became "marked men," and were obliged to remain in temporary concealment, or entirely to leave their country. It was Wilson's misfortune to have formed an intimacy with some of these, who, knowing his talents, prevailed upon him to revile and satirise the conduct of those who were most offensive to their views of liberty, or of propriety of conduct as masters. Wilson himself, well acquainted with all the circumstances regarding these matters in Paisley, and ever ready to redress what he imagined wrongs, was too easily wrought upon; and, entering into the feelings and prejudices of the operatives, at the solicitation of others, produced a number of poetical squibs which held up the subjects of popular dislike to contempt and ridicule. The principal ones went under the designations of "The Pedlar Insulted," "Hab's Door, or the Temple of Terror," "The Hollander, or Light Weights," and "The Shark, or Long Mills Detected;" the titles bearing reference to local circumstances of the times, and persons, which are now of no consequence. For one of these he underwent a trial, but the charge could not be proved, and for the last, written against a respectable and wealthy manufacturer, perhaps containing some truths, he was prosecuted, and sentenced to imprisonment in the jail of Paisley, and to burn the offensive poem with his own hand. It will be sufficient to say, that he deserved the punishment, having behaved with neither propriety nor honour in the transaction. Of this he was himself sensible, and deeply regretted his error; and such was the opinion even of his prosecutors, that the sentence would have been averted,

* When his satirical productions first gained notice, his father was repeatedly heard to say,—"Sandy, I see you have some talent about you, but my advice is, never to use it to wound the feelings of others."

if possible, after the verdict had been pronounced: it was, however, carried into execution in the most private and gentle manner, a very few of his intimate friends only seeing him burn the obnoxious pamphlet in question. In those his evil days, he seems to have been carried off by the persuasions of his companions, or, as he says, "led astray" by his too keen "imagination," without considering the pain or injury he inflicted on the objects of his satire, or the evil consequences it might bring upon himself. His future opinions were very different. In a copy of his poems, in possession of Mr Ord, his able editor, there is written by himself,—"I published these poems when only twenty-two, an age more abundant in *sail* than *ballast.* Reader, let this soften the rigour of criticism a little.—1804." In a letter to his father of a later date, he says, "In youth I had wrong ideas of life; imagination too often led me astray. You will find me much altered from the son you knew me in Paisley, more diffident of myself, and less precipitate, though often wrong." And when copies of these poems were afterwards brought to him, in America, by a friend, he threw them into the fire, saying, "Ay, David,* have you been at all this trouble? These were the follies of my youth, and I sincerely wish they had never seen the light. Had I taken the advice of our kind and excellent father, I should have done well, and saved myself many an uneasy hour."

After these unfortunate circumstances, we might easily judge, that the keen feelings of our author would not allow him to remain in Paisley, and might fancy him resuming for a time his travelling occupations. His thoughts, however, took a wider range. An honourable fear of ruining his friends actuated him. He could not trust to his own strength of mind to refrain from those satires, for which he was now under bail; and after using every argument to convince his father and favourite sister, Mary, of the propriety of his intentions, he at last observed, "I am bound now, and cannot ruin Thomas Witherspoon (his security), and I must get out my mind." He imagined, also, that his misfortunes would continue; and the ideal charms of a free land, and of liberty, drew his attention to a more distant country. His mind had become gradually, but firmly, reconciled to seek a new fortune in America.

* David Wilson, the poet's half brother.

By a little exertion at the loom, and some kind assistance from other quarters, he was enabled to earn the funds necessary to defray the expenses of his intended voyage; and, bidding a long farewell to his parents, and those companions, who had so often assisted him to the extent of their means—to the scenes where he had wandered from his boyhood— where every bush and tree had its story—every crag or bold feature in the landscape its associations and recollections, to be felt only by those who have been placed in similar circumstances, —he set out on foot from the land of his birth, and arrived at Belfast, where he had accidentally heard that a vessel was nearly ready to sail. His nephew, William Duncan, a lad of sixteen, was his companion. He had shared his confidence, and agreed to share his fortunes; and, on the morning of the 23rd May 1794, the young men set sail from Ireland; and, after a dangerous passage of twenty-two days, they arrived in safety at Newcastle, in the state of Delaware.

We now find Wilson in the land where he imagined all his wrongs would cease. "He had often," says Mr Ord, in his excellent "Memoirs," "cast a wistful look towards the western hemisphere; and his warm fancy had suggested the idea, that among that people only, who maintained the doctrine of an equality of rights, could political justice be found. He had become indignant at beholding the influence of the wealthy converted into the means of oppression, and had imputed the wrongs and sufferings of the poor, not to the condition of society, but to the nature and constitution of the government." The sequel will show how these opinions are borne out.

Upon landing in the New World, his funds were so scanty as to require an immediate exertion, and he set out on foot in search of work to Philadelphia; from thence he wrote to his parents, informing them of his safe arrival; and this letter being fortunately preserved,* I am enabled to give an account of his passage, and first opinions of America, in his own words:—

"PHILADELPHIA (UNITED STATES),
July 25, 1794.

"DEAR FATHER AND MOTHER,—You will see by this that I am at length landed in America, as is also my nephew, William Duncan

* This letter is now in possession of his sister, Mrs Bell, to whose kindness I am indebted for its use.

—both in good health. We sailed in the ship *Swift*, from Belfast Loch, on Friday the 23rd of May, about six in the morning, at which time I would have wrote you; but, hoping we would have a speedy passage, and feeling for the anxiety I feared you might be under in knowing we were at sea, I purposely omitted writing till our arrival in America. I fear that by this conduct I have given you more unhappiness than I am aware of; if I have, I hope you will forgive me, for I intended otherwise. We had 350 passengers, —a mixed multitude of men, women, and children. Each berth between decks was made to hold them all, with scarce a foot for each. At first sight, I own, it appeared to me almost impossible that the one-half of them could survive; but, on looking around, and seeing some whom I thought not much stouter than myself, I thought I might have a chance as well as the rest of some of them. I asked Willy if he was willing, and he saying he was, we went up to Belfast immediately for our clothes; and, in two days after we got on board, she sailed. We were very sick four days, but soon recovered; and having a good, steady, fair breeze for near a fortnight, had hopes of making an excellent voyage. On the third day, and just as we lost sight of land, we spoke the *Caledonia* of Greenock, a letter of marque, bound for the Bay of Fundy; and, on the Monday following, Dr Reynolds, who was tried and condemned by the Irish House of Lords, was discovered to be on board, and treated all the passengers and crew with rum-grog, which we drank to the confusion of despots, and the prosperity of liberty all the world over. Till the 17th of June, we had pretty good weather, and only buried an old woman and two children. On the 18th, we fell in with an amazing number of islands of ice; I counted at one time thirty-four in sight, some of whom that we nearly passed was more than twice as high as our main-topgallant mast head, and of great extent: we continued passing among them, with a good breeze, for two days, during which time we run at the least five knots an hour. On the 20th we had a storm of wind, rain, thunder and lightning, beyond anything I had ever witnessed. Next day a seaman dropped overboard; and, though he swam well, and made for the ship, yet the sea running high, and his clothes getting wet, he perished within six yards of a hen coop, which we had thrown over to him. On the 11th of July, we could plainly perceive land from the mast head; but a terrible gale of wind blowing all night from the shore, it was Sunday before we had again the satisfaction of seeing it, scarcely perceptible through the fog; but a pilot coming on board, and the sun rising, we found ourselves within the Capes of the Delaware, the shore on land having the appearance of being quite flat, and only a complete forest of trees. About seven at night, having had a good breeze all day, we cast anchor at a place called Reedy Island, where one of the cabin passengers and the first man who leapt ashore in the long boat was drowned in returning to the ship. We arrived at New-castle next day about mid-day, where we were all as happy as mortals could be; and being told that Wilmington was only five miles up the river, we set out on foot through a flat woody country, that looked in every respect like a new world to us from the great profusion of fruit that everywhere overhung our heads, the strange birds, shrubs, &c., and came at length to Wilmington, which lies on the side of a hill, about a mile from the Delaware, and may be about as large as Renfrew, or perhaps larger. We could hear of no employment here in our business, though I saw two silk looms going, and some jennies preparing for establishing some manufactory of cotton cloth; but they proceed with so little spirit, that I believe it may be some years before half-a-dozen of looms can be employed. From Wilmington we proceeded to Philadelphia, twenty-nine miles distant, where very little of the ground is cleared; the only houses we saw were made of large logs of wood, laid one over another; and what crops we could see consisted of Indian corn, potatoes, and some excellent oats. We made free to go in to a good many farm-houses on the road, but saw none of that kindness and hospitality so often told of them. We met with three weavers by the way, who live very quiet and well enough, but had no place for any of us. At length we came within sight of Philadelphia, which lies something like Glasgow, but on a much flatter piece of ground, extending in breadth along the Delaware for near three miles. Here we made a more vigorous search than ever for weavers, and found, to our astonishment, that, though the city contains between forty and fifty thousand people, there is not twenty weavers among the whole, and these had no conveniences for journeymen, nor seemed to wish for any; so, after we had spent every farthing we had, and saw no hopes of anything being done that way, we took the first offer of employment we could find, and have continued so since.

"The weather here is so extremely hot, that, even though writing in an open room, and dressed, according to the custom, in nothing but thin trousers and waistcoat, and though it is near eleven at night, I am wet with sweat. Judge, then, what it must be at noon with all kinds of tradesmen that come to this country, none with less encouragement than weavers; and those of that trade would do well to consider first, how they would agree with the spade or wheelbarrow under the almost intolerable heat of a scorching sun. I fear many of them never think of these. Necessities of life are here very high, owing to the vast numbers of emigrants from St Domingo and France. Flour, though you will scarce believe it, is near double the price to what it is in Scotland; beef, ninepence of their currency, which is about sixpence of ours; shoes, two dollars and a half; while house-rents are most exhorbitantly high. I was told yesterday, by a person who had come immediately from Washington, that that city does not contain above two dozen of houses, and if it come not faster on than they have done, it won't contain one thousand inhabitants these twenty years. As we passed through the woods, in our way to Philadelphia, I did not observe one bird such as those in Scotland, but all much richer in colours. We saw great numbers of squirrels, snakes about a yard long, and red birds, several of which I shot for our curiosity. I am sorry I have so little room. I beg once more you will write to me soon, and direct to the care of Mr William Young, bookseller, Chestnut Street, Philadelphia; and wishing you both as much happiness as this world can afford, I remain your affectionate son,

ALEXANDER WILSON."

Finding some difficulty to procure a livelihood in Philadelphia, by working at any of the occupations to which he had been accustomed, he introduced himself to a countryman, Mr John Aitken, who gave him a temporary employment at his own business, that of a copper-plate printer. From this period to about 1800, all his correspondence with friends in Scotland seems to have been destroyed, and we comparatively lose sight of him for nearly four years, getting only occasional glimpses as he from time to time settled, for a short period, in certain places. It is probable that he tried various schemes of bettering his fortune; and though none was very successful, he did not, at this time, lose all hope. In a letter to Mr Crichton, he says, "Let no man who is stout and healthy be discouraged. If he is a weaver, and cannot get employment at his own business, there are a thousand others which will offer, where he will save as much as he can in Scotland, and live ten times better."

He travelled again as a pedlar, on a sort of trading expedition, with considerable success; and during his journey through New Jersey, kept a journal, in which is sketched, with considerable spirit, the manner of the inhabitants, and the habits of the most remarkable quadrupeds and birds. We next hear of him in a school at Frankfort, Pennsylvania, whence he removed to Milestone, and taught in the village school-house. He had the merit, while there, of studying several branches of liberal education, and of reaching, by his exertions, to considerable proficiency. He advanced so far in mathematics as to bear some note in the science, and by his knowledge of surveying, was enabled to improve his income, during the leisure school hours. He also took a lead in a debating society, which the recollection of his essays in the Pantheon of Edinburgh, perhaps, gave him some claim to direct. It was conducted nearly in the same manner. The subjects proposed chiefly related to agriculture, such as, "Is the cultivation of the vine an object worthy of the attention of the American farmer?" This hard course of study, however, impaired his health, and we learn, from his letters to Mr Ord, that on this account he was obliged to resign his situation twice. He finally left Milestone, and taught and wrought his way to Bloomfield, New Jersey. Becoming more discouraged with the country and his prospects, on his establishment here, he thus writes to his friend Ord:—

"BLOOMFIELD, *near* NEWARK, NEW JERSEY,
12th *July* 1801.

"MY DEAR SIR,—If this letter reach you, it will inform you that I keep school at twelve shillings per quarter, York currency, with thirty-five scholars, and pay twelve shillings per week for board, and four shillings additional for washing, and four shillings per week for my horse. After I parted with Davidson, the Quakers not coming to any agreement about engaging me, I left Washington, and steered for New York, through a country entirely unknown to me; visited many wretched hovels of schools by the way; in two days reached York, and from every person who knew the Mitchells, received the most disagreeable accounts of them, viz., that James had, by too great a fondness for gaming, and sometimes taking his scholars along with him, entirely ruined his reputation, and lost his business; and, from his own mouth, I learned that he expected jail every day, for debts to a considerable amount. And William is lost for every good purpose in this world, and abandoned to the most shameful and excessive drinking, swearing, and wretched company: he called on me last Thursday morning, in company with a hocus-pocus-man, for whom he plays the clarionette. New York swarms with newly-imported Irishmen of all descriptions,—clerks, schoolmasters, &c. &c. The city is very sickly: Mitchell, and all the rest to whom I spoke of you, believed that your success here would be even more unsuccessful than in Philadelphia, and related so many stories to that purpose that I was quite discouraged. Mr Milne attempted it there, but was obliged to remove, and is now in Boston, wandering through the streets insane. I stayed only one night in York, and being completely run out except three elevenpenny bits, I took the first school, from absolute necessity, that I could find. I live six miles north from Newark, and twelve miles from New York, in a settlement of canting, preaching and praying, and snivelling, ignorant Presbyterians. They pay their minister £250 a-year for preaching twice a-week, and their teacher forty dollars a quarter, for the most spirit-sinking laborious work, six—I may say twelve—times weekly. I have no company, and live unknowing and unknown. I have lost all relish for this country; and, if Heaven spare me, I shall soon see the shores of old Caledonia. How happy I should be to have you beside me;—I am exceedingly uneasy to hear from you. Dear Ord, make no rash engagements that may bind you for ever to this unworthy soil. —I am, most sincerely, your affectionate friend,

"ALEXANDER WILSON."

"*P.S.*—Let's contrive a plan to leave this country, and try old Scotia once more in company."

Wilson remained only a little while at Bloomfield; for, hearing of a better situation, he applied for it, and obtained an engagement from the trustees of the Union School, a short way from Gray's Ferry on the Schuylkill, and about four miles from Philadelphia. Upon his first arrival in America, nothing appears to have struck him so much as the birds. The variety of their forms and rich colours had at once impressed his mind. The difference of the feathered race from those of his native country, is noticed in his first letter to his parents, written only a few days after his arrival; and his sensations on viewing the first bird that presented itself as he entered the forests of Delaware, were most vivid: it was a red-headed woodpecker, which he shot, and considered the most beautiful bird he had ever beheld. The acquisition of this situation, therefore, may be looked upon as the most important era in his whole life,—it commenced his acquaintance with the venerable Bartram. "His school-house and residence lay but a short distance from Bartram's botanic garden, situated on the western bank of the Schuylkill,—a sequestered spot, possessing attractions of no ordinary kind. An acquaintance was soon contracted with that venerable naturalist, which grew into an uncommon friendship, and continued, without the least abatement, until severed by death. Here it was that Wilson found himself translated, if we may so speak, into a new existence. He had long been a lover of nature, and had derived more happiness from the contemplation of her simple beauties, than from any other source of gratification. But he had hitherto been a mere novice; he was now about to receive instructions from one, whom the experience of a long life, spent in travel and rural retreat, had qualified to teach."*

Notwithstanding this improved condition in life, his mind, perhaps weakened by his late illness, was ill at rest, still brooding over his dependent situation, and, as we learn from his letters to Mr Ord, upon circumstances of a private nature, which it would be useless to introduce here. So much was he depressed that his anxious friends began even to dread the safety of his understanding; and Mr Lawson, the engraver, who enjoyed his confidence, succeeded in prevailing upon him, for a time, to lay aside music and poetry, in which he indulged during his solitary walks, and to

* Ord's Life, p. xxvii. 2d edit.

give his attention to drawing. He attempted some landscapes and sketches of the human figure, but turned from his first trials with disgust. At the suggestion of Mr Bartram, he was induced to make a second attempt, upon birds and other objects of natural history, and in this he succeeded beyond his anticipations. This, in some degree, diverted his mind from sadder thoughts, and he went on enthusiastically. His school-house, however, occupied much of his time. "The duties of my profession," he writes, "will not admit me to apply to this study with the assiduity and perseverance I could wish. Chief part of what I do is sketched by candle light; and for this I am obliged to sacrifice the pleasures of social life, and the agreeable moments which I might enjoy in company with you and your amiable friend.*

As Wilson acquired proficiency in this his new art, he began to examine such ornithological works as he could obtain, and conceived the idea of illustrating the ornithology of the *United States.* Over this wish he long pondered, before he could assume confidence to make it known to his friends. He at last resolved to entrust his venerable adviser, Bartram, with his views, who, zealous for everything that would promote his favourite science of nature, entered warmly into the plan, and freely expressed his confidence in the abilities and acquirements of Wilson.† The scheme was now unfolded to Mr Lawson, and met his approbation, though he began to make calculations which did not at all keep pace with the sanguine anticipations of our author, and even caused, for a little while, a sort of coolness between these friends. It was soon, however, forgotten, and Wilson, some time after, writes :—"I never was more wishful to spend an afternoon with you. In three weeks I shall have a few days' vacancy, and mean to be in town chief part of the time. I am most earnestly bent on pursuing my plan of making a collection of all the birds in this part of North America. Now I don't want you to throw cold water, as Shakespeare says, on this notion, Quixotic as it may appear. I have been so long accustomed to the building of airy castles, and brain windmills, that it has become one of my earthly comforts,—a sort of rough bone, that amuses me when sated with the dull drudgery of life." No

* Letter to Bartram.

† Ord's Memoirs, p. xxxix.

plan appears now to have been matured for commencing this undertaking, and Wilson went on with great zeal, improving his talents as a draughtsman, and adding to a rapidly increasing collection of birds. The following letter will best show the zeal with which he pursued this study, and the intimate terms of friendship which existed between him and Mr Bartram, who was now suffering under severe affliction :—

"*To Mr William Bartram.*

"KINGSESSING, *March* 31, 1804.

"I TAKE the first few moments I have had since receiving your letter, to thank you for your obliging attention to my little attempts at drawing, and for the very affectionate expressions of esteem with which you honour me. But sorry I am, indeed, that afflictions so severe as those you mention should fall where so much worth and sensibility reside, while the profligate, the unthinking, and unfeeling, so frequently pass through life strangers to sickness, adversity, or suffering. But God visits those with distress whose enjoyments he wishes to render more exquisite. The storms of affliction do not last for ever ; and sweet is the serene air and warm sunshine after a day of darkness and tempest. Our friend has, indeed, passed away in the bloom of youth and expectation ; but nothing has happened but what almost every day's experience teaches us to expect. How many millions of beautiful flowers have flourished and faded under your eye ! and how often has the whole profusion of blossoms, the hopes of a whole year, been blasted by an untimely frost ! He has gone only a little before us —we must soon follow ; but while the feelings of nature cannot be repressed, it is our duty to bow with humble resignation to the decisions of the Father of all, rather receiving with gratitude the blessings He is pleased to bestow, than repining at the loss of those He thinks proper to take from us. But allow me, my dear friend, to withdraw your thoughts from so melancholy a subject, since the best way to avoid the force of any overpowering passion is to turn its direction in another way.

"That lovely season is now approaching, when the garden, woods, and fields will again display their foliage and flowers. Every day we may expect strangers, flocking from the south to fill our woods with harmony. The pencil of Nature is now at work, and outlines, tints, and gradations of lights and shades, that baffle all description, will soon be spread before us by that great Master, our most benevolent Friend and Father. Let us cheerfully participate in the feast He is preparing for all our senses. Let us survey those millions of green strangers, just peeping into day, as so many happy messengers come to proclaim the power and munificence of the Creator. I confess that I was always an enthusiast in my admiration of the rural scenery of Nature; but, since your example and encouragement have set me to attempt to imitate her productions, I see new beauties in every bird, plant, or flower, I contemplate; and find my ideas of the incomprehensible First Cause still more exalted, the more minutely I examine His works.

"I sometimes smile to think, that while others are immersed in deep schemes of speculation and aggrandisement, in building towns and purchasing plantations, I am entranced in contemplation over the plumage of a lark, or gazing, like a despairing lover, on the lineaments of an owl. While others are hoarding up their bags of money, without the power of enjoying it, I am collecting, without injuring my conscience, or wounding my peace of mind, those beautiful specimens of Nature's works that are for ever pleasing. I have had live crows, hawks, and owls; opossums, squirrels, snakes, lizards, &c., so that my room has sometimes reminded me of Noah's ark; but Noah had a wife in one corner of it, and, in this particular, our parallel does not altogether tally. I receive every subject of natural history that is brought to me; and, though they do not march into my ark from all quarters, as they did into that of our great ancestor, yet I find means, by the distribution of a few fivepenny *bits*, to make them find the way fast enough. A boy, not long ago, brought me a large basketful of crows. I expect his next load will be bull frogs, if I don't soon issue orders to the contrary. One of my boys caught a mouse in school, a few days ago, and directly marched up to me with his prisoner. I set about drawing it that same evening; and all the while the pantings of its little heart showed it to be in the most extreme agonies of fear. I had intended to kill it, in order to fix it in the claws of a stuffed owl; but happening to spill a few drops of water near where it was tied, it lapped it up with such eagerness, and looked in my face with such an eye of supplicating terror, as perfectly overcame me. I immediately untied it, and restored it to life and liberty. The agonies of a prisoner at the stake, while the fire and instruments of torment are preparing, could not be more severe than the sufferings of that poor mouse; and, insignificant as the object was, I felt at that moment the sweet sensations that mercy leaves on the mind when she triumphs over cruelty."

He now found that he could not conscientiously discharge his duty to his pupils, and at the same time follow, as he wished, his pursuit after natural history; and being anxious to make excursions to a greater length than he had hitherto done, he sought after some literary employment. He applied to Mr Brown, the conductor of the *Literary Magazine*, and wrote for his work the "Rural Walk," and "Solitary Tutor." The former poem I have been unable to procure; the latter is a sort of epitome of his own history, where he describes the early intentions of his parents of bringing him up for the Church—his emigration—his school-house on the Schuylkill—and his favourite haunts in Bartram's woods.

In October 1804, Wilson, accompanied with two friends, set out on foot to visit the Falls of Niagara. The party were too late in starting, and, on this account, suffered many hardships on their return. These did not, however, discourage him; he soon after writes to Mr Bartram—

"GRAY'S FERRY, 15*th December* 1804.

"DEAR SIR,—Though now snugly at home, looking back in recollection on the long circuitous journey, which I have at length finished, through deep snows, and uninhabited forests—over stupendous mountains, and down dangerous rivers—passing over, in a course of 1300 miles, as great a variety of men and modes of living, as the same extent of country can exhibit in any part of the United States; though in this tour I have had every disadvantage of deep roads and rough weather—hurried marches, and many other inconveniences to encounter,—yet so far am I from being satisfied with what I have seen, or discouraged by the fatigues which every traveller must submit to, that I feel more eager than ever to commence some more extensive expedition, where scenes and subjects, entirely new and generally unknown, might reward my curiosity; and where, perhaps, my humble acquisitions might add something to the stores of knowledge. For all the hazards and privations incident to such an undertaking, I feel confident in my own spirit and resolution. With no family to enchain my affections—no ties but those of friendship—and the most ardent love to my adopted country—with a constitution which hardens amidst fatigues—and with a disposition sociable and open, which can find itself at home by an Indian fire in the depth of the woods, as well as in the best apartment of the civilised,—I have at present a real design of becoming a traveller. But I am miserably deficient in many acquirements absolutely necessary for such a character. Botany, mineralogy, and drawing, I most ardently wish to be instructed in. Can I yet make any progress in botany, sufficient to enable me to be useful? and what would be the most proper way to proceed? I have many leisure moments that should be devoted to this pursuit, provided I could have hopes of succeeding. Your opinion on this subject will confer an additional obligation on your affectionate friend."

This, his first journey after the acquirement of any knowledge of natural history, is described in the poem of the "Foresters," afterwards published in the "Portfolio;" and his visit to Niagara furnished the materials for his beautiful description and poem of the bald eagle and fish hawk. These expeditions destroyed the success of his school, which he yet retained; and the additional circumstance of the winter of 1805 being a very unfavourable one, from the severe and continued frosts, nearly exhausted the little emoluments produced by his labours. Writing to Mr W. Duncan, he says,—"This winter has been entirely lost to me, as well as to yourself. I shall, on the 12th of next month, be scarcely able to collect a sufficiency to pay my board, having not more than twenty-seven scholars. Five or six families who used to send me their children, have been almost in a state of starvation. The rivers Schuylkill and Delaware are still shut; and waggons are passing and repassing at this time on the ice." Wilson still remained at Union School; and, by perseverance, was enabled to maintain himself honestly. He could not, however, give up his design of illustrating the birds of the United States, though prudence, and the calculations of Mr Lawson, still forbade the scheme. On the 2d July of this year, he again writes to Mr Bartram: "I dare say you will smile at my presumption, when I tell you that I have seriously begun to make a collection of drawings of the birds to be found in Pennyslvania, or that occasionally pass through it. Twenty-eight, as a beginning, I send for your opinion. They are, I hope, inferior to what I shall produce, though as close copies of the original as I could make. One or two of these I cannot find either in your Nomenclature, or in the seven volumes of Edwards. Any hint for promoting my plan, or enabling me to execute better, I will receive from you with much pleasure. Criticise these, my dear friend, without fear of offending me. This will instruct, but not discourage me. To your advice and encouraging encomiums I am indebted for these few specimens, and for all that will follow. *They may yet tell posterity that I was honoured with your friendship, and that to your inspiration they owe their existence.*" In his examination of Edwards, Wilson perceived that that naturalist had etched his own plates. It appeared to him, that what one man had done, another might do; and he determined at least to try. Mr Lawson, of course, was applied to, and cheerfully supplied him with the copper and necessary tools. The sequel of this story is thus detailed by his American biographer:—The day after Wilson had parted from his preceptor, he was surprised to see him bouncing into his room, crying out, "*I have finished my plate! let us bite it in with aqua fortis at once, for I must have a proof before I leave town.*" Lawson burst into laughter at the ludicrous appearance of his friend, animated with impetuous zeal; and, to humour him, granted his request. A proof was taken, but fell far short of Wilson's expectations, or his ideas of correctness. However, he lost no time in conferring with Mr Bartram, to whom he wrote as follows:

"I have been amusing myself this some time in attempting to etch, and now send you a proof sheet of my first performance in this way. Be so kind as communicate to me your own corrections, and those of your young friend and pupil. I will receive them as a very kind and particular favour. My next attempt will perhaps be better, everything being new to me in this. I will send the first impression I receive after I finish the plate."

With the proof of his second plate he again writes—"Mr Wilson's affectionate compliments to Mr Bartram, and sends for his amusement and correction another proof of his *Birds of the United States.* The colouring being chiefly done last night, must soften criticism a little: will be thankful for my friend's advice and correction.

"Mr Wilson wishes his beloved friend a happy New Year, and every blessing."

This is written at the commencement of 1806, nearly two years after he first conceived the idea of such a work. These attempts seem to have convinced him that he could not himself attain sufficient proficiency to produce the effects he wished, and he proposed to Mr Lawson to embark in the work as a joint concern. These proposals Mr Lawson, from prudential motives, declined, and Wilson, with an enthusiasm similar to what had ever actuated him in like circumstances, declared that he would proceed alone in the publication, should it cost him his life. "*I shall at least leave a small beacon to point out where I perished.*"

A circumstance at this time occurred which prevented him from starting immediately with his design, and the delay was perhaps the indirect means of more completely forwarding his views; for it is more than probable, that a commencement of an undertaking of such extent, on his own narrow means alone, would have been crushed ere its merits could have spread. Mr Jefferson, the President of the United States, had it in contemplation to despatch an expedition to explore the country of the Mississippi, and Wilson, anxious to see these new regions, and to procure additional materials for his work, wished to be chosen as a naturalist to the party. He applied to Mr Bartram, who cheerfully wrote to the President, recommending his friend, and Wilson forwarded the communication with the following letter from himself:—

"To His Excellency THOMAS JEFFERSON, President of the United States.

"SIR,—Having been engaged, these several years, in collecting materials and furnishing drawings from nature, with the design of publishing a new Ornithology of the United States of America, so deficient in the works of Catesby, Edwards, and other Europeans, I have traversed the greater part of our northern and eastern districts, and have collected many birds undescribed by these naturalists.

Upwards of one hundred drawings are completed; and two plates in folio already engraved. But as many beautiful tribes frequent the Ohio, and the extensive country through which it passes, that probably never visit the Atlantic States; and as faithful representations of these can only be taken from living nature, or from birds newly killed,—I had planned an expedition down that river, from Pittsburg to the Mississippi, thence to New Orleans, and to continue my researches by land in return to Philadelphia. I had engaged, as a companion and assistant, Mr W. Bartram of this place, whose knowledge of botany, as well as zoology, would have enabled me to make the best of the voyage, and to collect many new specimens in both those departments. Sketches of these were to have been taken on the spot; and the subjects put in a state of preservation, to finish our drawings from, as time would permit. We intended to set out from Pittsburg about the beginning of May; and expected to reach New Orleans in September.

"But my venerable friend, Mr Bartram, taking into more serious consideration his advanced age, being near seventy, and the weakness of his eyesight, and apprehensive of his inability to encounter the fatigues and privations unavoidable in so extensive a tour; and having, to my extreme regret, and the real loss of science, been induced to decline the journey, I had reluctantly abandoned the enterprise, and all hopes of accomplishing my purpose; till, hearing that your Excellency had it in contemplation to send travellers this ensuing summer up the Red River, the Arkansaw, and other tributary streams of the Mississippi, and believing that my services might be of advantage to some of these parties, in promoting your Excellency's design, while the best opportunities would be afforded me of procuring subjects for the work which I have so much at heart,—under these impressions, I beg leave to offer myself for any of those expeditions; and can be ready at a short notice to attend your Excellency's orders.

"Accustomed to the hardships of travelling—without a family—and an enthusiast in the pursuit of natural history, I will devote my whole powers to merit your Excellency's approbation; and ardently wish for an opportunity of testifying the sincerity of my professions, and the deep veneration with which I have the honour to be, sir, your obedient servant, ALEX. WILSON.

"KINGSESS, *6th Feb.* 1806."

Wilson had been previously introduced to Jefferson, and entertained extravagant ideas of his talents and virtues. Writing to Bartram, he says—"My Dear Friend,—This day the heart of every republican, of every good man, within the immense limits of our happy country, will leap for joy.

"The reappointment and continuance of our beloved Jefferson to superintend our national concerns, is one of those distinguished blessings whose beneficent effects extend to posterity, and whose value our hearts may feel, but can never express.

"I congratulate you, my dear friend, on this happy event. The enlightened philosopher—the distinguished naturalist—the first statesman on earth—the friend, the ornament of science,—is the father of our country, the faithful guardian of our liberties.

"I am at present engaged in drawing the two birds which I brought from the Mohawk; and if I can finish them to your approbation, I intend to transmit them to our excellent President, as the child of an amiable parent presents to its affectionate father some little token of its esteem."

He did transmit these drawings to the President, whose answer in return will shew in what estimation he held the abilities of Wilson as a draughtsman and ornithologist:—

"MONTICELLO, *April* 7, 1805.

"SIR,—I received here yesterday your favour of March 18, with the elegant drawings of the new birds you found on your tour to Niagara, for which I pray you to accept my thanks. The jay is quite unknown to me. From my observations while in Europe, on the birds and quadrupeds of that quarter, I am of opinion that there is not in our continent a single bird or quadruped which is not sufficiently unlike all the members of its family there to be considered specifically different. On this general observation, I conclude with confidence that your jay is not a European bird.

"The first bird on the same sheet I judge to be a *Muscicapa*, from its bill, as well as from the following circumstance: Two or three days before my arrival here, a neighbour killed a bird, unknown to him, and never before seen here, as far as he could learn. It was brought to me soon after I arrived, but in the dusk of the evening, and so putrid, that it could not be approached but with disgust.

But I retain a sufficiently exact idea of its form and colours to be satisfied it is the same with yours. The only difference I find in yours is, that the white on the back is not so pure, and that the one I saw had a little of a crest. Your figure, compared with the white-bellied *gobe-mouche*, Buff. viii. 342, Pl. enl. 566, shews a near relation. Buffon's is dark on the back.

"As you are curious in birds, there is one well worthy your attention, to be found, or rather heard, in every part of America, and yet scarcely ever to be seen. It is in all the forests from spring to fall, and never but on the tops of the tallest trees, from which it perpetually serenades us with some of the sweetest notes, and as clear as those of the nightingale. I have followed it for miles, without ever, but once, getting a good view of it. It is of the size and make of the mocking bird, lightly thrush-coloured on the back, and a greyish white on the breast and belly. Mr Randolph, my son-in-law, was in possession of one which had been shot by a neighbour; he pronounces this also a *Muscicapa*, and I think it much resembling the *Moucherolle de la Martinique*, Buff. viii. 374, Pl. enl. 568. As it abounds in all the neighbourhood of Philadelphia, you may, perhaps, by patience and perseverance (of which much will be requisite), get a sight, if not possession, of it. I have for twenty years interested the young sportsmen of my neighbourhood to shoot me one; but, as yet, without success.—Accept my salutations and assurances of respect.

"TH. JEFFERSON."

After the encomiums bestowed upon him in this letter, and the apparent wish to forward his views, we are totally at a loss to conjecture the probable reason why no attention was paid to the application: neither Mr Bartram nor Wilson ever ascertained the cause. The latter again resumed his occupations, and was soon so fortunate as to obtain a situation of such importance and emolument as made him independent of his school, and proved the means of enabling him to commence his great work. Mr Samuel F. Bradford, bookseller in Philadelphia, being about to publish an improved edition of Rees' New Cyclopædia, Wilson was introduced to him as one qualified to superintend the work, and was engaged at a liberal salary as assistant editor. The agreement is dated 20th

April 1806, and, two days after, he writes to Mr Bartram, detailing his plans, and expressing diffidence in his ability for the superintendence of such varied subjects. "This engagement will, I hope, enable me, in more ways than one, to proceed with my intended Ornithology, to which all my leisure moments will be devoted. In the meantime, I anticipate with diffidence the laborious and very responsible situation I am soon to be placed in, requiring a much more general fund of scientific knowledge, and stronger powers of mind, than I am possessed of; but all these objections have been overruled, and I am engaged." He soon unfolded his darling project to Mr Bradford, who thought so favourably of the undertaking, and of Wilson's abilities, that he agreed to become the publisher, and to furnish the requisite funds. Now that all obstruction was removed, Wilson launched into the enterprise with his whole energies, devoting to it every moment that he could spare from his duties as editor. Mr Lawson was set to work; his prospectus was composed, and twenty-five hundred copies were to be thrown off; it was to be printed in all the newspapers; he already meditated the appointment of an agent "in every town of the Union." This hard study, however, again impaired his health, and he was forced sometimes to relax by an excursion to the country, which gave him a double portion of enjoyment, from being mostly in constant confinement, "immersed among musty books, and compelled to forego the harmony of the woods for the everlasting din of the city, the very face of the blessed heavens involved in soot, and interrupted by walls and chimney tops." *

"At length," writes Mr Ord, "in the month of September 1808, the first volume of the 'American Ornithology' made its appearance. From the date of the arrangement with the publisher, a prospectus had been issued, wherein the nature and intended execution of the work were specified. But yet no one appeared to entertain an adequate idea of the treat which was about to be afforded to the lovers of the fine arts and of elegant literature; and when the superb volume was presented to the public, their delight was equalled only by their astonishment that America, as yet in its infancy, should produce an original work in science, which could

* Letter to Mr Bartram.

vie in its essentials with the proudest productions of a similar nature of the European world."

The first hurry and excitement being over, and the work meeting with an approbation far beyond his hopes, he thought he might be able to increase the subscriptions at a distance, by his presence and personal exertions; he decided upon an expedition for this purpose, and set out in search of "birds and subscribers," through the eastern states, by Boston to Maine, and back through the state of Vermont. He set out on the 21st September 1808, and the incidents of this journey are detailed in letters addressed to Mr Miller and Mr Lawson. They bear the same evidence of his talent for observation, which marked his early journal in the east of Scotland: almost every incident worthy of notice is recorded; and whatever would promote the main object of his tour is followed with the warmest enthusiasm. They can be detailed in no language except his own; and though of considerable length, they are the best and only means by which a proper estimate of his character and disposition, at this time of his life, can be obtained; and besides, they form the commencement of letters descriptive of the various excursions he undertook during the remaining progress of his work:—

"Boston, *October* 12, 1808.

"Dear Sir,—I arrived here on Sunday last, after various adventures, the particulars of which, and the observations I have had leisure to make upon the passing scenery around me, I shall endeavour, as far as possible, to compress into this letter, for your own satisfaction, and that of my friends, who may be interested for my welfare. At Princeton I bade my fellow-travellers good-by, as I had to wait upon the reverend doctors of the college. I took my book under my arm, put several copies of the prospectus into my pocket, and walked up to this spacious sanctuary of literature. Dr Smith, the president, and Dr M'Lean, professor of natural history, were the only two I found at home. The latter invited me to tea, and both were much pleased and surprised with the appearance of the work. I expected to receive some valuable information from Mr M'Lean, on the ornithology of the country; but I soon found, to my astonishment, that he scarcely knew a sparrow from a woodpecker. I visited several other literary characters; and, at about half-past eight, the *Pilot* coming up, I took my passage in it to New Brunswick, which we reached about midnight.

"The next morning was spent in visiting the few gentlemen who were likely to patronise my undertaking. I had another task of the same kind at Elizabeth Town; and, without tiring you with details that would fill a volume, I shall only say that I reached Newark that day, having gratified the curiosity and feasted the eyes of a great number of people, who repaid me with the most extravagant compliments, which I would have very willingly exchanged for a few simple *subscriptions.* I spent nearly the whole of Saturday in Newark, where my book attracted as many starers as a bear or a mammoth would have done, and I arrived in New York the same evening. The next day I wrote a number of letters, enclosing copies of the prospectus, to different gentlemen in town, and, in the afternoon of Tuesday, I took my book and waited on each of those gentlemen to whom I had written the preceding day. Among these I found some friends, but more admirers. The professors of Columbia College expressed much esteem for my performance. The professor of languages being a Scotchman, and also a Wilson, seemed to feel all the pride of national partiality so common to his countrymen, and would have done me every favour in his power. I spent the whole of this week traversing the streets, from one particular house to another, till, I believe, I became almost as well known as the public crier or the clerk of the market; for I frequently could perceive gentlemen point me out to others as I passed with my book under my arm.

"On Sunday morning, October 2d, I went on board a packet for Newhaven, distant about ninety miles. The wind was favourable. The Sound here, between Longisland and the Maine, is narrowed to less than half a mile, and filled with small islands, and enormous rocks under water, among which the tide roars and boils violently, and has proved fatal to many a seaman. At high water it is nearly as smooth as any other flow, and can then be safely passed. The country, on the New York side, is ornamented with handsome villas, painted white, and surrounded by great numbers of Lombardy poplars. The breeze increasing to a gale, in eight hours from the time we set sail, the high red-fronted mountain of Newhaven rose to our view. In two hours more we landed, and, by the stillness and solemnity of the streets, recollected we were in New England, and that it was Sunday, which latter circumstance had been almost forgotten on board the packet-boat.

"This town is situated upon a sandy plain, and the streets are shaded with elm trees and poplars. In a large park, or common, covered with grass, and crossed by two streets and several footpaths, stands the church, the state house, and college buildings, which last are one hundred and eighty yards in front. From these structures rise four or five wooden spires, which, in former times, as one of the professors informed me, were so infested by woodpeckers, which bored them in all directions, that to preserve their steeples from destruction, it became necessary to set people with guns to watch and shoot these invaders of the sanctuary. About the town the pasture-fields and corn look well; but a few miles off, the country is poor and ill cultivated.

"The literati of Newhaven received me with politeness and respect; and, after making my usual rounds, which occupied a day and a half, I set off for Middleton, twenty-two miles distant. The country through which I passed was generally flat and sandy. In some places, whole fields were entirely covered with sand, not a blade of vegetation to be seen, like some parts of New Jersey. Round Middleton, however, the country is really beautiful; the soil rich; and here I first saw the river of Connecticut, stretching along the east side of the town, which consists of one very broad street, with rows of elms on each side. On entering, I found the streets filled with troops, it being muster-day. The sides of the street were choked up with waggons, carts, and wheelbarrows, filled with bread, roast beef, fowls, cheese, liquors, barrels of cider, and rum bottles. Some were singing out, 'Here's the best brandy you ever put into your head!' Others in dozens shouting, 'Here's the round and sound gingerbread! most capital gingerbread!' In one place I observed a row of twenty or thirty country girls drawn up, with their backs to a fence, and two young fellows supplying them with rolls of bread from a neighbouring stall, which they ate with a hearty appetite, keeping nearly as good time with their grinders as the militia did with their muskets. In another place, the crowd had formed a ring, within which they danced to the catgut scrapings of an old negro. The spectators looked on with as much gravity as if they were listening to a sermon, and the dancers laboured with such seriousness, that it seemed more like a penance imposed on the poor devils for past sins, than mere amusement.

"I waited on a Mr A—— of this town, and by him was introduced to several others. He also furnished me with a good deal of information respecting the birds of New England. He is a great sportsman, a man of fortune and education, and has a considerable number of stuffed birds, some of which he gave me, besides letters to several gentlemen of influence in Boston. I endeavoured to recompense him in the best manner I could, and again pursued my route to the north-east. The country between this and Hartford is extremely beautiful, much resembling that between Philadelphia and Frankfort. The road is a hard sandy soil; and, in one place, I had an immense prospect of the surrounding country, nearly equal to that which we saw returning from Easton, but less covered with woods. On reaching Hartford, I waited on Mr G——, a member of Congress, who recommended me to several others, particularly a Mr W——, a gentleman of taste and fortune, who was extremely obliging. The publisher of a newspaper here expressed the highest admiration of the work, and has since paid many handsome compliments to it in his publication, as three other editors did in New York. This is a species of currency that will neither purchase plates, nor pay the printer; but, nevertheless, it is gratifying to the vanity of an author, *when nothing* better can be got. My journey from Hartford to Boston, through Springfield, Worcester, &c., one hundred and twenty-eight miles, it is impossible for me to detail at this time. From the time I entered Massachussets, until within ten miles of Boston, which distance is nearly two-thirds the length of the whole state, I took notice that the principal features of the country were stony mountains, rocky pasture fields, and hills and swamps, adorned with pines. The fences, in every direction, are composed of strong stones; and, unless a few straggling, self-planted, stunted apple-trees, overgrown with moss, deserve the name, there is hardly an orchard to be seen in ten miles. Every six or eight miles you come to a meeting-house, painted white, with a spire. I could perceive little difference in the form or elevation of their steeples.

"The people here make no distinction between *town* and *town-*

ship; and travellers frequently asked the driver of the stage-coach, 'What town are we now in?' when, perhaps, we were upon the top of a miserable barren mountain, several miles from a house. It is in vain to reason with the people on the impropriety of this—custom makes every absurdity proper. There is scarcely any currency in this country but paper, and I solemnly declare, that I do not recollect having seen one hard dollar since I left New York. Bills even of twenty-five cents, of a hundred different banks, whose very names one has never heard of before, are continually in circulation, I say nothing of the jargon which prevails in the country. Their book schools, if I may judge by the state of their schoolhouses, are no better than our own.

"Lawyers swarm in every town like locusts; almost every door has the word *Office* painted over it, which, like the web of a spider, points out the place where the spoiler lurks for his prey. There is little or no improvement in agriculture; in fifty miles I did not observe a single grain or stubble field, though the country has been cleared and settled these one hundred and fifty years. In short, the *steady habits* of a great portion of the inhabitants of those parts of New England through which I passed, seem to be laziness, low bickerings, and———. A man here is as much ashamed of being seen walking the streets on a Sunday, unless in going and returning from church, as many would be of being seen going to a —— house.

"As you approach Boston, the country improves in its appearance; the stone fences give place to those of posts and rails; the road becomes wide and spacious; and everything announces a better degree of refinement and civilisation. It was dark when I entered Boston, of which I shall give you some account in my next. I have visited the celebrated Bunker's Hill, and no devout pilgrim ever approached the sacred tomb of his holy prophet with more awful enthusiasm, and profound veneration, than I felt in tracing the grassgrown entrenchments of this hallowed spot, made immortal by the bravery of those heroes who defended it—whose ashes are now mingled with its soil—and of whom a mean, beggarly *pillar of bricks* is all the *memento.*"

His next letter, to the same gentleman, is dated "Windsor, Vermont, October 26." He remained nearly a week at Boston, journeying through the streets with his book, and visiting all the literary characters he could meet with. He continues :—"The streets of Boston are a perfect labyrinth. The markets are dirty : the fish-market is so filthy, that I will not disgust you by a description of it. Wherever you walk, you hear the most hideous howling, as if some miserable wretch were expiring on the wheel at every corner ; this, however, is nothing but the draymen shouting to their horses. Their drays are twenty-eight feet long, drawn by two horses, and carry ten barrels of flour. From Boston I set out for Salem ; the country between, swampy, and, in some places, the most barren, rocky, and desolate in nature. Salem is a neat little town. The waters were crowded with vessels. One wharf here is twenty hundred and twenty-two feet long. I stayed here two days, and again set off for Newbury Port, through a rocky, uncultivated, sterile country.

"I travelled on through New Hampshire, stopping at every place where I was likely to do any business ; and went as far east as Portland, in Maine, where I stayed three days ; and, the supreme court being then sitting, I had an opportunity of seeing and conversing with people from the remotest boundaries of the United States in this quarter, and received much interesting information from them with regard to the birds that frequent these northern regions. From Portland, I directed my course across the country, among dreary, savage glens, and mountains covered with pines and hemlocks, amid whose black and half-burnt trunks the everlasting rocks and stones, that cover this country, 'grinned horribly.' One hundred and fifty-seven miles brought me to Dartmouth College, New Hampshire, on the Vermont line. Here I paid my addresses to the reverend fathers of literature, and met with a kind and obliging reception. Dr Wheelock, the President, made me eat at his table, and the professors vied with each other to oblige me.

"I expect to be in Albany in five days ; and, if the legislature be sitting, I shall be detained perhaps three days there. In eight days more, I hope to be in Philadelphia. I have laboured with the zeal of a knight-errant, in exhibiting this book of mine, wherever I went, travelling with it, like a beggar with his bantling, from town to town, and from one country to another. I have been loaded with praises, with compliments, and kindnesses ; shaken almost to pieces in stage-coaches; I have wandered among strangers, hearing the same Oh's and Ah's, and telling the same story, a thousand times over, and for what? Ay, that's it! You are very anxious to know, and you shall know the whole when I reach Philadelphia."

To Mr Alexander Lawson.

"ALBANY, *November* 3, 1808.

"DEAR SIR,—Having a few leisure moments at disposal, I will devote them to your service, in giving you a sketch of some circumstances in my long literary pilgrimage, not mentioned in my letters to Mr Miller. And, in the first place, I ought to thank you for the thousands of compliments I have received for my birds, from persons of all descriptions, which were chiefly due to the taste and skill of the engraver. In short, the book, in all its parts, so far exceeds the ideas and expectations of the first literary characters in the eastern section of the United States, as to command their admiration and respect. The only objection has been the price of *one hundred and twenty dollars*, which, in innumerable instances, has risen like an evil genius between me and my hopes. Yet I doubt not but when those copies subscribed for are delivered, and the book a little better known, the whole number will be disposed of; and, perhaps, encouragement given to go on with the rest. To effect this, to me, most desirable object, I have encountered the fatigues of a long, circuitous, and expensive journey, with a zeal which has increased with increasing difficulties; and sorry I am to say, that the whole number of subscribers which I have obtained amounts only to *forty-one*.

"While in New York, I had the curiosity to call on the celebrated author of the 'Rights of Man.' He lives in Greenwich, a short way from the city. In the only decent apartment of a small indifferent looking frame house, I found this extraordinary man, sitting wrapt in a night gown, the table before him covered with newspapers, with pen and ink beside him. Paine's face would have excellently suited the character of Bardolph; but the penetration and intelligence of his eye bespeak the man of genius and of the world. He complained to me of his inability to walk, an exercise he was formerly fond of; he examined my book, leaf by leaf, with great attention; desired me to put down his name as a subscriber;

and, after inquiring particularly for Mr P—— and Mr B——, wished to be remembered to both.

"My journey through almost the whole of New England has rather lowered the Yankees in my esteem. Except a few neat academies, I found their school-houses equally ruinous and deserted with ours; fields covered with stones; stone fences; scrubby oaks, and pine trees; wretched orchards; scarcely one grain field in twenty miles; the taverns along the road, dirty, and filled with loungers, brawling about law-suits and politics; the people snappish and extortioners, lazy, and two hundred years behind the Pennsylvanians in agricultural improvements. I traversed the country bordering the river Connecticut for nearly two hundred miles. Mountains rose on either side, sometimes three, six, or eight miles apart, the space between almost altogether alluvial; the plains fertile, but not half cultivated. From some projecting headlands I had immense prospects of the surrounding countries, everywhere clothed in pine, hemlock, and scrubby oak.

"It was late in the evening when I entered Boston, and, whirling through the narrow lighted streets, or rather lanes, I could form but a very imperfect idea of the town. Early the next morning, resolved to see where I was, I sought out the way to Beacon Hill, the highest part of the town, and whence you look down on the roofs of the houses—the bay, interspersed with islands—the ocean—the surrounding country, and distant mountains of New Hampshire; but the most singular objects are the long wooden bridges, of which there are five or six, some of them three quarters of a mile long, uniting the towns of Boston and Charleston with each other, and with the main land. I looked round, with an eager eye, for that eminence, so justly celebrated in the history of the revolution of the United States,—*Bunker's Hill;* but I could see nothing that I could think deserving of the name, till a gentleman, who stood by, pointed out a white monument upon a height beyond Charleston, which he said was the place. I explored my way thither without paying much attention to other passing objects; and, in tracing the streets of Charleston, was astonished and hurt at the indifference with which the inhabitants directed me to the place. I inquired if there were any person still living here who had been in the battle, and I was directed to a Mr Miller, who was a lieutenant in this memorable affair. He is a man about sixty, stout, remarkably fresh coloured, with a benign and manly countenance. I introduced myself without any ceremony—shook his hand with sincere cordiality—and said, with some warmth, that I was proud of the honour of meeting with one of the heroes of Bunker's Hill—the first unconquerable champions of their country. He looked at me, pressed my hand in his, and the tears instantly glistened in his eyes, which as instantly called up corresponding ones in my own. In our way to the place, he called on a Mr Carter, who, he said, was also in the action, and might recollect some circumstances which he had forgotten. With these two veterans I spent three hours, the most interesting to me of any of my life. As they pointed out to me the rout of the British—the American entrenchments—the place where the greatest slaughter was made —the spot where Warren fell, and where he was thrown amid heaps of the dead,—I felt as though I could have encountered a whole battalion myself in the same glorious cause. The old soldiers were highly delighted with my enthusiasm: we drank a glass of wine to the memory of the illustrious dead, and parted almost with regret.

"From Boston to Portland, in the district of Maine, you are almost always in the neighbourhood, or within sight, of the Atlantic. The country may be called a mere skeleton of rocks and fields of sand; in many places entirely destitute of wood, except a few low scrubby junipers; in others, covered with pines of a diminutive growth. On entering the tavern of Portland, I took up the newspaper of the day, in which I found my song of *Freedom and Peace*, which I afterwards heard read before a numerous company (for the Supreme Court was sitting), with great emphasis, as a most excellent song, but I said nothing on the subject.

"From Portland, I steered across the country for the northern parts of Vermont, among barren, savage, pine-covered mountains, through regions where nature and art had done infinitely less to make it a fit residence for men, than any country I ever traversed. Among these dreary tracts I found winter had already commenced, and the snow several inches deep. I called at Dartmouth College, the president of which, as well as all I visited in New England, subscribed. Though sick with a severe cold and great fatigue, I continued my route to this place, passing and calling at great numbers of small towns in the way.

"The legislature is at present in session. The newspapers have to-day taken notice of my book and inserted my advertisement. I shall call on the principal people—employ an agent among some of the booksellers in Albany—and return home by New York."

Wilson remained at home only for a few days, and though now winter, set out on another tour to the southward, visiting every town of importance as far as Savannah, in the state of Georgia, during which excursion he suffered considerably from the inclemency of the season,—the fatigue completely knocking up his horse. He was, however, gratified by it; and, in addition to a few subscribers, procured several friends, and some information useful to the future volumes. He had also an opportunity of renewing his acquaintance with the president at Washington, and the former misunderstanding regarding the expedition to the Mississippi, seems to have been mutually forgotten. He says—"The President received me very kindly. I asked for nobody to introduce me, but merely sent him a line that I was there, when he ordered me to be immediately admitted. He has given me a letter to a gentleman in Virginia, who is to introduce me to a person there, who, Mr Jefferson says, has spent his whole life in studying the manners of our birds, and from whom I am to receive a world of facts and observations."

He did not return till March 1809, having been absent above three months, and the fatigue and expense of travelling obliged him to return by sea. Immediately before going on board, he thus writes Mr Bartram from Savannah:—

"Three months, my dear friend, are passed since I started from you at Kingsess. This is the most arduous, expensive, and fatiguing expedition I ever undertook. I have, however, gained my point, in procuring two hundred and fifty subscribers, in all, for my Ornithology, and a great mass of information respecting the birds that winter in the southern states, and some that never visit the middle states; and this information I have derived personally, and can therefore the more certainly depend upon it.

"On the commons, near Charleston, I presided at a singular feast. The company consisted of two hundred and thirty-seven carrion crows (*Vultur atratus*), five or six dogs, and myself, though I only kept order, and left the eating part entirely to the others. I sat so near to the dead horse that my feet touched his, and yet, at one time, I counted thirty-eight vultures on and within him, so that hardly an inch of his flesh could be seen for them.

"As far north as Wilmington, in North Carolina, I met with the ivory-billed woodpecker. I killed two, and winged a male, who alarmed the whole town of Wilmington, by screaming exactly like a young child crying violently, so that everybody supposed I had a baby under the apron of my chair, till I took it out to prevent the people from stopping me. This bird I confined in the room I was to sleep in, and in less than half an hour he made his way through the plaster, the lath, and partly through the weather boards, and would have escaped if I had not accidentally come in."

This journey is more fully described in another letter to Mr Miller, the principal parts of which I shall insert, as too important to be omitted, and carrying us regularly on in the line of his various excursions.

To Mr D. H. Miller.

"CHARLESTON, *February* 22, 1809.

"DEAR SIR,—I have passed through a considerable extent of country since I wrote you last, and have met with a variety of adventures, some of which may perhaps amuse you. Norfolk turned out better than I expected: I left that place on one of the coldest mornings I have experienced since I left Philadelphia.

"I passed through a flat pine-covered country from Norfolk to Suffolk, twenty-four miles distant, and lodged, on the way, in the house of a planter, who informed me that every year, in August and September, almost all his family were laid up with the bilious fever; that at one time forty of his people were sick, and that of thirteen children, only three were living. Two of these, with their mother, appeared likely not to be long tenants of this world. Thirty miles farther I came to a place on the river Nottoway, called

Jerusalem. Here I found the river swelled to such an extraordinary height, that the oldest inhabitant had never seen the like.

"After passing along the bridge, I was conveyed in a boat, termed a *flat,* a mile and three quarters through the wood, where the torrent, sweeping along in many places, rendered this sort of navigation rather disagreeable. I proceeded on my journey, passing through solitary pine woods, perpetually interrupted by swamps, that covered the road with water two and three feet deep, frequently half a mile at a time, looking like a long river or pond. These, in the afternoon, were surmountable; but the weather being exceedingly severe, they were covered every morning with a sheet of ice, from half an inch to an inch thick, that cut my horse's legs and breast. After passing a bridge, I had many times to wade, and twice to swim my horse, to get to the shore. I attempted to cross the Roanoke at three different ferries, thirty-five miles apart, and at last succeeded at a place about fifteen miles below Halifax. A violent snow storm made the roads still more execrable.

"The productions of these parts of North Carolina are hogs, turpentine, tar, and apple brandy. A tumbler of toddy is usually the morning's beverage of the inhabitants, as soon as they get out of bed. So universal is the practice, that the first thing you find them engaged in, after rising, is preparing the brandy toddy. You can scarcely meet a man whose lips are not parched and chopped, or blistered with drinking this poison. Those who do not drink it, they say, are sure of the ague; I, however, escaped. The pine woods have a singular appearance, every tree being stripped, on one or more sides, of the bark, for six or seven feet up. The turpentine covers these parts in thick masses. I saw the people, in different parts of the woods, mounted on benches, chopping down the sides of the trees, leaving a trough or box for the turpentine to run into. Of hogs they have immense multitudes; one person will sometimes own five hundred. The leaders have bells round their necks; and every drove knows its particular call, whether it be a conch shell, or the bawling of a negro, though half a mile off. Their owners will sometimes drive them, for four or five days, to a market, without once feeding them.

"The taverns are the most desolate and beggarly imaginable,—bare, bleak and dirty walls: one or two old broken chairs, and a bench, form all the furniture. The white females seldom make their appearance, and everything must be transacted through the medium of negroes. At supper, you sit down to a meal, the very sight of which is sufficient to deaden the most eager appetite, and you are surrounded by half a dozen dirty, half naked blacks, male and female, whom any man of common scent might smell a quarter of a mile off. The house itself is raised upon props, four or five feet, and the space below is left open for the hogs, with whose charming vocal performance the wearied traveller is serenaded the whole night long, till he is forced to curse the hogs, the house, and everything about it.

"I crossed the river Taw at Washington, for Newbern, which stands upon a sandy plain, between the rivers Trent and Neuse, both of which abound with alligators. Here I found the shad fishery begun, on the 5th inst., and wished to have some of you with me to assist in dissecting some of the finest I ever saw. Thence to Wilmington was my next stage, one hundred miles, with only one house for the accommodation of travellers on the road,—two landlords having been broken up with the fever.

"The general features of North Carolina, where I crossed it, are immense, solitary pine savannas, through which the road winds among stagnant ponds, swarming with alligators; dark, sluggish creeks, of the colour of brandy, over which are thrown high wooden bridges, without railings, and so crazy and rotten, as not only to alarm one's horse, but also the rider, and to make it a matter of thanksgiving with both, when they get fairly over, without going through; enormous cypress swamps, which, to a stranger, have a striking, desolate, and ruinous appearance. Picture to yourself a forest of prodigious trees, rising as thick as they can grow, from a vast flat and impenetrable morass, covered for ten feet from the ground with reeds. The leafless limbs of the cypresses are clothed with an extraordinary kind of moss (*Tillandsia usneoides*), from two to ten feet long, in such quantities, that fifty men might conceal themselves in one tree. Nothing in this country struck me with such surprise as the prospect of several thousand acres of such timber, loaded, as it were, with many million tons of tow, waving in the wind. I attempted to penetrate several of these swamps with my gun, in search of something new; but, except in some chance places, I found it altogether impracticable. I coasted along their borders, however, in many places, and was surprised at the great profusion of evergreens, of numberless sorts, and a variety of berries that I knew nothing of. Here I found multitudes of birds, that never winter with us in Pennsylvania, living in abundance. Though the people told me the alligators were so numerous as to destroy many of their pigs, calves, dogs, &c., yet I have never been enabled to put my eye on one, though I have been several times in search of them with my gun. In Georgia, they tell me, they are ten times more numerous, and I expect some sport among them. I saw a dog at the river Santee, who swims across when he pleases, in defiance of these voracious animals. When he hears them behind him, he wheels round and attacks them, often seizing them by the snout. They generally retreat, and he pursues his route, seizing any one that attacks him in the same manner. He belongs to the boatmen, and when left behind always takes to the water.

"As to the character of the North Carolinians, were I to judge of it by the specimens which I met with in taverns, I should pronounce them to be the most ignorant, debased, indolent, and dissipated portion of the Union. But I became acquainted with a few such noble exceptions that for their sakes I am willing to believe they are all better than they seem to be.

"Wilmington contains about 3000 souls, and yet there is not one cultivated field within several miles of it. The whole country on this side of the river is a mass of sand, into which you sink up to the ankles, and hardly a blade of grass is to be seen. All about is pine barrens.

"From Wilmington I rode through solitary pine savannas and cypress swamps as before, sometimes thirty miles without seeing a hut or human being. On arriving at the Wackamaw, Pedee, and Black river, I made long zigzags among the rich nabobs, who live on their rice plantations, amidst large villages of negro huts. One of these gentlemen told me, that he had 'something better than six hundred head of blacks!'

"These excursions detained me greatly. The roads to the plantations were so long, so difficult to find, and so bad, and the hospitality of the planters was such, that I could scarcely get away again.

I ought to have told you, that the deep sands of South Carolina had so worn out my horse, that, with all my care, I found he would give up. Chance led me to the house of a planter, named V——, about forty miles north of the river Wackamaw, where I proposed to bargain with him, and to give up my young blood horse for another in exchange, giving him at least as good a character as he deserved. He asked twenty dollars to boot, and I thirty. We parted; but I could perceive that he had taken a liking to my steed, so I went on. He followed me to the sea beach, about three miles, under pretext of pointing out to me the road; and there on the sands, amidst the roar of the Atlantic, we finally bargained; and I found myself in possession of a large, well formed, and elegant sorrel horse, that ran off with me, at a canter, for fifteen miles along the sea shore; and travelled, the same day, forty-two miles, with nothing but a few mouthfuls of rice straw, which I got from a negro. If you have ever seen the rushes with which carpenters sometimes smooth their work, you may form some idea of the common fare of the South Carolina horses. I found now that I had got a very devil before my chair; the least round of the whip made him spring half a rod at a leap; no road, however long or heavy, could tame him. Two or three times he had nearly broke my neck, and chair to boot; and at Georgetown Ferry, he threw one of the boatmen into the river. But he is an excellent traveller, and for that one quality I forgive him all his sins, only keeping a close rein and sharp look out."

From the increasing subscriptions to the "Ornithology" it was thought expedient to throw off three hundred copies in addition to the first two hundred; and the second volume, published in January 1810, started with an impression of five hundred, and a fair proportion of subscribers,—the work gaining, as it advanced, fresh applause and support. A short time before the publication of the second volume, when he was relieved from the press of business by the completion of its various materials, he wrote the following letter to Mr Bartram, from which it would appear, that the extensive expedition, the events of which will be presently detailed in a series of excellent letters to Mr Lawson, was intended to have been undertaken in company:—

"PHILADELPHIA, *November* 11, 1809.

"DEAR SIR,—Since I parted from you yesterday evening, I have ruminated a great deal on my proposed journey; I have considered the advantages and disadvantages of the three modes of proceeding, —on horseback, in the stage coach, and on foot. Taking everything into view, I have at length determined to adopt the last, as being the cheapest; the best adapted for examining the country we pass through; the most favourable to health; and, in short, except for its fatigues, the best mode for a scientific traveller or naturalist, in every point of view. I have also thought, that, by this determination, I will be so happy as to secure your company, for which I would willingly sustain as much hardship, and as many deprivations as I am able to bear.

"If this determination should meet your approbation, and if you are willing to encounter the hardships of such a pedestrian journey, let me know as soon as is convenient. I think one dollar a-day each will be fully sufficient for our expenses, by a strict regard, at all times, to economy."

Bartram did not, however, accompany him, most probably prevented, or comparatively unable, from his increasing years; and Wilson set out alone for Pittsburgh, on his ornithological pilgrimage, in the end of January 1810. His adventures and successes are sketched in the following letters, some of which were published in the year following in the *Portfolio*:—

To Mr Alexander Lawson.

"PITTSBURGH, *February* 22, 1810.

"DEAR SIR,—From this first stage of my ornithological pilgrimage, I sit down with pleasure to give you some account of my adventures since we parted. On arriving at Lancaster, I waited on the governor, secretary of state, and such other great folks as were likely to be useful to me. The governor received me with civility, passed some good-natured compliments on the volumes, and readily added his name to my list. He seems an active man, of plain good sense, and little ceremony. By Mr L—— I was introduced to many members of both houses; but I found them in general such a pitiful, squabbling, political mob—so split up, and justling about the mere formalities of legislation, without knowing anything of its realities,—that I abandoned them in disgust. I must, however, except from this censure a few intelligent individuals, friends to science, and possessed of taste, who treated me with great kindness. I crossed the Susquehannah on Sunday forenoon with some difficulty, having to cut our way through the ice for several hundred yards; and, passing on to York, paid my respects to all the literati of that place without success. Five miles north of this town lives a very extraordinary character, between eighty and ninety years of age, who has lived by trapping birds and quadrupeds these thirty years. Dr F—— carried me out in a sleigh to see him; he has also promised to transmit to me such a collection of facts relating to this singular original, as will enable me to draw up an interesting narrative of him for the *Port-folio.* I carried him half a pound of snuff, of which he is insatiably fond, taking it by handfuls. I was much diverted with the astonishment he expressed on looking at the plates of my work: he could tell me anecdotes of the greater part of the subjects of the first volume, and some of the second. One of his traps, which he says he invented himself, is remarkable for ingenuity, and extremely simple. Having a letter from Dr Muhlenburg to a clergyman in Hanover, I passed on through a well-cultivated country, chiefly inhabited by Germans, to that place, where a certain judge took upon himself to say, that such a book as mine ought not to be encouraged, as it was not within the reach of the commonalty, and therefore inconsistent with our republican institutions! By the same mode of reasoning, which I did not dispute, I undertook to prove him a greater culprit than myself, in erecting a large, elegant, three-story brick house, so much beyond the reach of the commonalty, as he called them, and consequently grossly contrary to our republican institutions. I harangued this Solomon of the bench more seriously afterwards, pointing out to him the great influence of science on a young rising nation like ours, and particularly the science of natural history, till he began to shew such symptoms of intellect as to seem ashamed of what he had said.

"From Hanover I passed through a thinly inhabited country, and, crossing the North Mountain at a pass called Newman's Leap, arrived at Chambersburgh, whence I next morning returned to Carlisle, to visit the reverend doctors of the college.

"The town of Chambersburgh and Shippensburgh produced me nothing. On Sunday the 11th, I left the former of these places in a stage coach, and, in fifteen miles, began to ascend the alpine regions of the Alleghany Mountains, where above, around, and below us, nothing appeared but prodigious declivities covered with woods; and the weather being fine, such a profound silence prevailed among these aërial solitudes, as impressed the soul with awe and a kind of fearful sublimity. Something of this arose from my being alone, having left the coach several miles below. These high ranges continued for more than one hundred miles, to Greensburgh, thirty-two miles from Pittsburgh. Thence the country is nothing but an assemblage of steep hills and deep valleys, descending rapidly till you reach within seven miles of this place, where I arrived on the 15th instant. We were within two miles of Pittsburgh, when suddenly the road descends a long and very steep hill, where the Alleghany river is seen at hand, on the right, stretching along a rich bottom, and bounded by a high ridge of hills on the west. After following this road parallel with the river, and about a quarter of a mile from it, through a rich low valley, a cloud of black smoke at its extremity announced the town of Pittsburgh. On arriving at the town, which stands on a low flat, and looks like a collection of blacksmiths' shops, glass-houses, breweries, forges, and furnaces, the Monongahela opened to the view, on the left, running along the bottom of the range of hills, so high, that the sun, at this season, sets to the town of Pittsburgh at a little past four. This range continues along the Ohio, as far as the view reaches. The ice had just begun to give way in Monongahela, and came down in vast bodies for the three following days. It has now begun in the Alleghany, and, at the moment I write, the river presents a white mass of rushing ice.

"The country beyond the Ohio, to the west, appears a monotonous and hilly region. The Monongahela is lined with arks, usually called Kentucky boats, waiting for the rising of the river, and the absence of the ice, to descend. A perspective view of the town of Pittsburgh at this season, with the numerous arks and covered keel-boats preparing to descend the Ohio—its hills, its great rivers, the pillars of smoke rising from its furnaces and glass-works —would make a noble picture. I began a very diligent search in this place, the day after my arrival, for subscribers, and continued it for four days. I succeeded beyond expectation, having got nineteen names of the most wealthy and respectable part of the inhabitants. The industry of Pittsburgh is remarkable; everybody you see is busy; and as a proof of the prosperity of the place, an eminent lawyer told me, that there has not been one suit instituted against a merchant of the town these three years.

"Gentlemen here assure me, that the road to Chilocothe is impassable on foot, by reason of the freshes. I have, therefore, resolved to navigate myself in a small skiff which I have bought, and named the *Ornithologist*, down to Cincinnati, a distance of five hundred and twenty eight-miles, intending to visit five or six towns that lie in my way. From Cincinnati I will cross over to the opposite shore, and, abandoning my boat, make my way to Lexington, where I expect to be ere your letter can reach that place. Were I to go by Chilocothe, I should miss five towns as large as it. Some say that I ought not to attempt going down by myself—others think I may. I am determined to make the experiment, the expense of hiring a rower being considerable. As soon as the ice clears out of the Alleghany, and the weather will permit, I shall shove off, having everything in readiness. I have ransacked the woods and fields here, without finding a single bird new to me, or indeed anything but a few snow birds and sparrows. I expect to have something interesting to communicate in my next.

"*February* 23.—My baggage is on board; I have just to despatch this and set off. The weather is fine, and I have no doubt of piloting my skiff in safety to Cincinnati. Farewell! God bless you."

To Mr Alexander Lawson.

"LEXINGTON, *April* 4, 1810.

"MY DEAR SIR,—Having now reached the second stage of my bird-catching expedition, I willingly sit down to give you some account of my adventures and remarks since leaving Pittsburgh: by the aid of a good map, and your usual stock of patience, you will be able to listen to my story, and trace all my wanderings. Though generally dissuaded from venturing by myself on so long a voyage down the Ohio in an open skiff, I considered this mode, with all its inconveniences, as the most favourable to my researches, and the most suitable to my funds; and I determined accordingly. Two days before my departure, the Alleghany river was one wide torrent of broken ice, and I calculated on experiencing considerable difficulties on this score. My stock of provisions consisted of some biscuit and cheese, and a bottle of cordial, presented me by a gentleman of Pittsburgh; my gun, trunk, and greatcoat occupied one end of the boat; I had a small tin, occasionally to bail her, and to take my beverage from the Ohio with; and bidding adieu to the smoky confines of Pitt, I launched into the stream, and soon winded away among the hills that everywhere enclose this noble river. The weather was warm and serene, and the river, like a mirror, except where floating masses of ice spotted its surface, and which required some care to steer clear of; but these, to my surprise, in less than a day's sailing, totally disappeared. Far from being concerned at my new situation, I felt my heart expand with joy at the novelties which surrounded me; I listened with pleasure to the whistling of the red bird on the banks as I passed, and contemplated the forest scenery, as it receded, with increasing delight. The smoke of the numerous sugar camps, rising lazily among the mountains, gave great effect to the varying landscape; and the grotesque log cabins, that here and there opened from the woods, were diminished into mere dog-houses by the sublimity of the impending mountains. If you suppose to yourself two parallel ranges of forest-covered hills, whose irregular summits are seldom more than three or four miles apart, winding through an immense extent of country, and enclosing a river half a mile wide, which alternately washes the steep declivity on one side, and leaves a rich, forest-clad bottom on the other, of a mile or so in breadth, you will have a pretty correct idea of the appearance of the Ohio. The banks of these rich flats are from twenty to sixty and eighty feet high; and even these last were within a few feet of being overflowed in December 1808.

"I now stripped with alacrity to my new avocation. The current went about two and a half miles an hour, and I added about three and a half miles more to the boat's way with my oars. In the course of the day, I passed a number of arks, or, as they are usually called, Kentucky boats, loaded with what it must be acknowledged are the most valuable commodities of a country; viz., men, women, and children, horses and ploughs, flour, millstones, &c. Several of these floating caravans were loaded with store goods, for the supply of the settlements through which they passed; having a counter erected, shawls, muslins, &c., displayed, and everything ready for transacting business. On approaching a settlement, they blow a horn, or tin trumpet, which announces to the inhabitants their arrival. I boarded many of those arks, and felt much interested at the sight of so many human beings migrating, like birds of passage, to the luxuriant regions of the south and west. The arks are built in the form of a parallelogram, being from twelve to fourteen feet wide, and from forty to seventy feet long, covered above, rowed only occasionally by two oars before, and steered by a long and powerful one fixed above.

"The barges are taken up along shore by setting poles, at the rate of twenty miles or so a-day; the arks cost about one hundred and fifty cents per foot, according to their length; and when they reach their places of destination, seldom bring more than one-sixth their original cost. These arks descend from all parts of the Ohio and its tributary streams,—the Alleghany, Monongahela, Muskingum, Sciota, Miami, Kentucky, Wabash, &c., in the months of March, April, and May, particularly with goods, produce, and emigrants, the two former for markets along the river, or at New Orleans, the latter for various parts of Kentucky, Ohio, and the Indiana territory. I now return to my own expedition:

"I rowed twenty odd miles the first spell, and found I should be able to stand it perfectly well. About an hour after night, I put up at a miserable cabin, fifty-two miles from Pittsburgh, where I slept on what I supposed to be corn stalks, or something worse; so preferring the smooth bosom of the Ohio to this brush heap, I got up long before day, and, being under no apprehension of losing my way, I again pushed out into the stream. The landscape on each side lay in one mass of shade; but the grandeur of the projecting headlands and vanishing points, or lines, was charmingly reflected in the smooth glassy surface below. I could only discover when I was passing a clearing by the crowing of cocks, and, now and then, in more solitary places, the big horned owl made a most hideous hollowing, that echoed among the mountains. In this lonesome manner, with full leisure for observation and reflection, exposed to hardships all day, and hard berths all night, to storms of rain, hail, and snow,—for it froze severely almost every night,—I persevered, from the 24th of February to Sunday evening, March 17, when I moored my skiff safely in Bear Grass Creek, at the rapids of the Ohio, after a voyage of seven hundred and twenty miles. My hands suffered the most; and it will be some weeks yet before they recover their former feeling and flexibility. It would be the task of a month to detail all the particulars of my numerous excursions, in every direction, from the river. In Stubenville, Charlestown, and Wheeling, I found some friends. At Marietta, I visited the celebrated remains of Indian fortifications, as they are improperly called, which cover a large space of ground on the banks of the Muskingum. Seventy miles above this, at a place called Big Grave Creek, I examined some extraordinary remains of the same kind there. The Big Grave is three hundred paces round at the base, seventy feet perpendicular, and the top, which is about fifty feet over, has sunk in, forming a regular concavity, three or four feet deep. This tumulus is in the form of a cone, and the whole, as well as its immediate neighbourhood, is covered with a venerable growth of forest, four or five hundred years old, which gives it a most singular appearance. In clambering around its steep sides, I found a place where a large white oak had been lately blown down, and had torn up the earth to the depth of five or six feet. In this place I commenced digging, and continued to labour for about an hour, examining every handful of earth with great care; but except some shreds of earthenware, made of a coarse kind of gritty clay, and considerable pieces of charcoal, I found nothing else; but a person of the neighbourhood presented me with some beads fashioned out of a kind of white stone, which were found in digging on the opposite side of this gigantic mound, where I found the hole still remaining. The whole of an extensive plain, a short distance from this, is marked out with squares, oblongs, and circles, one of which comprehends several acres. The embankments by which they are distinguished are still two or three feet above the common level of the field. The Big Grave is the property of a Mr Tomlinson, or Tumblestone, who lives near, and who would not expend three cents to see the whole sifted before his face. I endeavoured to work on his avarice, by representing the probability that it might contain valuable matters, and suggested to him a mode by which a passage might be cut into it, level with the bottom, and by excavation and arching, a most noble cellar might be formed for keeping his turnips and potatoes. 'All the turnips and potatoes I shall raise this dozen years,' said he, 'would not pay the expense.' This man is no antiquary, or theoretical farmer, nor much of a practical one either, I fear: he has about two thousand acres of the best land, and just makes out to live. Near the head of what is called the Long Reach, I called on a certain Michael Cressop, son to the noted Colonel Cressop, mentioned in Jefferson's notes on Virginia. From him I received the head of a paddle fish, the largest ever seen in the Ohio, which I am keeping for Mr Peale, with various other curiosities. I took the liberty of asking whether Logan's accusation of his father having killed all his family, had any truth in it; but he replied that it had not. Logan, he said, had been misinformed. He detailed to me all the particulars, which are too long for repetition, and concluded by informing me that his father died early in the revolutionary war, of the camp fever, near New York.

"Marietta stands in a swampy plain, which has evidently once been the ancient bed of the Muskingum, and is still occasionally inundated to the depth of five or six feet. A Mr Putnam, son to the old general of Bunker's Hill memory, and Mr Gillman, and Mr Feering, are making great exertions here in introducing and multiplying the race of merinos. The two latter gentlemen are about establishing works by steam for carding and spinning wool, and intend to carry on the manufacture of broadcloth extensively. Mr Gillman is a gentleman of taste and wealth, and has no doubts of succeeding. Something is necessary to give animation to this place, for, since the numerous building of ships has been abandoned here, the place seems on the decline.

"The current of the Muskingum is very rapid, and the ferry boat is navigated across in the following manner:—A strong cable is extended from bank to bank, forty or fifty feet above the surface of the river, and fastened tight at each end. On this cable are two loose running blocks; one rope, from the bow of the boat, is fastened to the first of these blocks, and another from the after part of the boat to the second block, and by lengthening this last, a diagonal direction is given to the boat's head, a little up the stream, and the current striking forcibly and obliquely on her aft, she is hurried forward with amazing velocity, without any manual labour whatever. I passed Blannerhasset's island after night, but the people were burning brush, and by the light I had a distinct view of the mansion-house, which is but a plain frame of no great dimensions. It is now the property of a Mr Miller from Lexington, who intends laying it chiefly in hemp. It is nearly three miles long, and contains about three hundred acres, half of which is in cultivation; but, like all the rest of the numerous islands of the Ohio, is subject to inundations. At Galliopolis, which stands upon a high plain, and contains forty or fifty scattered houses, I found the fields well fenced and well cultivated, peach and apple orchards numerous, and a considerable appearance of industry. One half of the original French settlers have removed to a tract of land opposite to the mouth of Sandy River. This town has one shop and two taverns: the mountains press into within a short distance of the town. I found here another Indian mound planted with peach trees.

"On Monday, March 5, about ten miles below the mouth of the Great Scotia, where I saw the first flock of paroquets, I encountered a violent storm of wind and rain, which changed to hail and snow, blowing down trees and limbs in all directions, so that, for immediate preservation, I was obliged to steer out into the river, which rolled and foamed like a sea, and filled my boat nearly half full of water; and it was with the greatest difficulty I could make the least head way. It continued to snow violently until dusk, when I at length made good my landing, at a place on the Kentucky shore, where I had perceived a cabin; and here I spent the evening in learning the art and mystery of bear-treeing, wolf-trapping, and wild-cat-hunting, from an old professor. But, notwithstanding the skill of this great master, the country here is swarming with wolves and wild cats, black and brown: according to this hunter's own confession, he had lost sixty pigs from Christmas last, and all night long the distant howling of the wolves kept the dogs in a perpetual uproar of barking. This man was one of those people called *squatters*, who neither pay rent nor own land, but keep roving on the frontiers, advancing as the tide of civilised population approaches. They are the immediate successors of the savages, and far below them in good sense and good manners, as well as comfortable accommodations.

"Nothing adds more to the savage grandeur and picturesque effect of the scenery along the Ohio, than these miserable huts of human beings, lurking at the bottom of a gigantic growth of timber, that I have not seen equalled in any other part of the United States. And it is truly amusing to observe how dear and how familiar habit has rendered these privations, which must have been first the offspring of necessity; yet none pride themselves more on their possessions. The inhabitants of those forlorn sheds will talk to you with pride of the richness of their soil—of the excellence and abundance of their country—of the healthiness of their climate, and the purity of their waters; while the only bread you find among them is of Indian corn, coarsely ground in a horse mill, with half of the grains unbroken; even their cattle are destitute of stables and hay, and look like moving skeletons; their own houses worse than pig-styes; their clothes an assemblage of rags; their faces yellow and lank with disease, and their persons covered with filth, and frequently garnished with humours of the Scotch fiddle, from which dreadful disease, by the mercy of God, I have been most miraculously preserved. All this is the effect of laziness. The corn is thrown into the ground in the spring, and the pigs turned into the woods, where they multiply like rabbits. The labour of the *squatter* is now over till autumn, and he spends the winter in eating pork, cabbages, and hoe-cakes. What a contrast to the neat farm, and snug cleanly habitation, of the industrious settler, that opens his green fields, his stately barns, gardens and orchards, to the gladdened eye of the delighted stranger!

"At a place called Salt Lick I went ashore to see the salt works, and to learn whether the people had found any farther remains of an animal of the ox kind, one of whose horns, of a prodigious size, was discovered here some years ago, and is in the possession of Mr Peale. They make here about one thousand bushels weekly, which sell at one dollar and seventy-five cents per bushel. The wells are from thirty to fifty feet deep, but nothing very curious has lately been dug up. I landed at Maysville, or Limestone, where a considerable deal of business is done in importation for the interior of Kentucky. It stands on a high narrow plain, between the mountains and the river, which is fast devouring the bank, and encroaching on the town; part of the front street is gone already, and unless some effectual means are soon taken, the whole must go by piecemeal. This town contains about one hundred houses, chiefly log and frames. From this place I set out on foot for Washington. On the road, at the height of several hundred feet above the present surface of the river, I found prodigious quantities of petrified shells of the small cockle and fan-shaped kind, but whether marine remains or not, I am uncertain. I have since found these petrified concoctions of shells universal all over Kentucky, wherever I have been. The rocks look as if one had collected heaps of broken shells and wrought them up among clay, then hardened it into stone. These rocks lie universally in horizontal strata. A farmer, in the neighbourhood of Washington, assured me, that, from seven acres he reaped at once eight thousand weight of excellent hemp, fit for market.

"Amidst very tempestuous weather, I reached the town of Cincinnati, which does honour to the name of the old Roman, and is the neatest and handsomest situated place I have seen since I left Philadelphia. You must know, that, during an unknown series of ages, the river Ohio has gradually sunk several hundred feet below its former bed, and has left, on both sides, occasionally, what are called the first or nearest, and the second or next, high bank, the latter of which is never overflowed.

"The town of Cincinnati occupies two beautiful plains, one on the first, and the other on the second bank, and contains upwards of five hundred houses, the greater proportion of which are of brick. One block house is all that remains of Fort Washington. The river Licking comes in from the opposite shore, where the town of Newport, of forty or fifty houses, and a large arsenal and barracks, are lately erected. Here I met with Judge Turner, a man of extraordinary talents, well known to the literati of Philadelphia: he exerted himself in my behalf with all the ardour of an old friend. A large Indian mound, in the vicinity of this town, has been lately opened by Dr Drake, who showed me the collection of curiosities which he had found in that and others. In the centre of this mound he also found a large fragment of earthenware, such as I found at the *Big Grave*, which is a pretty strong proof that these works had been erected by a people, if not the same, differing little from the present race of Indians, whose fragments of earthenware, dug up about their late towns, correspond exactly with these. Twenty miles below this I passed the mouth of the great Miami, which rushes in from the north, and is a large and stately river, preserving its pure waters, uncontaminated for many miles with those of the Ohio, each keeping their respective sides of the channel. I rambled up the banks of this river for four or five miles, and in my return shot a turkey. I also saw five or six deer in a drove, but they were too light-heeled for me.

"In the afternoon of the 15th I entered Big Bone Creek, which being passable only about a quarter of a mile, I secured my boat, and left my baggage under the care of a decent family near, and set out on foot five miles through the woods for the Big Bone Lick, that great antediluvian rendezvous of the American elephants. This place, which lies 'far in the windings of a sheltered vale,' afforded me a fund of amusement in shooting ducks and paroquets (of which last I skinned twelve, and brought off two slightly wounded), and in examining the ancient buffalo roads to this great licking place. Mr Colquhoun, the proprietor, was not at home; but his agent and manager entertained me as well as he was able, and was much amused with my enthusiasm. This place is a low valley, everywhere surrounded by high hills; in the centre, by the side of the creek, is a quagmire of near an acre, from which, and another smaller one below, the chief part of these large bones have been taken; at the latter places, I found numerous fragments of large bones lying scattered about. In pursuing a wounded duck across this quagmire, I had nearly deposited my carcase among the grand congregation of mammoths below, having sunk up to the middle, and had hard struggling to get out.

"As the proprietor intends to dig in various places this season for brine, and is a gentleman of education, I have strong hopes that a more complete skeleton of that animal called the mammoth, than has yet been found, will be procured. I laid the strongest injunctions on the manager to be on the look out, and to preserve everything, I also left a letter for Mr Colquhoun to the same purport, and am persuaded that these will not be neglected. In this neighbourhood, I found the columbo plant in great abundance, and collected some of the seeds. Many of the old stalks were more than five feet high. I have since found it in various other parts of this country. In the afternoon of the next day, I returned to my boat, replaced my baggage, and rowed twenty miles to the Swiss settlement, where I spent the night. These hardy and industrious people have now twelve acres closely and cleanly planted with vines from the Cape of Good Hope. They last year made seven hundred gallons of wine, and expect to make three times as much the ensuing season. Their houses are neat and comfortable. They have orchards of peach and apple-trees, besides a great number of figs, cherries, and other fruit trees, of which they are very curious. They are of opinion, that this part of the Indiana territory is as well suited as any part of France to the cultivation of the vine; but the vines, they say, require different management here from what they are accustomed to in Switzerland. I purchased a bottle of their last vintage, and drank to all your healths, as long as it lasted, in going down the river. Seven miles below this, I passed the mouth of Kentucky river, which has a formidable appearance. I observed twenty or thirty scattered houses on its upper side, and a few below; many of the former seemingly in a state of decay. It rained on me almost the whole of this day, and I was obliged to row hard and drink healths to keep myself comfortable. My birds' skins were wrapt up in my greatcoat, and my own skin had to sustain a complete drenching, which, however, had no bad effects. This evening I lodged at the most wretched hovel I had yet seen. The owner, a meagre, diminutive wretch, soon began to let me know of how much consequence he had formerly been; that he had gone through the war with General Washington—had become one of his *life-guards*—and had sent many a British soldier to his long home. As I answered him with indifference, to interest me the more, he began to detail anecdotes of his wonderful exploits. 'One grenadier,' said he, 'had the impudence to get on the works, and to wave his cap in defiance. My commander (General Washington, I suppose) says to me, Dick, says he, can't you pepper that there fellow? says he. Please your honour, says I, I'll try at it; so I took a fair, cool, and steady aim, and touched my trigger; up went his heels like a turkey! down he tumbled! One buckshot had entered here, and another here' (laying a finger on each breast) 'and the bullet found the way to his brains, right through his forehead.' Though I believed every word of this to be a lie, yet I could not but look with disgust on the being who uttered it. This same miscreant pronounced a long prayer before supper; and, immediately after, called out, in a splutter of oaths, for the pine splinters to be held to let the gentleman see. Such a farrago of lies, oaths, prayers, and politeness, put me in good humour in spite of myself. The whole herd of this filthy kennel were in perpetual motion with the itch; so, having procured a large fire to be made, under pretence of habit, I sought for the softest plank, placed my trunk and greatcoat at my head, and stretched myself there till morning. I set out early, and passed several arks. A number of turkeys, which I observed from time to time on the Indiana shore, made me lose half the morning in search of them. On the Kentucky shore, I was also decoyed by the same temptations, but never could approach near enough to shoot one of them. These affairs detained me so, that I was dubious whether I should be able to reach Louisville that night. Night came on, and I could hear nothing of the Falls. About eight, I heard the first roaring of the Rapids; and, as it increased, I was every moment in hopes of seeing the lights of Louisville; but no lights appeared, and the noise seemed now within less than half a mile of me. Seriously alarmed, lest I might be drawn into the suction of the Falls, I cautiously coasted along shore, which was full of snags and sawyers, and at length, with great satisfaction, opened Bear Grass Creek, where I secured my skiff to a Kentucky boat, and, loading myself with my baggage, I groped my way through a swamp up to the town. The next day, I sold my skiff for exactly half what it cost me; and the man who bought it wondered why I gave it such a droll Indian name (*the Ornithologist*). 'Some old chief or warrior, I suppose?' said he. This day, I walked down along shore to Shipping Port, to take a view of these celebrated Rapids; but they fell far short of my expectation. I should have no hesitation in going down them in a skiff. The Falls of Oswego, in the state of New York, though on a smaller scale, are far more dangerous and formidable in appearance. Though the river was not high, I observed two arks and a barge run them with great ease and rapidity. The Ohio here is something more than a mile wide, with several islands interspersed ; the channel rocky, and the islands heaped with driftwood. The whole fall, in two miles, is less than twenty-four feet. The town of Louisville stands on a high *second* bank, and is about as large as Frankfort, having a number of good brick buildings and valuable shops. The situation would be as healthy as any on the river, but for the numerous swamps and ponds that intersect the woods in its neighbourhood. These, from their height above the river, might all be drained, and turned into cultivation ; but every man here is so intent on the immediate making of money, that they have neither time nor disposition for improvements, even where the article health is at stake. A man here told me, that last fall he had fourteen sick in his own family. On Friday the 24th, I left my baggage with a merchant of the place, to be forwarded by the first waggon; and set out on foot for Lexington, seventy-two miles distant. I passed through Middleton and Shelbyville, both inconsiderable places. Nine-tenths of the country is in forest, the surface undulating into gentle eminences and declivities, between each of which generally runs a brook, over loose flags of limestone. The soil, by appearance, is of the richest sort. I observed immense fields of Indian corn ; high, excellent fences ; few grain fields; many log-houses, and those of the meaner sort. I took notice of few apple orchards, but several very thriving peach ones. An appearance of slovenliness is but too general about their houses, barns, and barn-yards. Negroes are numerous ; cattle and horses lean, particularly the former, who appear as if struggling with starvation for their existence. The woods are swarming with pigs, pigeons, squirrels, and woodpeckers. The pigs are universally fat, owing to the great quantity of mast this year. Walking here in wet weather is most execrable, and is like travelling on soft soap: a few days of warm weather hardens this again into a stone. Want of bridges is the greatest inconvenience to a foot traveller here. Between Shelbyville and Frankfort, having gone out of my way to see a pigeon-roost (which, by the by, is the greatest curiosity I have seen since leaving home), I waded a deep creek, called Benson, nine or ten times. I spent several days in Frankfort, and in rambling among the stupendous cliffs of Kentucky river. On Thursday evening, I entered Lexington. But I cannot do justice to these subjects at the conclusion of a letter, which, in spite of all my abridgments, has far exceeded in length what I first intended. My next will be from Nashville. I shall then have seen a large range of Kentucky, and be more able to give you a correct delineation of the country and its inhabitants. In descending the Ohio, I amused myself with a poetical narrative of my expedition, which I have called the Pilgrim."

To Mr Alexander Lawson.

"NASHVILLE, TENNESSEE,
April 28, 1810.

"MY DEAR SIR,—Before setting out on my journey through the wilderness to Natchez, I sit down to give you, according to promise, some account of Lexington, and of my adventures through the state of Kentucky. These I shall be obliged to sketch as rapidly as possible. Neither my time nor my situation enables me to detail particulars with any degree of regularity, and you must condescend to receive them in the same random manner in which they occur, altogether destitute of fanciful embellishment, with nothing but their novelty and the simplicity of truth to recommend them.

"I saw nothing of Lexington till I had approached within half a mile of the place, when, the woods opening, I beheld the town before me on an irregular plain, ornamented with a small white spire, and consisting of several parallel streets, crossed by some others. Many of the houses are built of brick, others of stone, neatly painted; but a great proportion wore a more humble and inferior appearance. The fields around looked clean and well fenced; gently undulating, but no hills in view. In a hollow between two of these parallel streets, ran a considerable brook, that, uniting with a larger a little below the town, drives several mills. A large quarry of excellent building stone also attracted my notice as I entered the town. The main street was paved with large masses from this quarry, the foot-path neat, and guarded by wooden posts. The numerous shops filled with goods, and the many well dressed females I passed in the streets, the sound of social industry, and the gay scenery of 'the busy haunts of men,' had a most exhilarating effect upon my spirits, after being so long immured in the forest. My own appearance, I believe, was to many equally interesting; and the shopkeepers and other loungers interrogated me with their eyes as I passed, with symptoms of eager and inquisitive curiosity. After fixing my quarters, disposing of my arms, and burnishing myself a little, I walked out to have a more particular view of the place.

"This little metropolis of the western country is nearly as large as Lancaster, in Pennsylvania. In the centre of the town is a public square, partly occupied by the court-house and market-place, and distinguished by the additional ornament of the pillory and stocks. The former of these is so constructed as to serve well enough, if need be, occasionally for a gallows, which is not a bad thought; for as nothing contributes more to make *hardened villains* than the pillory, so nothing so effectually rids society of them as the gallows; and every knave may here exclaim,

'My *bane* and *antidote* are both before me.'

I peeped into the court-house as I passed; and, though it was court day, I was struck with the appearance its interior exhibited; for, though only a plain, square, brick building, it has all the gloom of the Gothic, so much admired of late by our modern architects. The exterior walls having, on experiment, been found too feeble for the superincumbent honours of the roof and steeple, it was found necessary to erect, from the floor a number of large, circular, and unplastered brick pillars, in a new order of architecture (the thick end uppermost), which, while they serve to impress the spectators with the perpetual dread that they will tumble about their ears, contribute also, by their number and bulk, to shut out the light, and to spread around a reverential gloom, producing a melancholy and chilling effect,—a very good disposition of mind, certainly, for a man to enter a court of justice in. One or two solitary individuals stole along the damp and silent floor; and I could just descry, elevated at the opposite extremity of the building, the judges sitting like spiders in a window corner, dimly distinguishable through the intermediate gloom. The market-place, which stands a little to the westward of this, stretches over the whole breadth of the square; is built of brick, something like that of Philadelphia, but is unpaved and unfinished. In wet weather, you sink over the shoes in mud at every step; and here, again, the wisdom of the police is manifest,—nobody, at such times, will wade in there unless forced by business or absolute necessity, by which means a great number of idle loungers are very properly kept out of the way of the market folks.

"I shall say nothing of the nature or quantity of the commodities which I saw exhibited there for sale, as the season was unfavourable to a display of their productions, otherwise something better than a few cakes of black maple sugar, wrapt in greasy saddle-bags, some cabbage, chewing tobacco, catmint and turnip tops, a few bags of meal, sassafras roots, and skinned squirrels, cut up into quarters, —something better than all this, I say, in the proper season, certainly covers the stalls of this market-place, in the metropolis of the fertile country of Kentucky.

"The horses of Kentucky are the hardiest in the world, not so much by nature as by education and by habit. From the commencement of their existence, they are habituated to every extreme of starvation and gluttony, idleness and excessive fatigue. In summer, they fare sumptuously every day. In winter, when not a blade of grass is to be seen, and when the cows have deprived them of the very bark and buds of every fallen tree, they are ridden into town fifteen or twenty miles, through roads and sloughs that would become the graves of any common animal, with a fury and celerity incomprehensible by you folks on the other side of the Alleghany. They are there fastened to the posts on the sides of the streets and around the public square, where hundreds of them may be seen, on a court-day, hanging their heads, from morning to night, in deep cogitation, ruminating, perhaps, on the long expected return of spring and green herbage. The country people, to their credit be it spoken, are universally clad in plain homespun. Soap, however, appears to be a scarce article, and Hopkins' *double cutters* would find here a rich harvest, and produce a very improving effect. Though religion here has its votaries, yet none can accuse the inhabitants of this flourishing place of bigotry, in shutting out from the pale of the church or churchyard any human being or animal whatever. *Some* of these sanctuaries are open at all hours, and to every visitor. The birds of heaven find a hundred passages through the broken panes, and the cows and hogs a ready access on all sides. The wall of separation is broken down between the living and the dead, and dogs tug at the carcass of the horse on the grave of his master. Lexington, however, with all its faults, which a few years will gradually correct, is an honourable monument of the enterprise, courage, and industry of its inhabitants. Within the memory of a middle-aged man, who gave me the information, there were only two log huts on the spot where the city is now erected; while the surrounding country was a wilderness, rendered hideous by skulking bands of bloody and ferocious Indians. Now, numerous excellent institutions for the education of youth, a public library, and a well-endowed university, under the superintendence of men of learning and piety, are in successful operation. Trade and manufactures are also rapidly increasing. Two manufactories for spinning cotton have lately been erected; one for woollen; several extensive ones for weaving sailcloth and bagging; and seven rope walks, which, according to one of the proprietors, export annually rope-yarn to the amount of 150,000 dollars. A taste for neat and even elegant buildings is fast gaining ground, and Lexington at present can boast of men who do honour to science, and of females whose beauty and amiable manners would grace the first circles of society. On Saturday, April 14th, I left this place for Nashville, distant about two hundred miles. I passed through Nicholasville, the capital of Jessamine county, a small village begun about ten years ago, consisting of about twenty houses, with three shops and four taverns. The woods were scarcely beginning to look green, which to me was surprising, having been led, by common report, to believe that spring here is much earlier than in the lower parts of Pennsylvania. I must farther observe, that, instead of finding the woods of Kentucky covered with a profusion of flowers, they were, at this time, covered with rotten leaves, and dead timber in every stage of decay and confusion; and I could see no difference between them and our own, but in the magnitude of the timber, and superior richness of the soil. Here and there the white blossoms of the *Sanguinaria Canadensis*, or red root, were peeping through the withered leaves; and the buds of the buckeye, or horse chestnut, and one or two more, were beginning to expand. Wherever the hackberry had fallen, or been cut down, the cattle had eaten the whole bark from the trunk, even to that of the roots.

"Nineteen miles from Lexington, I descended a long, steep, and rocky declivity, to the banks of Kentucky river, which is here about as wide as the Schuylkill, and winds away between prodigious perpendicular cliffs of solid limestone. In this deep and romantic valley, the sound of the boat horns from several Kentucky arks, which were at that instant passing, produced a most charming effect. The river, I was told, had already fallen fifteen feet, but was still high. I observed great numbers of uncommon plants and flowers growing among the cliffs, and a few solitary bank swallows were skimming along the surface. Reascending from this, and travelling for a few miles, I again descended a vast depth to another stream called Dick's River, engulfed among the same perpendicular masses of rock. Though it was nearly dark, I found some curious petrifactions, and some beautiful specimens of mother-of-pearl on the shore. The roaring of a mill dam, and the rattling of the mill, prevented the ferryman from hearing me till it was quite night, and I passed the rest of the road in the dark, over a rocky country abounding with springs, to Danville. This place stands on a slight eminence, and contains about eighty houses, chiefly log and frame buildings, disposed in two parallel streets crossed by several others. It has two rope works and a woollen manufactory, also nine shops and three taverns. I observed a great many sheep feeding about here, amidst fields of excellent pasture; it is, however, but a dull place. A Roman Catholic chapel has been erected here at the expense of one or two individuals. The shopkeepers trade from the mouth of Dick's River, down to New Orleans, with the common productions of the country,—flour, hemp, tobacco, pork, corn, and whisky. I was now one hundred and eighty miles from Nashville, and, as I was informed, not a town or village on the whole route. Every day, however, was producing wonders on the woods, by the progress of vegetation. The blossoms of the sassafras, dogweed, and red bud, contrasted with the deep green of the poplar and buckeye, enriched the scenery on every side, while the voices of the feathered tribes, many of which were to me new and unknown, were continually engaging me in the pursuit. Emerging from the deep solitude of the forest, the rich green of the grain fields, the farm-house, and cabins embosomed amidst orchards of glowing purple and white, gave the sweetest relief to the eye. Not far from the foot of a high mountain, called Mulder's Hill, I overtook one of those family caravans, so common in this country, moving to the westward; the procession occupied a length of road, and had a formidable appearance, though, as I afterwards understood, it was composed of the individuals of only a single family. In the front went a waggon drawn by four horses, driven by a negro, and filled with implements of agriculture; another heavy-loaded waggon, with six horses, followed, attended by two persons; after which came a numerous and mingled group of horses, steers, cows, sheep, hogs, and calves, with their bells; next followed eight boys, mounted double; also a negro wench, with a child before her; then the mother, with one child behind her, and another at the breast; ten or twelve colts brought up the rear, now and then picking herbage and trotting a-head. The father, a fresh good looking man, informed me that he was from Washington county, in Kentucky, and was going as far as Cumberland River; he had two ropes fixed to the top of the waggon, one of which he guided himself, and the other was entrusted to his eldest son, to keep it from oversetting in ascending the mountain. The singular appearance of this moving group, the mingled music of the bells, and the shouting of the drivers, mixed with the echoes of the mountains, joined to the picturesque solitude of the place, and various reflections that hurried through my mind, interested me greatly; and I kept company with them for some time, to lend my assistance, if necessary. The country now became mountainous, perpetually ascending and descending; and about forty-nine miles from Danville I passed through a pigeon roost, or rather breeding place, which continued for three miles, and, from information, exceeded in length more than forty miles. The timber was chiefly beech,—every tree loaded with nests; and I counted, in different places, more than ninety nests on a single tree. Beyond this I passed a large company of people engaged in erecting a horse mill for grinding grain. The few cabins I passed were generally poor, but much superior in appearance to those I met with on the shores of the Ohio. In the evening I lodged near the banks of the Green River. This stream, like all the rest, is sunk in a deep gulf, between high perpendicular walls of limestone; is about thirty yards wide at this place, and runs with great rapidity; but as it had fallen considerably, I was just able to ford it without swimming. The water was of a pale greenish colour, like that of the Licking and some other streams, from which circumstance I suppose it has its name. The rocky banks of the river are hollowed out, in many places, into caves of enormous size, and of great extent. These rocks abound with the same masses of petrified shells so universal in Kentucky.

"In the woods, a little beyond this, I met a soldier, on foot, from New Orleans, who had been robbed and plundered by the Choctaws, as he passed through their nation. 'Thirteen or fourteen Indians,' said he, 'surrounded me before I was aware, cut away my canteen, tore off my hat, took the handkerchief from my neck, and the shoes from my feet, and all the money I had from me, which was about forty-five dollars.' Such was his story. He was going to Chilocothe, and seemed pretty nearly done up. In the afternoon I crossed another stream, of about twenty-five yards in width, called Little Barren; after which, the country began to assume a new and very singular appearance. The woods, which had hitherto been stately, now degenerated into mere scrubby saplings, on which not a bud was beginning to unfold, and grew so open, that I could see for a mile through them. No dead timber or rotten leaves were to be seen, but the whole face of the ground was covered with rich verdure, interspersed with a variety of very beautiful flowers, altogether new to me. It seemed as if the whole country had once been one general level; but that, from some unknown cause, the ground had been undermined, and had fallen in, in innumerable places, forming regular funnel-shaped concavities, of all dimensions, from twenty feet in diameter, and six feet in depth, to five hundred by fifty, the surface or verdure generally unbroken. In some tracts, the surface was entirely destitute of trees, and the eye was presented with nothing but one general neighbourhood of these concavities, or, as they are usually called, sink-holes. At the centre, or bottom, of some of these, openings had been made for water. In several places these holes had broken in, on the sides, and even middle of the road, to an unknown depth; presenting their grim mouths as if to swallow up the unwary traveller. At the bottom of one of those declivities, at least fifty feet below the general level, a large rivulet of pure water issued at once from the mouth of a cave about twelve feet wide and seven high. A number of very singular sweet smelling lichens grew over the entrance, and a Peewee had fixed her nest, like a little sentry-box, on a projecting shelf of the rock above the water. The height and dimensions of the cave continued the same as far as I waded in, which might be thirty or forty yards; but the darkness became so great that I was forced to return. I observed numbers of small fish sporting about; and I doubt not but these abound even in its utmost subterranean recesses. The whole of this country, from Green to Red River, is hollowed out into these enormous caves; one of which, lately discovered in Warren county, about eight miles from the dripping spring, has been explored for upwards of six miles, extending under the bed of the Green River. The entrance to these caves generally commences at the bottom of a sink-hole, and many of them are used by the inhabitants as cellars, or spring houses, having generally a spring or brook of clear water running through them. I descended one of these belonging to a Mr Wood, accompanied by the proprietor, who carried the light. At first, the darkness was so intense that I could scarcely see a few feet beyond the circumference of the candle; but after being in for five or six minutes, the objects around me began to make their appearance more distinctly. The bottom, for fifteen or twenty yards at first, was so irregular that we had constantly to climb over large masses of wet and slippery rocks. The roof rose in many places to the height of twenty or thirty feet, presenting all the most irregular projections of surface, and hanging in gloomy and silent horror. We passed numerous chambers, or offsets, which we did not explore; and after three hours' wandering in these profound regions of gloom and silence, the particulars of which would detain me too long, I emerged, with a handkerchief filled with bats, including one which I have never seen described; and a number of extraordinary insects of the gryllus tribe, with antennæ upwards of six inches long, and which, I am persuaded, had never before seen the light of day, as they fled from it with seeming terror, and I believe were as blind in it as their companions, the bats. Great quantities of native glauber salts are found in these caves, and are used by the country people in the same manner, and with equal effect, as those of the shops. But the principal production is saltpetre, which is procured from the earth in great abundance. The cave in Warren county, above mentioned, has lately been sold for three thousand dollars to a saltpetre company; an individual of which informed me that, from every appearance, this cave had been known to the Indians many ages ago; and had evidently been used for the same purposes. At the distance of more than a mile from the entrance, the exploring party, on their first visit, found the roof blackened by smoke, and bundles of half-burnt canes scattered about. A bark mockasin, of curious construction, besides several other Indian articles, were found among the rubbish. The earth, also, lay piled in heaps, with great regularity, as if in preparation for extracting the saltpetre.

"Notwithstanding the miserable appearance of the timber in these barrens, the soil, to my astonishment, produced the most luxuriant fields of corn and wheat I had ever before met with. But one great disadvantage is the want of water; for the whole running streams, with which the surface of this country evidently once abounded, have been drained off to a great depth, and now murmur among these lower regions, secluded from the day. One forenoon I rode nineteen miles without seeing water; while my faithful horse looked round, but in vain, at every hollow, with a wishful and languishing eye, for that precious element. These barrens furnished me with excellent sport in shooting grouse, which abound here in great numbers; and in the delightful groves, that here and there rise majestically from these plains, I found many new subjects for my 'Ornithology.' I observed all this day, far to the right, a range of high, rocky, detached hills, or knobs, as they are called, that skirt the barrens, as if they had been once the boundaries of the great lake that formerly covered this vast plain. These, I was told, abound with stone, coal, and copperas. I crossed Big Barren river in a ferry boat, where it was about one hundred yards wide; and passed a small village called Bowling Green, near which I rode my horse up to the summit of one of these high insulated rocky hills, or knobs, which overlooked an immense circumference of country, spreading around bare and leafless, except where the groves appeared, in which there is usually water. Fifteen miles from this, induced by the novel character of the country, I put up for several days at the house of a pious and worthy presbyterian, whence I made excursions, in all directions, through the surrounding country. Between this and Red River, the country had a bare and desolate appearance. Caves continued to be numerous; and report made some of them places of concealment for the dead bodies of certain strangers who had disappeared there. One of these lies near the banks of the Red River, and belongs to a person of the name of ———, a man of notoriously bad character, and strongly suspected, even by his neighbours, of having committed a foul murder of this kind, which was related to me, with all its minutiæ of horrors. As this man's house stands by the roadside, I was induced by motives of curiosity to stop and take a peep of him. On my arrival I found two persons in conversation under the piazza, one of whom informed me that he was the landlord. He was a dark mulatto, rather above the common size, inclining to corpulency, with legs small in proportion to his size, and walked lame. His countenance bespoke a soul capable of deeds of darkness. I had not been three minutes in company, when he invited the other man (who I understood was a traveller) and myself to walk back and see his cave, to which I immediately consented. The entrance is in the perpendicular front of a rock, behind the house—has a door, with a lock and key to it, and was crowded with pots of milk, placed near the running stream. The roof and sides of solid rock were wet and dropping with water. Desiring —— to walk before with the lights, I followed, with my hand on my pistol, reconnoitring on every side, and listening to his description of its length and extent. After examining this horrible vault for forty or fifty yards, he declined going any further, complaining of a rheumatism; and I now first perceived that the other person had stayed behind, and that we two were alone together. Confident in my means of self-defence, whatever mischief the devil might suggest to him, I fixed my eye steadily on his, and observed to him, that he could not be ignorant of the reports circulated about the country relative to this cave. 'I suppose,' said I, 'you know what I mean?' 'Yes, I understand you,' returned he, without appearing the least embarrassed,—'that I killed somebody, and threw them into this cave. I can tell you the whole beginning of that damned lie,' said he; and, without moving from the spot, he detailed to me a long story, which would fill half my letter, to little purpose, and which, with other particulars, I shall reserve for your amusement when we meet. I asked him why he did not get the cave examined by three or four reputable neighbours, whose report might rescue his character from the suspicion of having committed so horrid a crime. He acknowledged it would be well enough to do so, but did not seem to think it worth the trouble; and we returned as we advanced, —— walking before with the lights. Whether this man be guilty or not of the transaction laid to his charge, I know not; but his manners and aspect are such as by no means to allay suspicion.

"After crossing Red River, which is here scarcely twenty yards broad, I found no more barrens. The timber was large, and the woods fast thicking with green leaves. As I entered the state of Tennessee, the face of the country became hilly, and even mountainous. After descending an immense declivity, and coursing along the rich valley of Manshas Creek, where I again met with large flocks of paroquets, I stopt at a small tavern to examine, for three or four days, this part of the country. Here I made some interesting additions to my stock of new subjects for the 'Ornithology.' On the fourth day, I crossed the Cumberland, where it is about two hundred and fifty yards wide, and of great depth, bounded, as usual, by high and precipitous banks, and reached the town of Nashville, which towers like a fortress above the river. Here I have been busily employed these eight days; and send you the enclosed parcel of drawings, the result of every moment of leisure and convenience I could obtain. Many of the birds are altogether new; and you will find, along with them, every explanation necessary for your purpose.

"You may rest assured of hearing from me by the first opportunity after my arrival at Natchez. In the meantime, I receive with much pleasure the accounts you give me of the kind inquiries of my friends. To me, nothing could be more welcome; for whether journeying in this world, or journeying to that which is to come, there is something of desolation and despair in the idea of being for ever forgotten in our absence, by those whom we sincerely esteem and regard."

To Mr Alexander Lawson.

"NATCHEZ, MISSISSIPPI TER.,
18th May 1810.

"DEAR SIR,—About three weeks ago, I wrote to you from Nashville, enclosing three sheets of drawings, which I hope you have received.* I was, at that time, on the point of setting out for St Louis; but, being detained a week by constant and heavy rains, and considering that it would add four hundred miles to my journey, and detain me at least a month, and the season being already far advanced, and no subscribers to be expected there, I abandoned the idea, and prepared for a journey through the wilderness. I was advised by many not to attempt it alone,—that the Indians were dangerous, the swamps and rivers almost impassable without assistance; and a thousand other hobgoblins were conjured up to dissuade me from going alone. But I weighed all these matters in my own mind; and, attributing a great deal of this to vulgar fears and exaggerated reports, I equipt myself for the attempt. I rode an excellent horse, on which I could depend. I had a loaded pistol in each pocket, a loaded fowling piece belted across my shoulders, a pound of gunpowder in my flask, and five pounds of shot in my belt. I bought some biscuit and dried beef, and, on Friday morning, May 4, I left Nashville. About half a mile from town, I observed a poor negro with two wooden legs, building himself a cabin in the woods. Supposing that this journey might afford you and my friends some amusment, I kept a particular account of the various occurrences, and shall transcribe some of the most interesting, omitting everything relative to my ornithological excursions and discoveries, as more suitable for another occasion. Eleven miles from Nashville, I came to the Great Harpath, a stream of about fifty yards wide, which was running with great violence. I could not discover the entrance of the ford, owing to the rains and inundations. There was no time to be lost; I plunged in, and almost immediately my horse was swimming. I set his head aslant the current, and, being strong, he soon landed me on the other side. As the weather was warm, I rode in my wet clothes without any inconvenience. The country to-day was a perpetual succession of steep hills and low bottoms; I crossed ten or twelve large creeks, one of which I swam with my horse, where he was near being entangled among some bad drift wood. Now and then a solitary farm opened from the woods, where the negro children were running naked about the yards. I also passed along the north side of a high hill, where the whole timber had been prostrated by some terrible hurri-

* These drawings never came to hand.—ORD.

cane. I lodged this night in a miner's, who told me he had been engaged in forming no less than thirteen companies for hunting mines, all of whom had left him. I advised him to follow his farm, as the surest vein of ore he could work. Next day (Saturday) I first observed the cane growing, which increased until the whole woods were full of it. The road this day winded along the high ridges of mountains that divide the waters of the Cumberland from those of the Tennessee. I passed few houses to-day; but met several parties of boatmen returning from Natchez and New Orleans, who gave me such an account of the road, and the difficulties they had met with, as served to stiffen my resolution to be prepared for everything. These men were as dirty as Hottentots; their dress, a shirt and trousers of canvass, black, greasy, and sometimes in tatters; the skin burnt wherever exposed to the sun; each with a budget, wrapt up in an old blanket; their beards, eighteen days old, added to the singularity of their appearance, which was altogether savage. These people came from the various tributary streams of the Ohio, hired at forty or fifty dollars a-trip, to return back on their own expenses. Some had upwards of eight hundred miles to travel. When they come to a stream that is unfordable, they coast it for fallen tree; if that cannot be had, they enter with their budget n their head, and, when they lose bottom, drop it on their shoulders, and take to swimming. They have sometimes fourteen or fifteen of such streams to pass in a day, and morasses of several miles in length, that I have never seen equalled in any country. I lodged this night at one Dobbin's, where ten or twelve of these men lay on the floor. As they scrambled up in the morning, they very generally complained of being unwell, for which they gave an odd reason,—lying within doors, it being the first of fifteen nights they had been so indulged. Next morning (Sunday) I rode six miles to a man's of the name of Grinder, where our poor friend Lewis perished.* In the same room where he expired, I took down from Mrs Grinder the particulars of that melancholy event, which affected me extremely. This house, or cabin, is seventy-two miles from Nashville, and is the last white man's as you enter

* "It is hardly necessary to state that this was the brave and enterprising traveller, whose journey across the Rocky Mountains, to the Pacific Ocean, has obtained for him well merited celebrity. The true cause of his committing the rash deed, so feelingly detailed here, is not yet known to the public."

the Indian country. Governor Lewis, she said, came thither about sunset, alone, and inquired if he could stay for the night; and, alighting, brought his saddle into the house. He was dressed in a loose gown, white, stripped with blue. On being asked if he came alone, he replied, that there were two servants behind, who would soon be up. He called for some spirits, and drank a very little. When the servants arrived, one of whom was a negro, he inquired for his powder, saying he was sure he had some powder in a canister. The servant gave no distinct reply, and Lewis, in the meanwhile, walked backwards and forwards before the door, talking to himself. Sometimes, she said, he seemed as if he were walking up to her, and would suddenly wheel round, and walk back as fast as he could. Supper being ready, he sat down, but had eaten only a few mouthfuls, when he started up, speaking to himself in a violent manner. At these times, she says, she observed his face to flush as if it had come on him in a fit. He lighted his pipe, and, drawing a chair to the door, sat down, saying to Mrs Grinder, in a kind tone of voice, 'Madam, this is a very pleasant evening.' He smoked for some time, but quitted his seat, and traversed the yard as before. He again sat down to his pipe, seemed again composed, and, casting his eyes wistfully towards the west, observed what a sweet evening it was. Mrs Grinder was preparing a bed for him; but he said he would sleep on the floor, and desired the servants to bring the bear skins and buffalo robe, which were immediately spread out for him; and, it being now dusk, the woman went off to the kitchen, and the two men to the barn, which stands about two hundred yards off. The kitchen is only a few paces from the room where Lewis was; and the woman, being considerably alarmed by the behaviour of her guest, could not sleep, but listened to him walking backwards and forwards, she thinks, for several hours, and talking aloud, as she said, 'like a lawyer.' She then heard the report of a pistol, and something fall heavily on the floor, and the words, 'O Lord!' Immediately afterwards she heard another pistol; and, in a few minutes, she heard him at her door, calling out, 'O Madam! give me some water, and heal my wounds.' The logs being open, and unplastered, she saw him stagger back, and fall against a stump that stands between the kitchen and the room. He crawled for some distance, raised himself by the side of a tree, where he sat about a minute. He once more got to the room; afterwards, he came to the kitchen door, but did not speak; she then heard him scraping the bucket with a gourd for water, but it appeared that this cooling element was denied the dying man! As soon as day broke, and not before, the terror of the woman having permitted him to remain for two hours in this most deplorable situation, she sent two of her children to the barn, her husband not being at home, to bring the servants; and, on going in, they found him lying on the bed. He uncovered his side, and showed them where the bullet had entered; a piece of the forehead was blown off, and had exposed the brains, without having bled much. He begged that they would take his rifle and blow out his brains, and he would give them all the money he had in his trunk. He often said, 'I am no coward; but I am so strong—so hard to die!' He begged the servant not to be afraid of him, for that he would not hurt him. He expired in about two hours, or just as the sun rose above the trees. He lies buried close by the common path, with a few loose rails thrown over his grave. I gave Grinder money to put a post fence round it, to shelter it from the hogs and from the wolves, and he gave me his written promise that he would do it. I left this place in a very melancholy mood, which was not much allayed by the prospect of the gloomy and savage wilderness which I was just entering alone.

"I was roused from this melancholy reverie by the roaring of Buffalo River, which I forded with considerable difficulty. I passed two or three solitary Indian huts in the course of the day, with a few acres of open land at each; but so wretchedly cultivated, that they just make out to raise maize enough to keep in existence. They pointed me out the distances by holding up their fingers. This is the country of the Chickasaws, though erroneously laid down in some maps as that of the Cherokees. I slept this night in one of their huts: the Indians spread a deer skin for me on the floor; I made a pillow of my portmanteau, and slept tolerably well: an old Indian laid himself down near me. On Monday morning I rode fifteen miles, and stopt at an Indian's to feed my horse. The sight of my paroquet brought the whole family around me. The women are generally naked from the middle upwards; and their heads, in many instances, being rarely combed, look like a large mop. They have a yard or two of blue cloth wrapt round by way of a petticoat, that reaches to their knees. The boys were generally naked, except a kind of bag of blue cloth, by way of a *fig leaf.* Some of the women have short jackets, with sleeves, drawn over their naked body, and the rag of a blanket is a general appendage. I met to-day two officers of the United States army, who gave me a better account of the road than I had received. I passed through many bad swamps to-day, and, at about five in the evening, came to the banks of the Tennessee, which was swelled by the rains, and is about half a mile wide thirty miles below the mussel shoals, and just below a long island laid down in your small map. A growth of canes, of twenty and thirty feet high, covers the low bottoms; and these cane swamps are the gloomiest and most desolate-looking places imaginable. I hailed for the boat as long as it was light without effect. I then sought for a place to encamp, kindled a large fire, stripped the canes for my horse, ate a bit of supper, and lay down to sleep, listening to the owls and the chuck-will's-widow, a kind of whip-poor-will that is very numerous here. I got up several times during the night to recruit my fire, and see how my horse did, and, but for the gnats, would have slept tolerably well. These gigantic woods have a singular effect by the light of a large fire, the whole scene being circumscribed by impenetrable darkness, except that in front, where every leaf is strongly defined, and deeply shaded. In the morning, I hunted until about six, when I again renewed my shouting for the boat, and it was not until it was near eleven that it made its appearance. I was so enraged with this delay, that had I not been cumbered with baggage, I believe I should have ventured to swim the river. I vented my indignation on the owner of the boat, who is a half-breed, threatening to publish him in the papers, and advise every traveller I met to take the upper ferry. This man charges one dollar for man and horse, and thinks, because he is a chief, he may do in the way what he pleases. The country now assumed a new appearance; no brushwood—no fallen or rotten timber; one could see a mile through the woods, which were covered with high grass, fit for mowing. These woods are burnt every spring, and thus are kept so remarkably clean, that they look like the most elegant nobleman's parks. A profusion of flowers altogether new to me, and some of them very elegant, presented themselves to my view as I rode along. This must be a heavenly place for the botanist. The most observable of these flowers was a kind of sweetwilliam of all tints, from white to the deepest crimson; a superb thistle, the most beautiful I had ever seen; a species of passion flower, very beautiful; a stately plant of the sun flower family, the button of the deepest orange, and the radiating petals bright carmine, the breadth of the flower is about four inches; a large white flower, like a deer's tail; great quantities of the sensitive plant, that shrunk instantly on being touched, covered the ground in some places. Almost every flower was new to me, except the Carolina pink root and Columbo, which grew in abundance on every side. At Bear Creek, which is a large and rapid stream, I first observed the Indian boys with their *blow-guns.* These are tubes of cane, seven feet long, and perfectly straight, when well made. The arrows are made of slender slips of cane, twisted, and straightened before the fire, and covered for several inches at one end with the down of thistles, in a spiral form, so as just to enter the tube. By a puff, they can send these with such violence, as to enter the body of a partridge twenty yards off. I set several of them a-hunting birds, by promises of reward, but not one of them could succeed. I also tried some of the blow-guns myself, but found them generally defective in straightness. I met six parties of boatmen to-day, and many straggling Indians, and encamped about sunset near a small brook, where I shot a turkey, and, on returning to my fire, found four boatmen, who stayed with me all night, and helped to pick the bones of the turkey. In the morning, I heard the turkeys gobbling all round me, but not wishing to leave my horse, having no great faith in my guest's honesty, I proceeded on my journey.

"This day (Wednesday) I passed through the most horrid swamps I had ever seen. These are covered with a prodigious growth of canes and high woods, which, together, shut out almost the whole light of day, for miles. The banks of the deep and sluggish creeks, that occupy the centre, are precipitous; where I had often to plunge my horse seven feet down, into a bed of deep clay, up to his belly, from which nothing but great strength and exertion could have rescued him; the opposite shore was equally bad, and beggars all description. For an extent of several miles, on both sides of these creeks, the darkness of night obscures every object around. On emerging from one of the worst of these, I met General Wade Hampton, with two servants and a pack-horse, going, as he said, towards Nashville. I told him of the mud campaign immediately before him; I was covered with mire and wet, and I thought he looked somewhat serious at the difficulties he was about to engage. He has been very sick lately. About half an hour before sunset, being within sight of the Indian's where I intended to lodge, the evening being perfectly clear and calm, I laid the reins on my horse's neck, to listen to a mocking bird, the first I had heard in the western country, which, perched on the top of a dead tree before the door, was pouring out a torrent of melody. I think I never heard so excellent a performer. I had alighted, and was fastening my horse, when, hearing the report of a rifle immediately beside me, I looked up, and saw the poor mocking bird fluttering to the ground: one of the savages had marked his elevation, and barbarously shot him. I hastened over to the yard, and, walking up to him, told him that was bad, very bad, that this poor bird had come from a far distant country to sing to him, and that, in return, he had cruelly killed him. I told him the Great Spirit was offended at such cruelty, and that he would lose many a deer for so doing. The old Indian, father-in-law to the bird killer, understanding, by the negro interpreter, what I said, replied, that, when these birds come singing and making a noise all day near the house, somebody will surely die, which is exactly what an old superstitious German, near Hampton, in Virginia, told me. This fellow has married the two eldest daughters of the old Indian, and presented one of them with the bird he had killed. The next day I passed through the Chickasaw *Bigtown,* which stands on the high open plain that extends through the country, three or four miles in breadth, by fifteen in length. Here and there you perceive little groups of miserable huts, formed of saplings, and plastered with mud and clay. About these are generally a few peach and plum trees. Many ruins of others stand scattered about, and I question whether there were twenty inhabited huts within the whole range of view. The ground was red with strawberries, and the boatmen were seen, in straggling parties, feasting on them. Now and then a solitary Indian, wrapt in his blanket, passed sullen and silent. On this plain are beds of shells, of a large species of clam, some of which are almost entire. I this day stopped at the house of a white man, who had two Indian wives, and a hopeful string of young savages, all in their fig leaves. Not one of them could speak a word of English. This man was by birth a Virginian, and had been forty years among the Chickasaws. His countenance and manners were savage, and worse than Indian. I met many parties of boatmen to-day, and crossed a number of bad swamps. The woods continue to exhibit the same open luxuriant appearance; and at night I lodged at a white man's, who has also two wives, and a numerous progeny. Here I met with a lieutenant of the United States army, anxiously inquiring for General Hampton. On Friday, the same open woods continued. I met several parties of Indians, and passed two or three of their hamlets. At one of these there were two fires in the yard, and at each eight or ten Indian men and women squat on the ground. In these hamlets there is generally one house built, of a circular form, and plastered thickly all over both without and within with clay. This they call a hot-house; and it is the general winter quarters of the hamlet in cold weather. Here they all kennel, and, having neither window, nor place for the smoke to escape, it must be a sweet place, while forty or fifty of them have it in occupancy. Round some of these hamlets were great droves of cattle, horses, and hogs. I lodged this night on the top of a hill, far from water, and suffered severely from thirst.

"On Saturday, I passed a number of most execrable swamps; the weather was extremely warm, and I had been attacked by something like the dysentery, which occasioned a constant burning thirst, and weakened me greatly. I stopt this day frequently to wash my head and throat in the water, to allay the burning thirst; and, putting on my hat without wiping, received considerable relief from it. Since crossing the Tennessee, the woods have been interspersed with pines, and the soil has become more sandy. This day I met a Captain Hughes, a traveller on his return from Santa Fee. My complaint increased so much, that I could scarcely sit on horseback; and all night my mouth and throat were parched with a burning thirst and fever. On Sunday, I bought some raw eggs, which I ate, and repeated the dose at mid-day, and towards evening, and found great benefit from this simple remedy. I inquired, all along the road, for fresh eggs, and, for nearly a week, made them almost my sole food, till I completed my cure. The water in these cane swamps is little better than poison; and, under the heat of a burning sun, and the fatigues of travelling, it is difficult to repress the urgent calls of thirst. On the Wednesday following, I was assailed by a tremendous storm of rain, wind, and lightning, until I and my horse were both blinded by the deluge, and unable to go on. I sought the first most open place, and, dismounting, stood for half an hour under the most profuse heavenly shower-bath I ever enjoyed. The roaring of the storm was terrible; several trees around were broken off, and torn up by the roots, and those that stood were bent almost to the ground; limbs of trees, of several hundred-weight, flew past, within a few yards of me, and I was astonished how I escaped. I would rather take my chance in a field of battle, than in such a tornado again.

"On the fourteenth day of my journey, at noon, I arrived at this place [Natchez, Mississippi territory], having overcome every obstacle, alone, and without being acquainted with the country; and, what surprised the boatmen more, without whisky. On an average, I met from forty to sixty boatmen every day, returning from this place and New Orleans. The Chickasaws are a friendly, inoffensive people; and the Chactaws, though more reserved, are equally harmless. Both of them treated me with civility, though I several times had occasion to pass through their camps, where many of them were drunk. The paroquet which I carried with me was a continual fund of amusement to all ages of these people; and, as they crowded around to look at it, gave me an opportunity of studying their physiognomies without breach of good manners.

"In thus hastily running over the particulars of this journey, I am obliged to omit much that would amuse and interest you; but my present situation,—a noisy tavern, crowded in every corner, even in the room where I write, with the sons of riot and dissipation,—prevents me from enlarging on particulars. I could also have wished to give you some account of this place, and of the celebrated Mississippi, of which you have heard so much. On these subjects, however, I can at present only offer you the following slight sketch, taken the morning after my arrival here:—

"The best view of this place and surrounding scenery, is from the old Spanish fort on the south side of the town, about a quarter of a mile distant. From this high point, looking up the river, Natchez lies on your right—a mingled group of green trees, and white and red houses, occupying an uneven plain, much washed into ravines, rising as it recedes from the bluff or high precipitous banks of the river. There is, however, neither steeple, cupola, nor distinguished object to add interest to its appearance. The country beyond it, to the right, is thrown up into the same irregular knolls; and, at the distance of a mile, in the same direction, you have a peep of some cultivated farms, bounded by the general forest. On your left, you look down, at a depth of two or three hundred feet, on the river, winding majestically to the south; the intermediate space exhibiting wild perpendicular precipices of brown earth. This part of the river and shore is the general rendezvous of all the arks, or Kentucky boats, several hundreds of which are at present lying moored there, loaded with the produce of the thousand shores of this noble river. The busy multitudes below present a perpetually varying picture of industry; and the noise and uproar, softened by the distance, with the continued crowing of the poultry with which many of those arks are filled, produce cheerful and exhilarating ideas. The majestic Mississippi, swelled by his ten thousand tributary streams, of a pale brown colour, half a mile wide, and spotted with trunks of trees, that show the different threads of the current, and its numerous eddies, bears his depth of water past in silent grandeur. Seven gun-boats, anchored at equal distances along the stream, with their ensigns displayed, add to the effect. A few scattered houses are seen on the low opposite shore, where a narrow strip of cleared land exposes the high gigantic trunks of some deadened timber that bounds the woods. The whole country beyond the Mississippi, from south round to west and north, presents to the eye one universal level ocean of forest, bounded only by the horizon. So perfect is this vast level, that not a leaf seems to rise above the plains, as if shorn by the hands of heaven. At this moment, while I write, a terrific thunder-storm, with all its towering assemblage of black alpine clouds, discharging lightning in every direction, overhangs this vast level, and gives a magnificence and sublime effect to the whole."

In June, Wilson reached New Orleans, and, sailing from thence to New York, again entered Philadelphia, after a very long and arduous journey; during which he experienced many difficulties from the season and climate, the wildness of the paths, and from a sickness which had nearly proved fatal, but which his still good constitution, and the simple prescription of an Indian, bore him safely through. He nevertheless procured much information, and new materials for his work, besides keeping up an extensive correspondence with his friends, and regularly entering in a diary the events of each day. From this, and the corresponding account of Mr Audubon, we learn that these ornithologists first met at Louisville; and have to regret that their intimacy and acquaintance had not a longer existence. Before this meeting, neither seemed to have had any idea of the other's existence, though both were eagerly pursuing the same object. Wilson thus notices it in his diary:—

"*March* 19.—Rambling round the town with my gun. Examined Mr ——'s drawings in crayons—very good. Saw two new birds he had—both *Motacillæ*.

"*March* 20.—Set out this afternoon with the gun—killed nothing new. People in taverns here devour their meals. Many shopkeepers board in taverns—also boatmen, land-speculators, merchants, &c. *No naturalist to keep me company.*

"March 21.—Went out this afternoon shooting with Mr A——. Saw a number of sandhill crows; pigeons numerous."

Till 1812, Wilson resided chiefly at Philadelphia, with Mr Bartram, writing and superintending his work, and making extensive excursions around the neighbouring country. The colouring of the plates cost him much trouble; and he even wrought at this department himself, anxious to render them as brilliant and true to nature as possible. This is a branch of the art yet understood in this country by a very few only, and is one of the greatest bars to the faithful representation of the more splendid forms of the feathered race.

His American biographer now, for the first time, introduces himself as the friend and companion of Wilson. He does not mention at what period this intimacy commenced, but tells us that, in 1812, a journey was undertaken by our author into the eastern states, for the purpose of visiting his subscribers and settling accounts with his agents. During this expedition Wilson writes to Mr Ord, giving a short account of the excursion :—

To Mr George Ord.

"BOSTON, *October* 13, 1812.

"DEAR SIR,—It is not in my power at present to give you anything more than a slight sketch of my rambles since leaving Philadelphia. My route up the Hudson afforded great pleasure, mingled with frequent regret that you were not along with me to share the enjoyment. About thirty miles south of Albany we passed within ten miles of the celebrated Catskil Mountains, a gigantic group, clothed with forest to the summits. In the river here I found our common reed (*Zizania aquatica*) growing in great abundance, in shoals extending along the middle of the river. I saw flocks of redwings, and some black ducks, but no rail or reed-birds. From this place my journey led me over a rugged mountainous country to Lake Champlain, along which I coasted as far as Burlington in Vermont. Here I found the little coot-footed tringa, or phalarope, that you sent to Mr Peale; a new and elegantly marked hawk; and observed some black ducks. The shores are alternate sandy bays and rocky headlands running into the lake. Every tavern was crowded with officers, soldiers, and travellers. Eight of us were left without a bed; but having an excellent greatcoat, I laid myself down in a corner, with a determination of sleeping in defiance of the uproar of the house, and the rage of my companions, who would not disgrace themselves by a prostration of this sort. From Lake Champlain I traversed a rude mountainous region to Connecticut River, one hundred miles above Dartmouth College. I spent several days with the gun in Graton and Ryegate townships, and made some discoveries. From this I coasted along the Connecticut to a place called Haverhill, ten miles from the foot of Moose-Hillock, one of the highest of the *White Mountains* of New Hampshire. I spent the greater part of a day in ascending to the peak of one of these majestic mountains, whence I had the most sublime and astonishing view that was ever afforded me. One immensity of forest lay below, extended on all sides to the farthest verge of the horizon ; while the only prominent objects were the columns of smoke from burning woods that rose from various parts of the earth beneath to the heavens ; for the day was beautiful and serene. Hence I travelled to Dartmouth, and thence in a direct course to Boston. From Boston I passed through Portsmouth to Portland, and got some things new. My return was by a different route. I have procured three new and beautiful hawks ; and have gleaned up a stock of materials that will be useful to me hereafter.

"I hope, my dear sir, that you have been well since I left you. I have myself been several times afflicted with a violent palpitation of the heart, and want to try whether a short voyage by sea will be beneficial or not.

"In New England, the rage of war, the virulence of politics, and the pursuit of commercial speculations, engross every faculty. The voice of science, and the charms of nature, unless these last present themselves in the form of prize sugars, coffee, or rum, are treated with contempt."

This letter concludes the series of those which were written Wilson's friends during his various excursions ; they have been given without abridgment, as the surest means to judge of his mind and disposition during his residence in America ; and we possess few additional records of the short remaining space of this ornithologist's life.

The Seventh Part of the "Ornithology" was far advanced, and soon after its publication, Wilson set out, accompanied by Mr Ord, on an expedition to Egg Harbour, to procure materials for the Eighth Volume, which would principally have contained the marine water fowl. This was his last expedition, and occupied nearly four months. On returning to Philadelphia, the anxiety to perfect the forthcoming volume, which he thought would bring his labours nearly to a conclusion, and would show him the end of a work upon which he had periled his dear reputation, brought on an attack of his old complaints, which had gradually become more frequent when his mind or body was harassed or agitated for the accomplishment of any favourite project. He was seized with dysentery, and yielded, after an attack of ten days, to a power which his constitution was unable longer to withstand.* Thus closed the life of Wilson, chequered from its commencement with changes of fortune very varied,—active in the extreme, and having, for its chief objects, the good or temporary relief of his fellow-men, with an ardent desire to communicate to them and illustrate the wonders and beauties in the works of his Creator. How far he has succeeded, must be judged of from the evidences he has left: of his wishes and anxiety for their perfection, there can be only one opinion.

As a poet Wilson is much esteemed by his townsmen, and perhaps his writings are there more highly appreciated, from the circumstance that many of the characters alluded to are still alive, and the scenes and incidents of their young days, which are often portrayed, are still vivid on their memories. Such recollection, and the retracing of their former companionship, will bear with them charms which another generation will not so highly prize, and who will judge of his productions with more impartial minds. I do not mean, on this account, to despise his poems; but they will not stand in that high place where his most enthusiastic admirers station them. They all bear the mark of truth in the manners, the scenes, and the incidents which they delineate, and show most correctly the view which their author took of his subject, and the feelings with which he was at the moment impressed; but they were avowedly written almost at the moment of conception, without bestowing very great care on their composition. He never appears really to have studied with a view to a fine poem, or to have attended to the rules of the art. He was passionately fond of poetry; and his ear was formed, and style taken, from the favourite authors he so constantly perused. If he had devoted himself to this branch with the same ardour that he engaged in ornithology, and bestowed an equal portion of time on it, he would probably have risen to considerable eminence, but it was only his relaxation and amusement. It was attended, how-

* The immediate cause of his last illness is stated by Captain Warnock to have been a cold, caught during a long chase after some much desired bird, in the course of which, and when overheated, he swam several rivers and small creeks.

ever, with one advantage,—the constant expression of his thoughts in this way gave him a facility of description, and flow of language, which destroyed the dry and stiff character so often and so unavoidably prevailing in scientific works.

It is as an ornithologist that Wilson's fame will last for after ages. There are two classes of naturalists,—who may be described as those who see and study the habits of the living creatures in their natural abodes, and describe or figure them under these circumstances; and those who receive the specimens in a state of preservation draw their conclusions from the little they have been able personally to observe, and learn the rest of their manners from the best authorities in their power. These two methods of study are, indeed, nearly indispensable to each other. To the general naturalist the first is impossible; and what the one class observes, the other generalises and brings to bear on the various facts seen and recorded by others. Wilson was an observing naturalist; and, perhaps, Nature never had a more ardent pursuer. His object was to illustrate the different birds in their various states, as closely to the truth as possible, and to describe those parts of their manners which he could from actual observation, throwing aside all hearsay evidence, and seldom indulging in any theories of classification, or the scale they hold in Nature. It is from these circumstances that his work derives its worth: the facts can be confidently quoted as authentic, and their value depended on in our reasonings upon their history—their migrations—their geographical distribution.

In his private character Wilson bore a very high station. As a youth, he was beloved by his acquaintances, and respected even by those whom ill advice had made his temporary enemies. In the New World, there existed an attachment of the warmest description between him and those friends whom his literary attainments had procured; and by them his memory continues to be fondly cherished, and his talents to be held in great respect. In his birthplace, a society has been formed by his admirers, who meet annually to talk over past recollections, where the merits of his works, and the remembrance of the deceased poet and naturalist, are commemorated in a speech or an ode. Among all his former friends who still enjoy life, his name is welcomed with an enthusiasm which I have never seen equalled. Paisley is justly proud of her late townsman.

AMERICAN ORNITHOLOGY.

1. Blue Jay. 2. Yellow-Bird or Goldfinch. 3. Baltimore Bird.
Corvus cristatus. Fringilla Tristis. Oriolus Baltimorus.

Drawn from Nature by A. Wilson — Engraved by W.H. Lizars

1.

WILSON'S

AMERICAN ORNITHOLOGY.

BLUE JAY. (*Corvus cristatus.*)

PLATE I.—FIG. 1.

Linn. Syst. i. p. 157, 158.—Garrulus Canadensis cœruleus, *Briss.* ii. p. 54, 2. t. 4. fig. 2.—Pica glandaria cristata, *Klein*, p. 61, 3.—Le geay bleu du Canada, *Buff.* iii. p. 120. *Pl. enl.* 529.—Blue Jay, *Catesb. Car.* i. 15.—*Edw.* 239.—*Arct. Zool.* ii. No. 38.—*Lath. Syn.* i. p. 386, 20.—*Bartram*, p. 290.—*Peale's Museum*, No. 1290.

GARRULUS CRISTATUS.—VIEILLOT.

Garrulus cristatus, *Vieill. Gal. des Ois.* pl. 102.—*North. Zool.* ii. p. 293.—*Bonap. Synop.* No. 63.—Pica cristata, *Wagl.* No. 8.

THIS elegant bird, which, as far as I can learn, is peculiar to North America, is distinguished as a kind of beau among the feathered tenants of our woods, by the brilliancy of his dress; and, like most other coxcombs, makes himself still more conspicuous by his loquacity, and the oddness of his tones and gestures. The jay measures eleven inches in length; the head is ornamented with a crest of light blue or purple feathers, which he can elevate or depress at pleasure; a narrow line of black runs along the frontlet, rising on each side higher than the eye, but not passing over it, as Catesby has represented, and as Pennant and many others have described it; back and upper part of the neck, a fine light purple, in which the blue predominates; a collar of black, proceeding from the hind head, passes with a graceful curve down each side of the neck to the upper part of the breast, where it forms a crescent; chin, cheeks, throat, and belly, white, the three former slightly tinged with blue; greater wing-coverts, a rich blue; exterior sides of the primaries, light blue, those of the secondaries, a deep purple, except the three feathers next the body, which are of a splendid light blue; all these, except the primaries, are beautifully barred with crescents of black, and tipt with white; the interior sides of the wing-feathers are dusky black; tail long and cuneiform, composed of twelve feathers of a glossy light blue, marked at half inches with transverse curves of black, each feather being tipt with white, except the two middle ones, which deepen into a dark purple at the extremities. Breast and sides under the wings, a dirty white, faintly stained with purple; inside of the mouth, the tongue, bill, legs, and claws, black; iris of the eye, hazel.

The blue jay is an almost universal inhabitant of the woods, frequenting the thickest settlements as well as the deepest recesses of the forest, where his squalling voice often alarms the deer, to the disappointment and mortification of the hunter; one of whom informed me, that he made it a point, in summer, to kill every jay he could meet with. In the charming season of spring, when every thicket pours forth harmony, the part performed by the jay always catches the ear. He appears to be among his fellow-musicians what the trumpeter is in a band, some of his notes having no distant resemblance to the tones of that instrument. These he has the faculty of changing through a great variety of modulations, according to the particular humour he happens to be in. When disposed for ridicule, there is scarce a bird whose peculiarities of song he cannot tune his notes to. When engaged in the blandishments of love, they resemble the soft chatterings of a duck, and, while he nestles among the thick branches of the cedar, are scarce heard at a few paces' distance; but he no sooner discovers your approach than he sets up a sudden and vehement outcry, flying off, and screaming with all his might, as if he called the whole feathered tribes of the neighbourhood to witness some outrageous usage he had received. When he hops undisturbed among the high branches of the oak and hickory, they become soft and musical; and his calls of the female, a stranger would readily mistake for the repeated screakings of an ungreased wheelbarrow. All these he accompanies with various nods, jerks, and other gesticulations, for which the whole tribe of jays are so remarkable, that, with some other peculiarities, they might have very well justified the great Swedish naturalist in forming them into a separate genus by themselves.*

The blue jay builds a large nest, frequently in the cedar,

* This has now been done; and modern ornithologists adopt the title *Garrulus*, of Brisson, for this distinct and very well defined group, containing many species, which agree intimately in their general form and habits, and are dispersed over every quarter of the world, New Holland excepted. The colours of their plumage are brown, gray, blue, and black; in some distributed with sober chastity, while, in others, the deep tints and decided markings rival the richest gems.

> Proud of cœrulean stains,
> From heaven's unsullied arch purloined, the jay
> Screams hoarse. GISBORNE'S *Walks in a Forest.*

In geographical distribution, we find those of splendid plumage following the warmer climates, and associating there with our ideas of Eastern magnificence; while the more sober dressed, and, in our opinion, not the least pleasing, range through more temperate and northern regions, or those exalted tracts in tropical countries where all the productions in some manner receive the impress of an alpine or northern station. This is nowhere better exemplified than in the specimens lately sent to this country from the lofty and extensive plains of the Himalaya, where we have already met with prototypes of the European jay, black and green woodpeckers, greater titmouse, and nutcracker. They inhabit woody districts; in their dispositions are cunning, bold, noisy, active, and restless, but docile and easily tamed when introduced to the care of man, and are capable of being taught tricks and various sounds. The following instance of the latter propensity is thus related by Bewick :—"We have heard one imitate the sound made by the action of a saw, so exactly, that though it was on a Sunday, we could hardly be persuaded that the person who kept it had not a carpenter sometimes on an apple tree, lines it with dry fibrous roots, and lays five eggs of a dull olive, spotted with brown. The male is particularly careful of not being heard near the place, making his visits as silently and secretly as possible. His favourite food is chestnuts, acorns, and Indian-corn. He occasionally feeds on bugs and caterpillars, and sometimes pays a plundering visit to the orchard, cherry rows, and potato patch; and has been known, in times of scarcity, to venture into the barn, through openings between the weather boards. In these cases he is extremely active and silent, and, if surprised in the fact, makes his escape with precipitation, but without noise, as if conscious of his criminality.

Of all birds, he is the most bitter enemy to the owl. No sooner has he discovered the retreat of one of these, than he

at work in the house. Another, at the approach of cattle, had learned to hound a cur dog upon them, by whistling and calling upon him by his name. At last, during a severe frost, the dog was, by that means, excited to attack a cow big with calf, when the poor animal fell on the ice, and was much hurt: the jay was complained of as a nuisance; and its owner was obliged to destroy it." They feed indiscriminately, and, according to circumstances, on either animal or vegetable substances; plundering nests of their eggs and young, and even, in the more exposed farmyards, disappointing the hopes of the mistress, in the destruction of a favourite brood. They are also robbers of orchards and gardens of their finest fruits; but, when without the reach of these luxuries, they will be content to satisfy their hunger with Nature's own productions, the wild berries or fruits and seeds of the forest and the field.

Several new species have been added to the North American list, some of which are described by the Prince of Musignano in our third volume; and, in addition, we may mention one new species, published by Dr Richardson and Mr Swainson in the *Arctic Zoology*. The only specimen brought home was killed on the roof of the dwelling-house at Fort Franklin, and was so similar to the Canada jay, that it was not then recognised as a distinct species. The chief distinctions mentioned in the above work are the shorter bill, broader at the base, and narrower on the ridge. The plumage looser than in *G. Canadensis;* the secondaries proportionally longer, and all end in slender, but very distinct points, scarcely discernible in the blue jay, and not nearly so much developed in the whisky-jack. Tail is shorter than the latter, the tarsus is more robust.—ED.

summons the whole feathered fraternity to his assistance, who surround the glimmering *solitaire*, and attack him from all sides, raising such a shout as may be heard, in a still day, more than half a mile off. When, in my hunting excursions, I have passed near this scene of tumult, I have imagined to myself that I heard the insulting party venting their respective charges with all the virulency of a Billingsgate mob; the owl, meanwhile returning every compliment with a broad goggling stare. The war becomes louder and louder, and the owl at length, forced to betake himself to flight, is followed by his whole train of persecutors, until driven beyond the boundaries of their jurisdiction.

But the blue jay himself is not guiltless of similar depredations with the owl, and becomes in his turn the very tyrant he detested, when he sneaks through the woods, as he frequently does, and among the thickets and hedgerows, plundering every nest he can find of its eggs, tearing up the callow young by piecemeal, and spreading alarm and sorrow around him. The cries of the distressed parents soon bring together a number of interested spectators (for birds in such circumstances seem truly to sympathise with each other), and he is sometimes attacked with such spirit as to be under the necessity of making a speedy retreat.

He will sometimes assault small birds, with the intention of killing and devouring them; an instance of which I myself once witnessed, over a piece of woods near the borders of Schuylkill; where I saw him engaged for more than five minutes pursuing what I took to be a species of *Motacilla*, wheeling, darting, and doubling in the air, and at last, to my great satisfaction, got disappointed, in the escape of his intended prey. In times of great extremity, when his hoard or magazine is frozen up, buried in snow, or perhaps exhausted, he becomes very voracious, and will make a meal of whatever carrion or other animal substance comes in the way, and has been found regaling himself on the bowels of a robin in less than five minutes after it was shot.

There are, however, individual exceptions to this general character for plunder and outrage, a proneness for which is probably often occasioned by the wants and irritations of necessity. A blue jay, which I have kept for some time, and with whom I am on terms of familiarity, is in reality a very notable example of mildness of disposition and sociability of manners. An accident in the woods first put me in possession of this bird, while in full plumage, and in high health and spirits; I carried him home with me, and put him into a cage already occupied by a golden-winged woodpecker, where he was saluted with such rudeness, and received such a drubbing from the lord of the manor, for entering his premises, that, to save his life, I was obliged to take him out again. I then put him into another cage, where the only tenant was a female orchard oriole. She also put on airs of alarm, as if she considered herself endangered and insulted by the intrusion; the jay, meanwhile, sat mute and motionless on the bottom of the cage, either dubious of his own situation, or willing to allow time for the fears of his neighbour to subside. Accordingly, in a few minutes, after displaying various threatening gestures (like some of those Indians we read of in their first interviews with the whites), she began to make her approaches, but with great circumspection, and readiness for retreat. Seeing, however, the jay begin to pick up some crumbs of broken chestnuts, in a humble and peaceable way, she also descended, and began to do the same; but, at the slightest motion of her new guest, wheeled round, and put herself on the defensive. All this ceremonious jealousy vanished before evening; and they now roost together, feed, and play together, in perfect harmony and good humour. When the jay goes to drink, his messmate very impudently jumps into the water to wash herself, throwing the water in showers over her companion, who bears it all patiently; venturing now and then to take a sip between every splash, without betraying the smallest token of irritation. On the contrary, he seems to take pleasure in his little fellow-prisoner, allowing her to pick (which she does very gently) about his whiskers, and to clean his claws from the minute fragments of chestnuts which happen to adhere to them. This attachment on the one part, and mild condescension on the other, may, perhaps, be partly the effect of mutual misfortunes, which are found not only to knit mankind, but many species of inferior animals, more closely together; and shows that the disposition of the blue jay may be humanised, and rendered susceptible of affectionate impressions, even for those birds which, in a state of nature, he would have no hesitation in making a meal of.

He is not only bold and vociferous, but possesses a considerable talent for mimicry, and seems to enjoy great satisfaction in mocking and teasing other birds, particularly the little hawk (*F. sparverius*), imitating his cry wherever he sees him, and squealing out as if caught: this soon brings a number of his own tribe around him, who all join in the frolic, darting about the hawk, and feigning the cries of a bird sorely wounded, and already under the clutches of its devourer; while others lie concealed in bushes, ready to second their associates in the attack. But this ludicrous farce often terminates tragically. The hawk, singling out one of the most insolent and provoking, sweeps upon him in an unguarded moment, and offers him up a sacrifice to his hunger and resentment. In an instant the tune is changed; all their buffoonery vanishes, and loud and incessant screams proclaim their disaster.

Wherever the jay has had the advantage of education from man, he has not only shown himself an apt scholar, but his suavity of manners seems equalled only by his art and contrivances; though it must be confessed that his itch for thieving keeps pace with all his other acquirements. Dr Mease, on the authority of Colonel Postell, of South Carolina, informs me, that a blue jay which was brought up in the family of the latter gentleman, had all the tricks and loquacity of a parrot; pilfered everything he could conveniently carry off, and hid them in holes and crevices; answered to his name with great sociability, when called on; could articulate a number of words pretty distinctly; and, when he heard any uncommon noise or loud talking, seemed impatient to contribute his share to the general festivity (as he probably thought it) by a display of all the oratorical powers he was possessed of.

Mr Bartram relates an instance of the jay's sagacity worthy of remark. "Having caught a jay in the winter season," says he, "I turned him loose in the greenhouse, and fed him with corn (zea, maize), the heart of which they are very fond of. This grain being ripe and hard, the bird at first found a difficulty in breaking it, as it would start from his bill when he struck it. After looking about, and, as if considering for a moment, he picked up his grain, carried and placed it close up in a corner on the shelf, between the wall and a plant box, where, being confined on three sides, he soon effected his purpose, and continued afterwards to make use of this same practical expedient. The jay," continues this judicious observer, "is one of the most useful agents in the economy of nature for disseminating forest trees, and other ruciferous and hard-seeded vegetables on which they feed. Their chief employment, during the autumnal season, is foraging to supply their winter stores. In performing this necessary duty, they drop abundance of seed in their flight over fields, hedges, and by fences, where they alight to deposit them in the post-holes, &c. It is remarkable what numbers of young trees rise up in fields and pastures after a wet winter and spring. These birds alone are capable, in a few years' time, to replant all the cleared lands." *

The blue jays seldom associate in any considerable numbers, except in the months of September and October, when they hover about, in scattered parties of from forty to fifty, visiting the oaks in search of their favourite acorns. At this season they are less shy than usual, and keep chattering to each other in a variety of strange and querulous notes. I have counted fifty-three, but never more, at one time; and these generally

* Letter of Mr William Bartram to the author.

following each other in straggling irregularity from one range of woods to another. Yet we are told by the learned Dr Latham—and his statement has been copied into many respectable European publications—that the blue jays of North America "often unite into flocks of twenty thousand at least! which, alighting on a field of ten or twelve acres, soon lay waste the whole."* If this were really so, these birds would justly deserve the character he gives them, of being the most destructive species in America. But I will venture the assertion, that the tribe *Oriolus phœniceus*, or red-winged blackbirds, in the environs of the river Delaware alone, devour and destroy more Indian-corn than the whole blue jays of North America. As to their assembling in such immense multitudes, it may be sufficient to observe, that a flock of blue jays of twenty thousand would be as extraordinary an appearance in America, as the same number of magpies or cuckoos would be in Britain.

It has been frequently said, that numbers of birds are common to the United States and Europe; at present, however, I am not certain of many. Comparing the best descriptions and delineations of the European ones with those of our native birds said to be of the same species, either the former are very erroneous, or the difference of plumage and habits in the latter justifies us in considering a great proportion of them to be really distinct species. Be this, however, as it may, the blue jay appears to belong exclusively to North America. I cannot find it mentioned by any writer or traveller among the birds of Guiana, Brazil, or any other part of South America. It is equally unknown in Africa. In Europe, and even in the eastern parts of Asia, it is never seen in its wild state. To ascertain the exact limits of its native regions would be difficult. These, it is highly probable, will be found to be bounded by the extremities of the temperate zone. Dr Latham has indeed asserted, that the blue jay of America is not found

* Synopsis of Birds, vol. i. p. 387. See also Encyclopædia Britannica, art. Corvus.

farther north than the town of Albany.* This, however, is a mistake. They are common in the eastern States, and are mentioned by Dr Belknap in his enumeration of the birds of New Hampshire.† They are also natives of Newfoundland. I myself have seen them in Upper Canada. Blue jays and yellow birds were found by Mr M'Kenzie, when on his journey across the continent, at the head waters of the Unjigah, or Peace River, in N. lat. 54°, W. lon. 121°, on the west side of the great range of Stony Mountains.‡ Steller, who, in 1741, accompanied Captain Behring in his expedition for the discovery of the north-west coast of America, and who wrote the journal of the voyage, relates, that he himself went on shore near Cape St Elias, in N. lat. 58° 28′, W. lon. 141° 46′, according to his estimation, where he observed several species of birds, *not known in Siberia;* and one, in particular, described by Catesby under the name of the blue jay.§ Mr William Bartram informs me, that they are numerous in the peninsula of Florida, and that he also found them at Natchez, on the Mississippi. Captain Lewis and Clark, and their intrepid companions, in their memorable expedition across the continent of North America to the Pacific Ocean, continued to see blue jays for six hundred miles up the Missouri.‖ From these accounts it follows, that this species occupies, generally or partially, an extent of country stretching upwards of seventy degrees from east to west, and more than thirty degrees from north to south; though, from local circumstances, there may be intermediate tracts, in this immense range, which they seldom visit.

* Synopsis, vol. i. p. 387.
† History of New Hampshire, vol. iii. p. 163.
‡ Voyages from Montreal, &c., p. 216, 4to, London, 1801.
§ See Steller's Journal, apud Pallas.
‖ This fact I had from Captain Lewis.

YELLOW BIRD, OR GOLDFINCH. (*Fringilla tristis.*)

PLATE I.—FIG. 2.

Linn. Syst. i. p. 320.—Carduelis Americana, *Briss.* iii. p. 6, 3.—Le Chardonnerat jaune, *Buff.* iv. p. 112. *Pl. enl.* 202, fo. 2.—American Goldfinch, *Arct. Zool.* ii. No. 242.—*Edw.* 274.—*Lath. Syn.* iii. p. 288, 57. *Id. Sup.* p. 166.—*Bartram*, p. 290.—*Peale's Museum*, No. 6344.

CARDUELIS AMERICANA.—EDWARDS.

New York Siskin, *Penn. Arct. Zool.* p. 372. (Male changing his plumage, and the male in his winter dress taken for female, auct. *Swains.*)—Fringilla tristis, *Bonap. Syn.* p. 111, No. 181.—Carduelis Americana, *North. Zool.* ii. p. 268.

THIS bird is four inches and a half in length, and eight inches in extent, of a rich lemon yellow, fading into white towards the rump and vent. The wings and tail are black, the former tipt and edged with white, the interior webs of the latter are also white; the fore part of the head is black, the bill and legs of a reddish cinnamon colour. This is the summer dress of the male; but in the month of September the yellow gradually changes to a brown olive, and the male and female are then nearly alike. They build a very neat and delicately formed little nest, which they fasten to the twigs of an apple tree, or to the strong branching stalks of hemp, covering it on the outside with pieces of lichen, which they find on the trees and fences; these they glue together with their saliva, and afterwards line the inside with the softest downy substances they can produre. The female lays five eggs, of a dull white, thickly marked at the greater end; and they generally raise two broods in a season. The males do not arrive at their perfect plumage until the succeeding spring; wanting, during that time, the black on the head, and the white on the wings being of a cream colour. In the month of April they begin to change their winter dress, and, before the middle of May, appear in brilliant yellow; the whole plumage towards its roots is of a dusky bluish black.

The song of the yellow bird resembles that of the goldfinch of Britain; but is in general so weak as to appear to proceed from a considerable distance, when perhaps the bird is perched on the tree over your head. I have, however, heard some sing in cages with great energy and animation. On their first arrival in Pennsylvania, in February, and until early in April, they associate in flocks, frequently assembling in great numbers on the same tree to bask and dress themselves in the morning sun, singing in concert for half an hour together; the confused mingling of their notes forming a kind of harmony not at all unpleasant.*

About the last of November, and sometimes sooner, they generally leave Pennsylvania, and proceed to the south; some, however, are seen even in the midst of the severest winters. Their flight is not direct, but in alternate risings and sinkings; twittering as they fly, at each successive impulse of the wings.†

* *Carduelis* of Brisson, having types in the common goldfinch and siskin of this country, is now generally used as the generic appellation for the group to which our present species belongs. It contains several American and European species. They are closely allied to the true linnets; and the lesser red-poll (the *Fringilla linaria auctorum*), has even by some been ranked with them. They also much resemble the latter group in their manners, their haunts, their breeding, and feeding. Every one who has lived much in the country, must have often remarked the common European gray linnets, in the manner above described of the American goldfinch, congregating towards the close of a fine winter's evening, perched on the summit of some bare tree, pluming themselves in the last rays of the sun, chirruping the commencement of their evening song, and then bursting simultaneously into one general chorus; again resuming their single strains, and again joining, as if happy, and rejoicing at the termination of their day's employment. Mr Audubon has remarked the same trait in their manners, and con firms the resemblance of their notes: "So much does the song of our goldfinch resemble that of the European species, that, whilst in France and England, I have frequently thought, and with pleasure thought, that they were the notes of our own bird which I heard."—Ed.

† The flight of the American goldfinch, and its manners during it, are described by Mr Audubon with greater minuteness: it is exactly similar to the European bird of the same name, being performed in deep curved lines, alternately rising and falling, after each propelling motion

During the latter part of summer they are almost constant visitants in our gardens in search of seeds, which they dislodge from the husk with great address, while hanging frequently head downwards, in the manner of the titmouse. From these circumstances, as well as from their colour, they are very generally known, and pass by various names expressive of their food, colour, &c., such as thistle bird, lettuce bird, salad bird, yellow bird, &c. The gardeners, who supply the city of Philadelphia with vegetables, often take them in trap-cages, and expose them for sale in market. They are easily familiarised to confinement, and feed with seeming indifference a few hours after being taken.

The great resemblance which the yellow bird bears to the canary has made many persons attempt to pair individuals of the two species together. An ingenious French gentleman, who resides in Pottsgrove, Pennsylvania, assured me that he

of the wings. It scarcely ever describes one of those curves, without uttering two or three notes whilst ascending, such as its European relative uses on similar occasions. In this manner its flight is prolonged to considerable distances, and it frequently moves in a circling direction before alighting. Their migration is performed during the day. They seldom alight on the ground, unless to procure water, in which they wash with great liveliness and pleasure ; after which they pick up some particles of gravel and sand. So fond of each other's company are they, that a party of them soaring on the wing will alter their course at the calling of a single one perched on a tree. This call is uttered with much emphasis : the bird prolongs its usual note, without much alteration ; and as the party approaches, erects its body, and moves to the right and left, as if turning on a pivot, apparently pleased at showing the beauty of its plumage and elegance of its manners.

This natural group has been long celebrated for their docility and easy instruction, whether in music or to perform a variety of tricks They are, consequently, favourites with bird-fanciers, and often doomed to undergo a severe and cruel discipline. The goldfinch, canary, the various linnets, the siskin, and chaffinch, are principally used for this purpose ; and it is often astonishing, and almost incredible, with what correctness they will obey the voice or motions of their masters. Mr Syme, in his "History of British Song Birds," when speaking of the Sieur Roman, who some years since exhibited goldfinches, linnets, and had tried the male yellow bird with the female canary, and the female yellow bird with the male canary, but without effect, though he kept them for several years together, and supplied them with proper materials for building. Mr Hassey of New York, however, who keeps a great number of native as well as foreign birds, informed me that a yellow bird paired with a canary in his possession, and laid eggs, but did not hatch, which he attributed to the lateness of the season.

These birds were seen by Mr M'Kenzie, in his route across the continent of North America, as far north as lat. 54°; they are numerous in all the Atlantic States north of the Carolinas; abound in Mexico, and are also found in great numbers in the savannahs of Guiana.

The seeds of the lettuce, thistle, hemp, &c., are their favourite food, and it is pleasant to observe a few of them at work in a calm day, detaching the thistle down, in search of

canaries, wonderfully trained, relates, that "one appeared dead, and was held up by the tail or claw without exhibiting any signs of life; a second stood on its head with its claws in the air: a third imitated a Dutch milkmaid going to market with pails on its shoulders; a fourth mimicked a Venetian girl looking out at a window; a fifth appeared as a soldier, and mounted guard as a sentinel; and the sixth acted as a cannonier, with a cap on its head, a firelock on its shoulder, and a match in its claw, and discharged a small cannon. The same bird also acted as if it had been wounded. It was wheeled in a barrow, to convey it, as it were, to the hospital; after which it flew away before the company: a seventh turned a kind of windmill; and the last bird stood in the midst of some fireworks, which were discharged all round it, and this without exhibiting the least symptom of fear." The American goldfinch is no less docile than its congeners. Mr Audubon relates, that they are often caught in trap-cages; and that he knew one, which had undergone severe training, draw water for its drink from a glass, by means of a little chain fastened to a soft leathern belt round its body, and another, equally light, fastened to a little bucket, which was kept by its weight in the water: it was also obliged to supply itself with food, by being obliged to draw towards its bill a little chariot filled with seeds.

Female is represented on Plate VI. of Vol. III., in Bonaparte's continuation.—ED.

the seeds, making it fly in clouds around them. The figure on the plate represents this bird of its natural size.

The American goldfinch has been figured and described by Mr Catesby,* who says, that the back part of the head is a dirty green, &c. This description must have been taken while the bird was changing its plumage. At the approach of fall, not only the rich yellow fades into a brown olive, but the spot of black on the crown and forehead becomes also of the same olive tint. Mr Edwards has also erred in saying, that the young male bird has the spot of black on the forehead; this it does not receive until the succeeding spring.† The figure in Edwards is considerably too large; and that by Catesby has the wings and tail much longer than in nature, and the body too slender—very different from the true form of the living bird. Mr Pennant also tells us that the legs of this species are black; they are, however, of a bright cinnamon colour; but the worthy naturalist, no doubt, described them as he found them in the dried and stuffed skin, shrivelled up and blackened with decay; and thus too much of our natural history has been delineated.

* Nat. Hist. Car., vol. i. p. 43.

† These changes take place in the common siskin of this country: indeed, changes, and, in many cases, similar to those alluded to, are common, according to season, among all our *Fringillidæ;* the common chaffinch loses the pale gray of his forehead, which becomes deep bluish purple; the head and back of the brambling, or mountain finch, becomes a deep glossy black; and the forehead and breasts of the different linnets, from a russet brown, assume a rich and beautiful crimson. They are chiefly produced by the falling off of the ends of the plumules of each feather, which before concealed the richer tints of its lower parts; at other times, by the entire change of colour. The tint itself, however, is always much increased in beauty and gloss as the season for its display advances; at its termination the general moult commences, when the feathers are replaced with their new elongated tips, of a more sombre hue, which, no doubt, adds to the heat of the winter clothing, and remain until warmer weather and desires promote their dispersion.—Ed.

BALTIMORE ORIOLE. (*Oriolus Baltimore.*)

PLATE I.—FIG. 3.

Linn. Syst. i. p. 162, 10.—Icterus minor, *Briss.* ii. p. 109, 19. t. 12. fig. 1.—Le Baltimore, *Buff.* iii. p. 231. *Pl. enl.* 506. fig. 1.—Baltimore Bird, *Catesb. Car.* i. 48.—*Arct. Zool.* ii. p. 142.—*Lath. Syn.* ii. p. 432, 19.—*Bartram*, p. 290.—*Peale's Museum*, No. 1506.

ICTERUS BALTIMORE.—DAUDIN.

Yphantes Baltimore, *Vieill. Gal. des Ois.* pl. 87.—Icterus Baltimore, *Bonap. Syn.* p. 51.—*North. Zool.* ii. p. 284.—Baltimore Oriole, pl. 12. and *Orn. Biog.* p. 66.

THIS is a bird of passage, arriving in Pennsylvania, from the south, about the beginning of May, and departing towards the latter end of August, or beginning of September.* From the singularity of its colours, the construction of its nest, and its preferring the apple trees, weeping willows, walnut and tulip trees, adjoining the farmhouse, to build on, it is generally known, and, as usual, honoured with a variety of names, such as hang-nest, hanging bird, golden robin, fire bird (from the bright orange seen through the green leaves, resembling a flash of fire), &c., but more generally the Baltimore bird, so named, as Catesby informs us, from its colours, which are black and orange, being those of the arms or livery of Lord Baltimore, formerly proprietary of Maryland.

The baltimore oriole is seven inches in length; bill, almost straight, strong, tapering to a sharp point, black, and sometimes lead coloured, above, the lower mandible light blue towards the base. Head, throat, upper part of the back and wings, black; lower part of the back, rump, and whole under parts, a bright orange, deepening into vermilion on the breast; the

* During migration, the flight of the baltimore is high above all the trees, and is straight and continuous; it is mostly performed during the day, as I have usually observed them alighting, always singly, about the setting of the sun, uttering a note or two, and darting into the lower branches to feed, and afterwards to rest.—*Audubon.*—ED.

black on the shoulders is also divided by a band of orange; exterior edges of the greater wing-coverts, as well as the edges of the secondaries, and part of those of the primaries, white; the tail-feathers under the coverts, orange; the two middle ones, from thence to the tips, are black; the next five, on each side, black near the coverts, and orange towards the extremities, so disposed that, when the tail is expanded, and the coverts removed, the black appears in the form of a pyramid, supported on an arch of orange. Tail, slightly forked, the exterior feather on each side, a quarter of an inch shorter than the others; legs and feet, light blue, or lead colour; iris of the eye, hazel.

The female has the head, throat, upper part of the neck and back, of a dull black, each feather being skirted with olive yellow; lower part of the back, rump, upper tail-coverts, and whole lower parts, orange yellow, but much duller than that of the male; the whole wing-feathers are of a deep dirty brown, except the quills, which are exteriorly edged, and the greater wing-coverts, and next superior row, which are broadly tipt with a dull yellowish white; tail, olive yellow; in some specimens, the two middle feathers have been found partly black, in others wholly so; the black on the throat does not descend so far as in the male, is of a lighter tinge, and more irregular; bill, legs, and claws, light blue.*

Buffon and Latham have both described the male of the bastard baltimore (*Oriolus spurius*) as the female baltimore. Mr Pennant has committed the same mistake; and all the ornithologists of Europe, with whose works I am acquainted,

* The change of the plumage of this bird, according to age, is beautifully represented on one of Mr Audubon's gigantic plates, together with its favourite tulip tree, and curious pensile nest. According to that gentleman, the male does not receive his full plumage until the third spring. In the male of one year, the bill is dark brown above, pale blue beneath; the iris, brown; feet, light blue. The general colour is dull brownish yellow, tinged with olive on the head and back; the wings, blackish brown; the quills and large coverts margined and tipped with white; the lesser coverts are olivaceous; the

who have undertaken to figure and describe these birds, have mistaken the proper males and females, and confounded the two species together in a very confused and extraordinary manner, for which, indeed, we ought to pardon them, on account of their distance from the native residence of these birds, and the strange alterations of colour which the latter are subject to.

This obscurity I have endeavoured to clear up in the present volume of this work, Plate IV., by exhibiting the male and female of the *Oriolus spurius* in their different changes of dress, as well as in their perfect plumage; and by introducing representations of the eggs of both, have, I hope, put the identity of these two species beyond all future dispute or ambiguity.

Almost the whole genus of orioles belong to America, and, with a few exceptions, build pensile nests.* Few of them, however, equal the baltimore in the construction of these receptacles for their young, and in giving them, in such a superior degree, convenience, warmth, and security. For these purposes he generally fixes on the high bending extremities of the branches, fastening strong strings of hemp or

tail, destitute of black; and the under parts paler than in the adult, without any approach to the vivid orange tints displayed on it. In that of the second spring, the distribution of colour has become the same as in the adult male, but the yellow is less vivid; the upper mandible is brownish black above, and the iris is light brown: in the third spring, they receive the rich and brilliant plumage described by our author.—Ed.

* The true orioles, having the *Oriolus galbula* of Europe and Africa, with *O. melanocephalus* of India, as typical, are entirely excluded from the New World; nevertheless Wilson was perfectly correct, meaning the *Icteri* of Brisson, which are nearly confined to North and South America, represent the orioles in that country, and have now been arranged into several genera. These contain many species remarkable as well for their elegant form and bright and beautiful plumage, as for the singular and often matchless workmanship of their nests. The materials of the latter are woven and entwined in such a way as would defy the skill of the most expert sempstress, and unite all the requisites of dryness, security, and warmth. They are mostly pendulous from the ends of

flax round two forked twigs, corresponding to the intended width of the nest: with the same materials, mixed with quantities of loose tow, he interweaves or fabricates a strong firm kind of cloth, not unlike the substance of a hat in its raw state, forming it into a pouch of six or seven inches in depth, lining it substantially with various soft substances, well interwoven with the outward netting, and, lastly, finishes with a layer of horse-hair; the whole being shaded from the sun and rain by a natural pent-house, or canopy of leaves. As to a hole being left in the side for the young to be fed and void their excrements through, as Pennant and others relate, it is certainly an error: I, have never met with anything of the kind in the nest of the baltimore.

Though birds of the same species have, generally speaking, a common form of building, yet, contrary to the usually received opinion, they do not build exactly in the same manner. As much difference will be found in the style, neatness, and finishing of the nests of the baltimores, as in their voices. Some appear far superior workmen to others: and probably age may improve them in this, as it does in their colours. I have a number of their nests now before me, all completed, and with eggs. One of these, the neatest, is in

branches, and form thus a security from snakes or other depredators, which could easily reach them if placed on a more solid foundation. They are formed of the different grasses, of dry roots, lichens, long and slender mosses, and, in the present instances mentioned by our author, of substances which could not occur in the early or really natural state of the country, but had been adopted either from necessity, or "*with the sagacity of a good architect*," improving every circumstance to the best advantage. Among the different species, they vary in shape, from being round or resembling a compact ball, to nearly every bottle-shaped gradation of form, until they exceed three or four feet in length. Many species being gregarious, they breed numerously on the same tree, and their nests, suspended from the pensile branches, and waving in the wind, render the landscape and woods singular to an unaccustomed eye, and present appearances which those only who have had the good fortune to witness them in their native wilds can appreciate.

The female is given by Wilson in Plate LIII. in our second volume. —ED.

the form of a cylinder, of five inches diameter, and seven inches in depth, rounded at bottom. The opening at top is narrowed, by a horizontal covering, to two inches and a half in diameter. The materials are flax, hemp, tow, hair, and wool, woven into a complete cloth; the whole tightly sewed through and through with long horse-hairs, several of which measure two feet in length. The bottom is composed of thick tufts of cow-hair, sewed also with strong horse-hair. This nest was hung on the extremity of the horizontal branch of an apple tree, fronting the southeast; was visible a hundred yards off, though shaded from the sun; and was the work of a very beautiful and perfect bird. The eggs are five, white, slightly tinged with flesh colour, marked on the greater end with purple dots, and on the other parts with long hair-like lines, intersecting each other in a variety of directions. I am thus minute in these particulars, from a wish to point out the specific difference between the true and bastard baltimore, which Dr Latham, and some others, suspect to be only the same bird in different stages of colour.

So solicitous is the baltimore to procure proper materials for his nest, that, in the season of building, the women in the country are under the necessity of narrowly watching their thread that may chance to be out bleaching, and the farmer to secure his young grafts; as the baltimore, finding the former, and the strings which tie the latter, so well adapted for his purpose, frequently carries off both; or, should the one be too heavy, and the other too firmly tied, he will tug at them a considerable time before he gives up the attempt. Skeins of silk and hanks of thread have been often found, after the leaves were fallen, hanging round the baltimore's nest; but so woven up and entangled as to be entirely irreclaimable. Before the introduction of Europeans, no such material could have been obtained here; but, with the sagacity of a good architect, he has improved this circumstance to his advantage; and the strongest and best materials are uniformly found in those parts by which the whole is supported.

Their principal food consists of caterpillars, beetles, and bugs, particularly one of a brilliant glossy green, fragments of which I have almost always found in their stomach, and sometimes these only.

The song of the baltimore is a clear mellow whistle, repeated at short intervals as he gleans among the branches. There is in it a certain wild plaintiveness and *naïveté* extremely interesting. It is not uttered with the rapidity of the ferruginous thrush (*Turdus rufus*), and some other eminent songsters; but with the pleasing tranquillity of a careless ploughboy, whistling merely for his own amusement. When alarmed by an approach to his nest, or any such circumstance, he makes a kind of rapid chirruping, very different from his usual note. This, however, is always succeeded by those mellow tones which seem so congenial to his nature.

> High on yon poplar, clad in glossiest green,
> The orange black-capped baltimore is seen;
> The broad extended boughs still please him best,
> Beneath their bending skirts he hangs his nest;
> There his sweet mate, secure from every harm,
> Broods o'er her spotted store, and wraps them warm;
> Lists to the noontide hum of busy bees,
> Her partner's mellow song, the brook, the breeze;
> These day by day the lonely hours deceive,
> From dewy morn to slow descending eve.
> Two weeks elapsed, behold! a helpless crew
> Claim all her care, and her affection too;
> On wings of love the assiduous nurses fly,
> Flowers, leaves, and boughs, abundant food supply;
> Glad chants their guardian, as abroad he goes,
> And waving breezes rock them to repose.

The baltimore inhabits North America, from Canada to Mexico, and is even found as far south as Brazil. Since the streets of our cities have been planted with that beautiful and stately tree, the Lombardy poplar, these birds are our constant visitors during the early part of summer; and, amid the noise and tumult of coaches, drays, wheelbarrows, and the din of the multitude, they are heard chanting "their native wood notes wild;" sometimes, too, within a few yards of an oysterman, who stands bellowing, with the lungs of a Stentor, under the shade of the same tree; so much will habit reconcile even birds to the roar of the city, and to sounds and noises that, in other circumstances, would put a whole grove of them to flight.

These birds are several years in receiving their complete plumage. Sometimes the whole tail of a male individual in spring is yellow, sometimes only the two middle feathers are black, and frequently the black on the back is skirted with orange, and the tail tipt with the same colour. Three years, I have reason to believe, are necessary to fix the full tint of the plumage, and then the male bird appears as already described.

WOOD THRUSH. (*Turdus melodus.*)

PLATE II.—FIG. 1.

Bartram, p. 290.—*Peale's Museum*, No. 5264.

TURDUS MUSTELINUS.—GMELIN.

Turdus mustelinus, *Gm. Linn.* ii. 817, No. 57.—*Bonap. Synop.* p. 75.—*Penn. Arct. Zool.* ii. p. 337.—The Wood Thrush, *Aud.* p. 372.

THIS bird is represented on the plate of its natural size, and particular attention has been paid to render the figure a faithful likeness of the original. It measures eight inches in length, and thirteen from tip to tip of the expanded wings; the bill is an inch long; the upper mandible, of a dusky brown, bent at the point, and slightly notched; the lower, a flesh colour towards the base; the legs are long, and, as well as the claws, of a pale flesh colour, or almost transparent. The whole upper parts are of a brown fulvous colour, brightening into reddish on the head, and inclining to an olive on the rump and tail; chin, white; throat and breast, white, tinged with a light buff colour, and beautifully marked with pointed spots of black or dusky, running in chains from the sides of the mouth, and intersecting each other all over the breast to the belly, which, with the vent, is of a pure white; a narrow circle of white surrounds the eye, which is large, full, the pupil black,

Drawn from Nature by A. Wilson.　　Engraved by W. H. Lizars.

1. Wood Thrush. 2. Red-breasted Thrush or Robin. 3. White-breasted-black-capped Nuthatch. 4. Red-bellied-black-capped Nuthatch.

Turdus Melodus　*Turdus Migratorius.*　*Sitta Carolinensis.*　*Sitta Varia.*

2

and the iris of a dark chocolate colour; the inside of the mouth is yellow. The male and female of this species, as, indeed, of almost the whole genus of thrushes, differ so little, as scarcely to be distinguished from each other. It is called by some the wood robin, by others the ground robin, and by some of our American ornithologists *Turdus minor*, though, as will hereafter appear, improperly. The present name has been adopted from Mr William Bartram, who seems to have been the first and almost only naturalist who has taken notice of the merits of this bird.*

* Almost every country has its peculiar and favourite songsters, and even among the rudest nations the cries and songs of birds are listened to, and associated with their general occupations, their superstitions, or religion. In America, the wood thrush appears to hold a rank equal to the nightingale and song thrush of Europe: like the latter, he may be oftentimes seen perched on the summit of a topmost branch, during a warm and balmy evening or morning, pouring forth in rich melody his full voice, and will produce associations which a foreigner would assimilate with the warblers of his own land.

"The song of the wood thrush," says Mr Audubon, "although composed of but few notes, is so powerful, distinct, clear, and mellow, that it is impossible for any person to hear it without being struck with the effect it produces on the mind. I do not know to what instrumental sounds I can compare these notes, for I really know none so melodious and harmonical. They gradually rise in strength, and then fall in gentle cadence, becoming at length so low as to be scarcely audible." They are easily reared from the nest, and sing nearly as well in confinement as when free.

Prince C. L. Bonaparte, in his "Nomenclature of Wilson's North American Ornithology," remarks, that our author was the first to distinguish the three closely allied species of North American thrushes by decided characters, but that he has nevertheless embroiled the nomenclature of this and his *T. mustelinus*:—"This bird being evidently the *T. mustelinus* of Gmelin and Latham, Wilson's new name, which is not modelled agreeably to any language, must be rejected."

The title for our present species, allowing Bonaparte to be correct, and of which there appears little doubt, will therefore now stand, *Wood Thrush*, Wilson; *Turdus mustelinus*, Gmelin; and *T. melodus* will come in as a synonym; while Wilson's *T. mustelinus*, being without a name, has been most deservedly dedicated to the memory of the great American ornithologist himself.—Ed.

This sweet and solitary songster inhabits the whole of North America, from Hudson's Bay to the peninsula of Florida. He arrives in Pennsylvania about the 20th of April, or soon after, and returns to the south about the beginning of October. The lateness or earliness of the season seems to make less difference in the times of arrival of our birds of passage than is generally imagined. Early in April the woods are often in considerable forwardness, and scarce a summer bird to be seen. On the other hand, vegetation is sometimes no further advanced on the 20th of April, at which time (*e.g.*, this present year, 1807) numbers of wood thrushes are seen flitting through the moist woody hollows, and a variety of the *Motacilla* genus chattering from almost every bush, with scarce an expanded leaf to conceal them. But at whatever time the wood thrush may arrive, he soon announces his presence in the woods. With the dawn of the succeeding morning, mounting to the top of some tall tree that rises from a low thick shaded part of the woods, he pipes his few, but clear and musical notes, in a kind of ecstasy; the prelude or symphony to which strongly resembles the double-tonguing of a German flute, and sometimes the tinkling of a small bell; the whole song consists of five or six parts, the last note of each of which is in such a tone as to leave the conclusion evidently suspended; the finalé is finely managed, and with such charming effect as to soothe and tranquillise the mind, and to seem sweeter and mellower at each successive repetition. Rival songsters, of the same species, challenge each other from different parts of the wood, seeming to vie for softer tones and more exquisite responses. During the burning heat of the day, they are comparatively mute; but in the evening the same melody is renewed, and continued long after sunset. Those who visit our woods, or ride out into the country at these hours, during the months of May and June, will be at no loss to recognise, from the above description, this pleasing musician. Even in dark, wet, and gloomy weather, when scarce a single chirp is heard from any other bird, the clear notes of the wood thrush thrill through the dropping woods, from morning to night; and it may truly be said, that the sadder the day the sweeter is his song.

The favourite haunts of the wood thrush are low, thick shaded hollows, through which a small brook or rill meanders, overhung with alder bushes, that are mantled with wild vines. Near such a scene he generally builds his nest, in a laurel or alder bush. Outwardly it is composed of withered beech leaves of the preceding year, laid at bottom in considerable quantities, no doubt to prevent damp and moisture from ascending through, being generally built in low, wet situations; above these are layers of knotty stalks of withered grass, mixed with mud, and smoothly plastered, above which is laid a slight lining of fine black fibrous roots of plants. The eggs are four, sometimes five, of a uniform light blue, without any spots.

The wood thrush appears always singly or in pairs, and is of a shy, retired, unobtrusive disposition. With the modesty of true merit, he charms you with his song, but is content, and even solicitous, to be concealed. He delights to trace the irregular windings of the brook, where, by the luxuriance of foliage, the sun is completely shut out, or only plays in a few interrupted beams on the glittering surface of the water. He is also fond of a particular species of lichen which grows in such situations, and which, towards the fall, I have uniformly found in their stomachs: berries, however, of various kinds, are his principal food, as well as beetles and caterpillars. The feathers on the hind head are longer than is usual with birds which have no crest; these he sometimes erects; but this particular cannot be observed but on a close examination.*

Those who have paid minute attention to the singing of birds know well that the voice, energy, and expression, in the same tribe, differ as widely as the voices of different indi-

* In addition to the above picture of the manners of this thrush, Mr Audubon remarks, that it performs its migrations during the day, gliding swiftly through the woods, without appearing in the open country;

viduals of the human species, or as one singer does from another. The powers of song, in some individuals of the wood thrush, have often surprised and delighted me. Of these I remember one, many years ago, whose notes I could instantly recognise on entering the woods, and with whom I had been, as it were, acquainted from his first arrival. The top of a large white oak that overhung part of the glen, was usually the favourite pinnacle from whence he poured the sweetest melody; to which I had frequently listened till night began to gather in the woods, and the fireflies to sparkle among the branches. But, alas! in the pathetic language of the poet—

> One morn I missed him on the accustomed hill,
> Along the vale, and on his favourite tree—
> Another came, nor yet beside the rill,
> Nor up the glen, nor in the wood was he.

A few days afterwards, passing along the edge of the rocks, I found fragments of the wings and broken feathers of a wood thrush killed by the hawk, which I contemplated with unfeigned regret, and not without a determination to retaliate on the first of these murderers I could meet with.

That I may not seem singular in my estimation of this bird, I shall subjoin an extract of a letter from a distinguished American gentleman, to whom I had sent some drawings, and whose name, were I at liberty to give it, would do honour to my humble performance, and render any further observations on the subject from me unnecessary.

"As you are curious in birds, there is one well worthy your attention, to be found, or rather heard, in every part of America, and yet scarcely ever to be seen. It is in all the forests from spring to fall, and never but on the tops of the tallest trees, from which it perpetually serenades us with some of the

that, on alighting upon a branch, it gives its tail a few jets, uttering at each motion a low chuckling note, peculiar to itself; it then stands still for a while, with the feathers of the hind part a little raised. It walks and hops along the branches with much ease, and bends down its head to peep at the objects around.—Ed.

sweetest notes, and as clear as those of the nightingale. I have followed it for miles, without ever but once getting a good view of it. It is of the size and make of the mocking bird, lightly thrush coloured on the back, and a grayish white on the breast and belly. Mr ——, my son-in-law, was in possession of one, which had been shot by a neighbour; he pronounced it a *Muscicapa,* and I think it much resembles the *Mouche rolle de la Martinique,* 8 Buffon, 374, *Pl. enlum,* 568. As it abounds in all the neighbourhood of Philadelphia, you may, perhaps, by patience and perseverance (of which much will be requisite), get a sight, if not a possession, of it. I have, for twenty years, interested the young sportsmen of my neighbourhood to shoot me one, but, as yet, without success."

It may seem strange that neither Sloane,* Catesby, Edwards, nor Buffon, all of whom are said to have described this bird, should say anything of its melody; or rather, assert that it had only a single cry or scream. This I cannot account for in any other way than by supposing, what I think highly probable, that this bird has never been figured or described by any of the above authors.

Catesby has, indeed, represented a bird which he calls *Turdus minimus,*† but it is difficult to discover, either from the figure or description, what particular species is meant; or whether it be really intended for the wood thrush we are now describing. It resembles, he says, the English thrush; but is less, never sings, has only a single note, and abides all the year in Carolina. It must be confessed that, except the first circumstance, there are few features of the wood thrush in this description. I have searched the woods of Carolina and Georgia in winter for this bird in vain, nor do I believe it ever winters in these States. If Mr Catesby found his bird mute during spring and summer, it was not the wood thrush, otherwise he must have changed his very nature. But Mr Edwards has also described and delineated the little thrush,‡

* Hist. Jam. ii. 305. † Catesby's Nat. Hist. Car., i. 31.

‡ Edwards, 296.

and has referred to Catesby as having drawn and engraved it before. Now this thrush of Edwards I know to be really a different species; one not resident in Pennsylvania, but passing to the north in May, and returning the same way in October, and may be distinguished from the true song thrush (*Turdus melodus*) by the spots being much broader, brown, and not descending so far below the breast. It is also an inch shorter, with the cheeks of a bright tawny colour. Mr William Bartram, who transmitted this bird, more than fifty years ago, to Mr Edwards, by whom it was drawn and engraved, examined the two species in my presence; and on comparing them with the one in Edwards, was satisfied that the bird there figured and described is not the wood thrush (*Turdus melodus*), but the tawny-cheeked species above mentioned. This I have never seen in Pennsylvania but in spring and fall. It is still more solitary than the former, and utters, at rare times, a single cry, similar to that of a chicken which has lost its mother. This very bird I found numerous in the myrtle swamps of Carolina in the depth of winter, and I have not a doubt of its being the same which is described by Edwards and Catesby.

As the Count de Buffon has drawn his description from those above mentioned, the same observations apply equally to what he has said on the subject; and the fanciful theory which this writer had formed to account for its want of song, vanishes into empty air; viz., that the song thrush of Europe (*Turdus musicus*), had, at some time after the creation, rambled round by the northern ocean, and made its way to America; that, advancing to the south, it had there (of consequence) become degenerated by change of food and climate, so that its cry is now harsh and unpleasant, "as are the cries of all birds that live in wild countries inhabited by savages." *

* Buffon, vol. iii. 289. The figure in Pl. enl. 398, has little or no resemblance to the wood thrush, being of a deep green olive above, and spotted to the tail below with long streaks of brown.

ROBIN. (*Turdus migratorius.*)

PLATE II.—FIG. 2.

Linn. Syst. i. p. 292, 6.—Turdus Canadensis, *Briss.* ii. p. 225, 9.—La Litorne de Canada, *Buff.* iii. p. 307.—Grive de Canada, *Pl. enl.* 556, 1.—Fieldfare of Carolina, *Cat. Car.* i. 29.—Red-breasted Thrush, *Arct. Zool.* ii. No. 196.—*Lath. Syn.* ii. p. 26.—*Bartram*, p. 290.—*Peale's Museum*, No. 5278.

*TURDUS MIGRATORIUS.**—LINNÆUS.

Turdus migratorius, *Bonap. Synop.* p. 75.—Merula migratoria, *North. Zool.* ii. p. 177.

THIS well-known bird, being familiar to almost everybody, will require but a short description. It measures nine inches and a half in length; the bill is strong, an inch long, and of a full yellow, though sometimes black or dusky near the tip

* In the beautifully wrought-out arrangement of the *Merulidæ*, by Mr Swainson, in the second volume of the "Northern Zoology," that family will form the second among the *Dentirostres* or the subtypical group; including, for its five principal divisions, the families *Merulinæ*, *Myotherinæ*, *Brachypodinæ*, *Oriolinæ*, and *Crateropodinæ*; among these, however, two, or at most three, only come within the range of the northern continent of America—the first and third. The first, *Merulinæ*, or more properly the typical form, will now claim our attention.

In all the members taken collectively, and in adaptation to their general habits, they show considerable perfection, though their form as a part of the *Dentirostres* does not come up to the typical perfections of that group. The parts are adapted for extensive locomotion, either in walking or perching, and in flight; many perform very considerable migrations, and long and rapid flights are often taken in those countries even where the climate does not seem to render this necessary. They are nearly omnivorous. A great part of their sustenance is sought for upon the ground, particularly during that season when insects are not indispensable for the welfare of their broods; and their feet and tarsi are admirably formed for walking and inspecting the various places where their food is then chiefly to be found. At other times they live principally upon fruits and some vegetables, with the larvæ of insects, and the abundant supply of large and succulent caterpillars; but during winter, the harder grains, and more fleshy insects common to low meadows and moist woods, such as the various snails, flies, and worms, are nearly their only food; for after the first month of the inclement season has passed, most of the winter wild fruits and berries have either of the upper mandible; the head, back of the neck, and tail, is black; the back and rump, an ash colour; the wings are black, edged with light ash; the inner tips of the two exterior tail-feathers are white; three small spots of white border the eye; the throat and upper part of the breast is black, the former streaked with white; the whole of the rest of the breast, down as far as the thighs, is of a dark orange; belly and vent, white, slightly waved with dusky ash; legs, dark brown; claws, black and strong. The colours of the female are more of the light ash, less deepened with black; and the orange on the breast is much paler, and more broadly skirted with white. The name of this bird bespeaks him a bird of passage, as are all the different species of thrushes we have; but the one we are now describing being more unsettled, and continually roving about from one region to another, during fall and winter, seems particularly entitled to the appellation. Scarce

fallen from their stocks, or have been already consumed by these and many other tribes that subsist upon them. Very few are quite solitary: during the breeding season they all separate, but after the broods have been raised, they congregate either in very large flocks or in groups of five or six. Those of smaller numbers generally either become more domestic, and approach dwellings and cultivated districts on the approach of winter, or retire entirely to the depths of solitary forests. Those that congregate in large flocks are always remarkably shy, suffer persons to approach with difficulty, and have a sentinel or watch on the look out, to warn them of danger. Their cry is harsh and sharp, or shrill and monotonous, except during the season of incubation, when they all produce strains of more interest. Some possess great melody, and in others the notes are remarkably pensive and melancholy. On this account they are universal favourites; and the early song of the *mavis* is watched for, by those residing much in the country, as the harbinger of a new season and brighter days. The true thrushes are all inhabitants of woods, and only from the necessity of procuring food resort to the open countries. In distribution, they range over the world, and the proportion seems pretty equal; India and Southern Europe may, perhaps, have the most extensive list, and North America will rank in the least proportion. They are often used as articles of food, and the immense havoc made among the *Northern robins* of our author will show the estimation in which they are held as luxuries for the table; in Spain and Italy, great numbers are taken for the same purpose, with a winter passes but innumerable thousands of them are seen in the lower parts of the whole Atlantic States, from New Hampshire to Carolina, particularly in the neighbourhood of our towns; and, from the circumstance of their leaving, during that season, the country to the northwest of the great range of the Alleghany, from Maryland northward, it would appear that they not only migrate from north to south, but from west to east, to avoid the deep snows that generally prevail on these high regions for at least four months in the year.

The robin builds a large nest, often on an apple tree, plasters it in the inside with mud, and lines it with hay or fine grass. The female lays five eggs of a beautiful sea-green. Their principal food is berries, worms, and caterpillars. Of the first, he prefers those of the sour gum (*Nyssa sylvatica*). So fond are they of gum-berries, that, wherever there is one of these trees covered with fruit, and flocks of robins in the

nets and various kinds of snares; with the severity of the season, however, and the difference of food, the flesh acquires a bitter flavour, which renders them unfit for culinary purposes, and affords a temporary respite from their merciless persecutions.

The title *Merula*, which Mr Swainson and several of our modern ornithologists have adopted, was used by Ray only as a subgenus among his "*Turdinum genus*," and contained that division to which the blackbird and ringousel would belong; *Turdus* being confined to those with spotted breasts. I do not consider the very trifling difference in form between the plain and spotted species to be of sufficient importance, and prefer retaining the generic name of *Turdus*, as one well known and long accepted.

Robin seems to be applied in America generally to several of the thrushes, some expletive going before to designate the species by its habits, as *wood robin*, *swamp robin*, *ground robin*, &c. Our present species is THE ROBIN; and, as the preceding was a favourite on account of its song, this is no less so from the unassuming and dependent familiarity of its manners: it was most probably this, joined with the colour of the breast, which first suggested the name of our own homely bird to the earlier British settlers, and along with it part of the respect with which its namesake is treated in this country.

An African species, *Turdus olivaceus* (*le Griveron*, Vieill.) is nearly allied in the distribution of the markings. I have another, I believe, from South America, which approaches both nearly.—ED.

neighbourhood, the sportsman need only take his stand near it, load, take aim, and fire; one flock succeeding another, with little interruption, almost the whole day: by this method, prodigious slaughter has been made among them with little fatigue. When berries fail, they disperse themselves over the fields, and along the fences, in search of worms and other insects. Sometimes they will disappear for a week or two, and return again in greater numbers than before; at which time the cities pour out their sportsmen by scores, and the markets are plentifully supplied with them at a cheap rate. In January 1807, two young men, in one excursion after them, shot thirty dozen. In the midst of such devastation, which continued many weeks, and, by accounts, extended from Massachusetts to Maryland, some humane person took advantage of a circumstance common to these birds in winter, to stop the general slaughter. The fruit called poke-berries (*Phytolacca decandra*, Linn.) is a favourite repast with the robin, after they are mellowed by the frost. The juice of the berries is of a beautiful crimson, and they are eaten in such quantities by these birds, that their whole stomachs are strongly tinged with the same red colour. A paragraph appeared in the public papers, intimating that, from the great quantities of these berries which the robins had fed on, they had become unwholesome, and even dangerous food, and that several persons had suffered by eating of them. The strange appearance of the bowels of the birds seemed to corroborate this account. The demand for and use of them ceased almost instantly; and motives of self-preservation produced at once what all the pleadings of humanity could not effect.* When fat, they are in considerable esteem for the table, and probably not inferior

* Governor Drayton, in his "View of South Carolina," p. 86, observes, that "the robins in winter devour the berries of the bead tree (*Melia azedarach*) in such large quantities, that, after eating of them, they are observed to fall down, and are readily taken. This is ascribed more to distension from abundant eating, than from any deleterious qualities of the plant." The fact, however, is, that they are literally choked, many of the berries being too large to be swallowed.

to the *Turdi* of the ancients, which they bestowed so much pains on in feeding and fattening. The young birds are frequently and easily raised, bear the confinement of the cage, feed on bread, fruits, &c., sing well, readily learn to imitate parts of tunes, and are very pleasant and cheerful domestics. In these I have always observed that the orange on the breast is of a much deeper tint, often a dark mahogany or chestnut colour, owing, no doubt, to their food and confinement.

The robin is one of our earliest songsters; even in March, while snow yet dapples the fields, and flocks of them are dispersed about, some few will mount a post or stake of the fence, and make short and frequent attempts at their song.* Early in April, they are only to be seen in pairs, and deliver their notes with great earnestness, from the top of some tree detached from the woods.

This song has some resemblance to, and indeed is no bad imitation of, the notes of the thrush or thrasher (*Turdus rufus*); but, if deficient in point of execution, he possesses more simplicity, and makes up in zeal what he wants in talent; so that the notes of the robin, in spring, are universally known, and as universally beloved. They are, as it were, the prelude

* "The male is one of the loudest and most assiduous of the songsters that frequent the fur countries, beginning his chant immediately on his arrival. Within the arctic circle, the woods are silent in the bright light of noonday; but, towards midnight, when the sun travels near the horizon, and the shades of the forest are lengthened, the concert commences, and continues till six or seven in the morning." Thus speaks Dr Richardson, in the "Northern Zoology," regarding the song of this bird; and he further adds, regarding the breeding and geographical range :—"Its nests were observed, by the last Northern expedition, conducted by Captain Sir J. Franklin, as high as the 67th parallel of latitude. It arrives on the Missouri, in lat. 41½°, from the eastward, on the 11th of April; and in the course of its northerly movement, reaches Severn River, in Hudson's Bay, about a fortnight later. Its first appearance at Carlton House, in the year 1827, in lat. 53°, was on the 22d April. In the same season it reached Fort Chippewyan, in lat. 55¾°, on the 7th of May; and Fort Franklin, in lat. 65°, on the 20th of that month. Those that build their nests in the 54th parallel of latitude, begin to hatch in the end of May; but 11° farther to the north, that

to the grand general concert that is about to burst upon us from woods, fields, and thickets, whitened with blossoms, and breathing fragrance. By the usual association of ideas, we therefore listen with more pleasure to this cheerful bird than to many others possessed of far superior powers, and much greater variety. Even his nest is held more sacred among schoolboys than that of some others; and, while they will exult in plundering a jay's or a cat bird's, a general sentiment of respect prevails on the discovery of a robin's. Whether he owes not some little of this veneration to the well-known and long-established character of his namesake in Britain, by a like association of ideas, I will not pretend to determine. He possesses a good deal of his suavity of manners; and almost always seeks shelter for his young in summer, and subsistence for himself in the extremes of winter, near the habitations of man.

The robin inhabits the whole of North America, from Hudson's Bay to Nootka Sound, and as far south as Georgia, though they rarely breed on this side the mountains farther south than Virginia. Mr Forster says, that about the beginning of May they make their appearance in pairs at the settle-

event is deferred till the 11th of June. The snow, even then, partially covers the ground; but there are, in those high latitudes, abundance of the berries of *Vaccinium uliginosum* and *Vitis idea*, *Arbutus alpina*, *Empetrum nigrum*, and of some other plants, which, after having been frozen up all winter, are exposed to the first melting of the snows, full of juice, and in high flavour: shortly after, the parents obtain abundance of grubs for their callow young."

We thus see the extreme regularity with which the migrations are performed, and cannot too much admire the power which enables them to perceive, and calculate so exactly, the time required for their journey to the climates best suited to their duties at that season. We also see another wonderful provision, both for the migratory species and those which subsist as they best can during the winter, in the preservation of the berries and fruits fresh and juicy under the snow. Were it not for this, the ground, on the melting of its covering, would present a more desolate appearance than in the extremest storms of winter, and all animal life would inevitably perish for want of food before the various and abundant plants could flower and perfect their fruits.—Ed.

ments of Hudson's Bay, at Severn River ; and adds a circumstance altogether unworthy of belief, viz., that, at Moose Fort, they build, lay, and hatch, in fourteen days ! but that at the former place, four degrees more north, they are said to take twenty-six days.* They are also common in Newfoundland, quitting these northern parts in October. The young, during the first season, are spotted with white on the breast, and at that time have a good deal of resemblance to the fieldfare of Europe.

Mr Herne informs us that the red-breasted thrushes are commonly called, at Hudson's Bay, the red birds—by some, the blackbirds, on account of their note—and by others, the American fieldfares; that they make their appearance at Churchill River about the middle of May, and migrate to the south early in the fall. They are seldom seen there but in pairs; and are never killed for their flesh, except by the Indian boys.†

Several authors have asserted that the red-breasted thrush cannot brook the confinement of the cage, and never sings in that state. But, except the mocking bird (*Turdus polyglottus*), I know of no native bird which is so frequently domesticated, agrees better with confinement, or sings in that state more agreeably than the robin. They generally suffer severely in moulting time; yet often live to a considerable age. A lady, who resides near Tarrytown, on the banks of the Hudson, informed me that she raised and kept one of these birds for seventeen years, which sung as well, and looked as sprightly, at that age as ever; but was at last unfortunately destroyed by a cat. The morning is their favourite time for song. In passing through the streets of our large cities on Sunday, in the months of April and May, a little after daybreak, the general silence which usually prevails without at that hour will enable you to distinguish every house where one of these songsters resides, as he makes it then ring with his music.

* Phil. Trans. lxii. 399.

† Journey to the Northern Ocean, p. 418, 4to, Lond. 1795.

Not only the plumage of the robin, as of many other birds, is subject to slight periodical changes of colour, but even the legs, feet, and bill ; the latter, in the male, being frequently found tipt and ridged for half its length with black. In the depth of winter their plumage is generally best; at which time the full-grown bird, in his most perfect dress, appears as exhibited in the plate.

WHITE-BREASTED, BLACK-CAPPED NUTHATCH.
(*Sitta Carolinensis.*)

PLATE II. FIG. 3.

Catesb. i. 22, fig. 2.—*Lath.* i. 650, B.—*Briss.* iii. p. 596, 4.—Sitta Carolinensis, *Turton.*—Sitta Europea, Gray Black-capped Nuthatch, *Bartram*, p. 289.—*Peale's Museum*, No. 20, 36.

*SITTA CAROLINENSIS.**

Sitta Carolinensis, *Bonap. Synop.* 96.—Sitta melanocephala, *Vieill. Gal. des Ois.* p. 280, pl. 174.

THE bill of this bird is black, the upper mandible straight, the lower one rounded upwards towards the point, and white near the base; the nostrils are covered with long curving black hairs; the tongue is of a horny substance, and ending in

* The true nuthatches, *Sittæ* (for I would not admit *S. velata* of Horsfield, and some allied species, nor the *S. chrysoptera* from New Holland), are all natives of Europe and South America. With this restriction of geographical distribution, the genus will contain only four species, three of which, *S. Carolinensis*, *Canadensis*, and *pusilla*, figured and described by our author, are confined to North America ; and the fourth, *S. Europea*, has been only found in Europe. With regard to their situation in our systems, I would prefer placing them near to *Certhia*, *Xeops*, *Anabates*, *Dendrocolaptes*, and not far distant from the titmice ; with the former, they seem intimately connected, and there appears little in their structure in common with the woodpeckers, except the act of running up the trunks of trees. In habit and general economy they resemble the titmice, always actively employed in turning or twisting round the branches, or in running up or down the trunks, for they do both with equal facility, searching after the insects, or their eggs and larvæ, which lie concealed under the moss or loose bark ; but occasionally also, like them, feeding upon different grains, on the seeds of the pine cones, as mentioned by our author in his description of the red-

several sharp points; the general colour above is of a light blue or lead; the tail consists of twelve feathers, the two middle ones lead colour, the next three are black, tipt with white for one-tenth, one-fourth, and half of an inch; the two next are also black, tipt half an inch or more with white, which runs nearly an inch up their exterior edges, and both have the white at the tips touched with black; the legs are of a purple or dirty flesh colour; the hind claw is much the largest; the inside of the wing at the bend is black; below this is a white spot spreading over the roots of the first five primaries; the whole length is five inches and a half; extent, eleven.

Mr Pennant considers this bird as a mere variety of the European nuthatch; but if difference in size, colour, and habits, be sufficient characteristics of a distinct species, this bird is certainly entitled to be considered as such. The head and back of the European species is of a uniform bluish gray; the upper parts of the head, neck, and shoulders of ours, are a deep black glossed with green; the breast and belly of the former is a dull orange, with streaks of chestnut; those parts in the latter are pure white. The European has a line of

bellied species; or, according to Montagu, like the *S. Europea* frequenting the orchards during the cider season, and picking the seeds from the refuse of the pressed apples. In a state of confinement they will thrive well upon raw meat or fat, and if taken at a proper age, become extremely familiar and amusing; if not, they will most likely destroy themselves in their endeavours to get free from confinement, as mentioned by the anonymous writer of an interesting account of this bird in Loudon's "Magazine of Natural History." I had lately an opportunity of observing a nest of our native species which had been taken young. They became remarkably tame; and, when released from their cage, would run over their owner in all directions, up or down his body and limbs, *poking* their bills into seams or holes, as if in search of food upon some old and rent tree, and uttering, during the time, a low and plaintive cry. When running up or down, they rest upon the back part of the whole tarsus, and make great use as a support of what may be called the real heel, and never use the tail. Their bills are comparatively strong, and the power they possess of using them great, equal apparently to that of a woodpecker of like size. They breed in hollow trees, and black passing through the eye, half way down the neck; the present species has nothing of the kind, but appears with the inner webs of the three shortest secondaries and the primaries of a jet black; the latter tipt with white, and the vent and lower parts of the thighs of a rust colour: the European, therefore, and the present, are evidently two distinct and different species.*

This bird builds its nest early in April, in the hole of a tree, in a hollow rail in the fence, and sometimes in the wooden cornice under the eaves; and lays five eggs of a dull white, spotted with brown at the greater end. The male is extremely attentive to the female while sitting; supplying her regularly with sustenance, stopping frequently at the mouth of the hole, calling and offering her what he has brought, in the most endearing manner. Sometimes he seems to stop merely to inquire how she is, and to lighten the tedious moments with his soothing chatter. He seldom rambles far from the spot; and when danger appears, regardless of his own safety, he flies

produce a rather numerous brood. The male attends carefully during the time. According to Montagu, our British species chooses the deserted habitation of some woodpecker. "The hole is first contracted by a plaster of clay, leaving only sufficient room for itself to pass out and in; the nest is made of dead leaves, chiefly those of the oak, which are heaped together without much order. If the barrier of plaster at the entrance is destroyed when they have eggs, it is speedily replaced,—a peculiar instinct to prevent their nest being destroyed by the woodpecker, and other birds of superior size, which build in the same manner." Or, as Mr Rennie, in his late edition of the same work, thinks probable, the wall may be to prevent the unfledged young from tumbling out of the nest when they begin to stir about. It is probable that the nuthatch does not look forward to any of these considerations; and although the effects above mentioned may be in reality the consequence, I should conceive the hole contracted as being really too large, and as increasing the heat and apparent comfort within. When roosting, they sleep with the head and back downwards, in the manner of several titmice.—ED.

* Wilson is perfectly correct in considering this species as distinct from that of Europe; he has marked out the distinctions well in the description. It is described by Vieillot as *Sitta melanocephala*.—ED.

instantly to alarm her. When both are feeding on the trunk of the same tree, or of adjoining ones, he is perpetually calling on her; and, from the momentary pause he makes, it is plain that he feels pleased to hear her reply.

The white-breasted nuthatch is common almost everywhere in the woods of North America, and may be known, at a distance, by the notes, *quank, quank,* frequently repeated, as he moves, upward and down, in spiral circles, around the body and larger branches of the tree, probing behind the thin scaly bark of the white oak, and shelling off considerable pieces of it, in search after spiders, ants, insects, and their larvæ. He rests and roosts with his head downwards, and appears to possess a degree of curiosity not common to many birds; frequently descending, very silently, within a few feet of the root of the tree where you happen to stand, stopping, head downward, stretching out his neck in a horizontal position, as if to reconnoitre your appearance; and, after several minutes of silent observation, wheeling round, he again mounts, with fresh activity, piping his unisons as before. Strongly attached to his native forests, he seldom forsakes them; and, amidst the rigours of the severest winter weather, his note is still heard in the bleak and leafless woods, and among the howling branches. Sometimes the rain, freezing as it falls, encloses every twig, and even the trunk of the tree, in a hard transparent coat or shell of ice. On these occasions I have observed his anxiety and dissatisfaction at being with difficulty able to make his way along the smooth surface; at these times generally abandoning the trees, gleaning about the stables, around the house, mixing among the fowls, entering the barn, and examining the beams and rafters, and every place where he may pick up a subsistence.

The name nuthatch has been bestowed on this family of birds from their supposed practice of breaking nuts by repeated hatchings, or hammerings with their bills. Soft-shelled nuts, such as chestnuts, chinkopins, and hazel nuts, they may, probably, be able to demolish, though I have never yet seen them so engaged; but it must be rather in search of maggots, that sometimes breed there, than for the kernel. It is, however, said, that they lay up a large store of nuts for winter; but as I have never either found any of their magazines, or seen them collecting them, I am inclined to doubt the fact. From the great numbers I have opened at all seasons of the year, I have every reason to believe that ants, bugs, small seeds, insects, and their larvæ, form their chief subsistence, such matters alone being uniformly found in their stomachs. Neither can I see what necessity they could have to circumambulate the trunks of trees with such indefatigable and restless diligence, while bushels of nuts lay scattered round their roots. As to the circumstance mentioned by Dr Plott, of the European nuthatch "putting its bill into a crack in the bough of a tree, and making such a violent sound as if it was rending asunder," this, if true, would be sufficient to distinguish it from the species we have been just describing, which possesses no such faculty.* The female differs little from the male in colour, chiefly in the black being less deep on the head and wings.

* When the nuthatch cracks or splits nuts, or stones of fruit, it is for the kernels alone; it is seen, from our various accounts, to be both a seed and grain eater. The very curious manner in which our own nuthatch splits nuts seems perfectly proved by several observers; and it is no less curious, that the same place is often resorted to different times in succession, as if it were more fit than another, or required less labour than to seek a new situation. Montagu says, that the most favourite position for breaking a nut is with the head downwards; and that in autumn it is no uncommon thing to find in the crevices of the bark of an old tree a great many broken nutshells, the work of this bird, who repeatedly returns to the same spot for this purpose: when it has fixed the nut firm in a chink, it turns on all sides to strike it with most advantage; this, with the common hazel nut, is the work of some labour; but it breaks a filbert with ease.—ED.

RED-BELLIED, BLACK-CAPPED NUTHATCH.
(*Sitta varia.*)

PLATE II.—FIG. 4.

Sitta varia, *Bart.* p. 289.—Sitta Canadensis, *Turton.*—Small Nuthatch, *Lath.* i. 651.

SITTA CANADENSIS.—LINNÆUS.

Sitta Canadensis, *Bonap. Synop.* p. 96.

THIS bird is much smaller than the last, measuring only four inches and a half in length, and eight inches in extent. In the form of its bill, tongue, nostrils, and in the colour of the back and tail-feathers, it exactly agrees with the former; the secondaries are not relieved with the deep black of the other species; and the legs, feet, and claws are of a dusky greenish yellow; the upper part of the head is black, bounded by a stripe of white passing round the frontlet; a line of black passes through the eye to the shoulder; below this is another line of white; the chin is white; the other under parts a light rust colour, the primaries and whole wings a dusky lead colour. The breast and belly of the female are not of so deep a brown, and the top of the head is less intensely black.

This species is migratory, passing from the north, where they breed, to the southern States, in October, and returning in April. Its voice is sharper, and its motions much quicker, than those of the other, being so rapid, restless, and small, as to make it a difficult point to shoot one of them. When the two species are in the woods together, they are easily distinguished by their voices, the note of the least being nearly an octave sharper than that of its companion, and repeated more hurriedly. In other respects, their notes are alike unmusical and monotonous. Approaching so near to each other in their colours and general habits, it is probable that their mode of building, &c., may be also similar.

Buffon's *Torchepot de la Canada* (Canada nuthatch of other European writers) is either a young bird of the present species, in its imperfect plumage, or a different sort, that rarely visits the United States. If the figure (Pl. enl. 623) be correctly coloured, it must be the latter, as the tail and head appear of the same bluish grey or lead colour as the back. The young birds of this species, it may be observed, have also the crown of a lead colour during the first season; but the tail-feathers are marked nearly as those of the old ones. Want of precision in the figures and descriptions of these authors makes it difficult to determine; but I think it very probable that *Sitta Jamaicensis minor*, Briss., the least loggerhead of Brown, *Sitta Jamaicensis var. t. st.*, Linn., and *Sitta Canadensis* of Linnæus, Gmelin, and Brisson, are names that have been originally applied to different individuals of the species we ar now describing.

This bird is particularly fond of the seeds of pine trees. You may traverse many thousand acres of oak, hickory, and chestnut woods, during winter, without meeting with a single individual; but no sooner do you enter among the pines than, if the air be still, you have only to listen for a few moments, and their note will direct you where to find them. They usually feed in pairs, climbing about in all directions, generally accompanied by the former species, as well as by the titmouse, *Parus atricapillus*, and the crested titmouse, *Parus bicolor*, and not unfrequently by the small spotted woodpecker, *Picus pubescens;* the whole company proceeding regularly from tree to tree through the woods like a corps of pioneers; while, in a calm day, the rattling of their bills, and the rapid motions of their bodies, thrown, like so many tumblers and rope dancers, into numberless positions, together with the peculiar chatter of each, are altogether very amusing; conveying the idea of hungry diligence, bustle, and activity.* Both these

* It is curious to remark the similarity, as it were, in the feeling and disposition of some species. In this country, during winter, when the different kinds have laid aside those ties which connected them by sexual intercourse, nothing is more common than to see a whole troop of the blue, marsh, cole, and long-tailed titmice, accompanied with a host of golden-crested wrens, and perhaps a solitary creeper, proceed in the

Drawn from Nature by A. Wilson.

Engraved by W. H. Lizars.

1. Gold winged Woodpecker. *Picus Auratus.* 2. Black throated Bunting. *Emberiza Americana.* 3. Blue Bird. *Motacilla Sialis.*

3.

little birds, from the great quantity of destructive insects and larvæ they destroy, both under the bark and among the tender buds of our fruit and forest trees, are entitled to, and truly deserving of, our esteem and protection.

GOLD-WINGED WOODPECKER. (*Picus auratus.*)

PLATE III.—FIG. 1.

Le pic aux ailes dorees, *Buffon*, vii. 39. *Pl. enl.* 693.—Picus auratus, *Linn. Syst.* 174.—Cuculus alis de auratis, *Klein*, p. 30.—*Catesby*, i. 18.—*Latham*, ii. 597.—*Bartram*, p. 289.—*Peale's Museum*, No. 1938.

COLAPTES AURATUS.—SWAINSON.*

Picus auratus, *Penn. Arct. Zool.* ii. p. 270.—*Wagler*, No. 84—*Bonap. Synop.* p. 44.—Golden-winged Woodpecker, *Aud.* i. p. 191.—Colaptes auratus, *North. Zool.* ii. 314.

THIS elegant bird is well known to our farmers and junior sportsmen, who take every opportunity of destroying him ; the former, for the supposed trespasses he commits on their Indian-corn, or the trifle he will bring in market; and the latter for

manner here mentioned, and regularly follow each other, as if in a laid-out path. An alarm may cause a temporary digression of some of the troop ; but these are soon perceived making up their way to the main body. The whole may be found out and traced by their various and constantly reiterated cries.—ED.

* This beautiful species is typical of one form among the *Picianæ*, and has been designated under the above title by Mr Swainson. The form appears to range in North and South America, the West Indian Islands, and in Africa ; our present species is confined to North America alone. They are at once distinguished from the true woodpeckers and the other groups by the curved and compressed bill, the broad and strong shafts of the quills, which are also generally brightly coloured, and appear very conspicuous during flight when the wings are expanded. In the typical species they are of a bright golden yellow, whence the common name ; and in one closely allied, the *C. Mexicanus*, Sw., of a bright reddish orange ; in a third, *C. Brasiliensis*, they are of a pale straw yellow. The upper parts of the plumage are, in general, barred, and the feathers on the hindhead are of a uniform length, never crested. A difference in form will always produce a difference in habit ; and we accordingly find that these birds more frequently perch on the branches,

the mere pleasure of destruction, and perhaps for the flavour of his flesh, which is in general esteem. In the State of Pennsylvania, he can scarcely be called a bird of passage, as, even in severe winters, they may be found within a few miles of the city of Philadelphia; and I have known them exposed for sale in market every week during the months of November, December, and January, and that, too, in more than commonly rigorous weather. They no doubt, however partially, migrate even here, being much more numerous in spring and fall than in winter. Early in the month of April they begin to prepare their nest, which is built in the hollow body or branch of a tree, sometimes, though not always, at a considerable height from the ground; for I have frequently known them fix on the trunk of an old apple tree, at not more than six feet from the root. The sagacity of this bird in discovering, under a sound bark, a hollow limb or trunk of a tree, and its perseverance in perforating it for the purpose of incubation, are

and feed a great deal upon the ground; they seem also to possess more of the activity of the nuthatch and titmice than the regular *climb* of the typical woodpeckers. The golden-winged woodpecker is known to feed a great deal upon ants, seeking them about the hills, and, according to Mr Audubon, also picks up grains and seed from the ground. In a Brazilian species, *Picus campestris* of Spix and Martius, we have analogous habits; and, as the name implies, it is often seen upon the ground, frequenting the ordure of cattle, and turning it over in search of insects; or in the neighbourhood of anthills, where they find an abundant and very favourite food. We find also the general development of form joined to habit in the typical form of another group, the common green and grey-headed woodpeckers of Europe, which feed much on ants, and of course seek them on the ground.

M. Lesson, in his "Manual d'Ornithologie," has given it the title of *Cucupicus*, making the African species typical. He of course was not aware of its having been previously characterised; and in that of America, all the forms are more clearly developed.

The *C. Mexicanus*, mentioned before, was met with in the last overland expedition, and will form an addition to the North American species; it was killed by Mr David Douglas to the westward of the Rocky Mountains. The more common country is Mexico, whence it extends along the shores of the Pacific some distance northward of the Columbia River, and to New California.—Ed.

truly surprising; the male and female alternately relieving and encouraging each other, by mutual caresses, renewing their labours for several days, till the object is attained, and the place rendered sufficiently capacious, convenient, and secure. At this employment they are so extremely intent, that they may be heard till a very late hour in the evening, thumping like carpenters. I have seen an instance where they had dug first five inches straight forward, and then downward more than twice that distance, through a solid black oak. They carry in no materials for their nest, the soft chips and dust of the wood serving for this purpose. The female lays six white eggs, almost transparent, very thick at the greater end, and tapering suddenly to the other. The young early leave the nest, and, climbing to the higher branches, are there fed by their parents.

The food of this bird varies with the season. As the common cherries, bird cherries, and berries of the sour gum successively ripen, he regales plentifully on them, particularly on the latter; but the chief food of this species, or that which is most usually found in his stomach, is wood lice, and the young and larvæ of ants, of which he is so immoderately fond, that I have frequently found his stomach distended with a mass of these, and these only, as large nearly as a plumb. For the procuring of these insects, nature has remarkably fitted him: the bills of woodpeckers, in general, are straight, grooved, or channelled, wedge-shaped, and compressed to a thin edge at the end, that they may the easier penetrate the hardest wood; that of the gold-winged woodpecker is long, slightly bent, ridged only on the top, and tapering almost to a point, yet still retaining a little of the wedge form there. Both, however, are admirably adapted for the peculiar manner each has of procuring its food. The former, like a powerful wedge, to penetrate the dead and decaying branches, after worms and insects; the latter, like a long and sharp pickaxe, to dig up the hillocks of pismires that inhabit old stumps in prodigious multitudes. These beneficial services would entitle him to some regard from the husbandman, were he not accused, and perhaps not without just cause, of being too partial to the Indian-corn when in that state which is usually called *roasting-ears.* His visits are indeed rather frequent about this time; and the farmer, suspecting what is going on, steals through among the rows with his gun, bent on vengeance, and forgetful of the benevolent sentiment of the poet, that

> Just as wide of justice he must fall,
> Who thinks all made for one, not one for all.

But farmers, in general, are not much versed in poetry, and pretty well acquainted with the value of corn, from the hard labour requisite in raising it.

In rambling through the woods one day, I happened to shoot one of these birds, and wounded him slightly in the wing. Finding him in full feather, and seemingly but little hurt, I took him home, and put him into a large cage, made of willows, intending to keep him in my own room, that we might become better acquainted. As soon as he found himself inclosed on all sides, he lost no time in idle fluttering, but, throwing himself against the bars of the cage, began instantly to demolish the willows, battering them with great vehemence, and uttering a loud piteous kind of cackling, similar to that of a hen when she is alarmed and takes to wing. Poor Baron Trenck never laboured with more eager diligence at the walls of his prison than this son of the forest in his exertions for liberty; and he exercised his powerful bill with such force, digging into the sticks, seizing and shaking them so from side to side, that he soon opened for himself a passage; and, though I repeatedly repaired the breach, and barricaded every opening, in the best manner I could, yet, on my return into the room, I always found him at large, climbing up the chairs, or running about the floor, where, from the dexterity of his motions, moving backward, forward, and sidewise, with the same facility, it became difficult to get hold of him again. Having placed him in a strong wire cage, he seemed to give up all hopes of making his escape, and soon became very tame; fed on young ears of Indian-corn, refused apples, but ate the berries of the sour gum greedily, small winter grapes, and several other kinds of berries; exercised himself frequently in climbing, or rather hopping perpendicularly along the sides of the cage; and, as evening drew on, fixed himself in a high hanging or perpendicular position, and slept with his head in his wing. As soon as dawn appeared, even before it was light enough to perceive him distinctly across the room, he descended to the bottom of the cage, and began his attack on the ears of Indian-corn, rapping so loud as to be heard from every room in the house. After this, he would sometimes resume his former position, and take another nap. He was beginning to become very amusing, and even sociable, when, after a lapse of several weeks, he became drooping, and died, as I conceived, from the effects of his wound.*

* Mr Audubon says they live well in confinement. "The golden-winged woodpecker never suffers its naturally lively spirit to droop. It feeds well; and by way of amusement will continue to destroy as much furniture in a day as can well be mended by a different kind of workman in a week." The same gentleman, when speaking of their flight, again adds, that it is more "strong and prolonged, being performed in a straighter manner, than any other of our woodpeckers. They propel themselves by numerous beats of the wings, with short intervals of sailing, during which they scarcely fall from the horizontal. When passing from one tree to another, they also fly in a straight line, until within a few yards of the spot on which they intend to alight, when they suddenly raise themselves a few feet, and fasten themselves to the bark of the trunk by their claws and tail. Their migrations, although partial (as many remain even in the middle districts during the severest winters), are performed under night, as is known by their note and the whistling of their wings, which are heard from the ground." Of its movement he also speaks: "It easily moves sidewise on a small branch, keeping itself as erect as other birds usually do; but with equal care does it climb by leaps along the trunks of trees or their branches, descend, and move sidewise or spirally, keeping at all times its head upwards, and its tail pressed against the bark, as a support."

I have thus at length transcribed Mr Audubon's minuter details, as tending to show the differences of habit in this form, which will be still better observed when compared with those we have yet to describe.

There is another peculiarity in these birds, and some others of the

Some European naturalists (and, among the rest, Linnæus himself, in his tenth edition of *Systema Naturæ*) have classed this bird with the genus *Cuculus*, or cuckoo, informing their readers that it possesses many of the habits of the cuckoo; that it is almost always on the ground; is never seen to climb trees like the other woodpeckers, and that its bill is altogether unlike theirs; every one of which assertions, I must say, is incorrect, and could have only proceeded from an entire unacquaintance with the manners of the bird. Except in the article of the bill—and that, as has been before observed, is still a little wedge-formed at the point—it differs in no one

genus mentioned by Mr Audubon, which does not seem to have been noticed before, though I am not sure that it is confined to the *Pici* only. In many of our sandpipers—the purre, for instance—the first plumage is that of the adult female in the nuptial dress; and, in those which have black breasts, an occasional tinge of that colour may be traced. A great portion of these also receive at least a part of the winter dress during the first year. What I have alluded to is as follows, and it may be well that it is attended to in the description of the different species of woodpeckers; Mr Audubon, however, uses the word "*frequently*," as if it were not a constant appearance in the young:—"In this species, as in a few others, there is a singular arrangement in the colouring of the feathers of the upper part of the head, which I conceive it necessary for me to state, that it may enable persons better qualified than myself to decide as to the reasons of such arrangement. The young of this species frequently have the whole upper part of the head tinged with red, which, at the approach of winter disappears, when merely a circular line of that colour is to be observed on the hind part, becoming of a rich silky vermilion tint. The hairy, downy, and red-cockaded woodpeckers are subject to the same extraordinary changes, which, as far as I know, never reappear at any future period of their lives. I was at first of opinion that this change appeared only on the head of the male birds; but, on dissection, I found it equally affecting both sexes. I am induced to believe, that, in consequence of this, many young woodpeckers, of different species, have been described and figured as forming distinct species themselves. I have shot dozens of young woodpeckers in this peculiar state of plumage, which, on being shown to other persons, were thought by them to be of different species from what the birds actually were. This occurrence is the more worthy of notice, as it is exhibited on all the species of this genus, on the heads of which, when in full plumage, a very narrow line exists."—Ed.

characteristic from the rest of its genus. Its nostrils are covered with tufts of recumbent hairs, or small feathers; its tongue is round, worm-shaped, flattened towards the tip, pointed, and furnished with minute barbs; it is also long, missile, and can be instantaneously protruded to an uncommon distance. The os hyöides, or internal parts of the tongue, like those of its tribe, is a substance for strength and elasticity resembling whalebone, divided into two branches, each the thickness of a knitting-needle, that pass, one on each side of the neck, to the hind head, where they unite, and run up along the skull in a groove, covered with a thin membrane, or sheath, descend into the upper mandible by the right side of the right nostril, and reach to within half an inch of the point of the bill, to which they are attached by another extremely elastic membrane, that yields when the tongue is thrown out, and contracts as it is retracted. In the other woodpeckers we behold the same apparatus, differing a little in different species. In some, these cartilaginous substances reach only to the top of the cranium; in others, they reach to the nostril; and in one species they are wound round the bone of the right eye, which projects considerably more than the left for its accommodation.

The tongue of the gold-winged woodpecker, like the others, is also supplied with a viscid fluid, secreted by two glands that lie under the ear on each side, and are at least five times larger in this species than in any other of its size; with this the tongue is continually moistened, so that every small insect it touches instantly adheres to it. The tail, in its strength and pointedness, as well as the feet and claws, prove that the bird was designed for climbing; and in fact I have scarcely ever seen it on a tree five minutes at a time without climbing; hopping not only upward and downward, but spirally; pursuing and playing with its fellow in this manner round the body of the tree. I have also seen them a hundred times alight on the trunk of the tree, though they more frequently alight on the branches; but that they climb, construct like nests, lay the same number and the like coloured eggs, and have the manners and habits of the woodpeckers, is notorious to every American naturalist; while neither in the form of their body, nor any other part, except in the bill being somewhat bent, and the toes placed two before and two behind, have they the smallest resemblance whatever to the cuckoo.

It may not be improper, however, to observe, that there is another species of woodpecker, called also gold-winged,* which inhabits the country near the Cape of Good Hope, and resembles the present, it is said, almost exactly in the colour and form of its bill, and in the tint and markings of its plumage, with this difference, that the mustaches are red, instead of black, and the lower side of the wings, as well as their shafts, are also red, where the other is golden yellow. It is also considerably less. With respect to the habits of this new species, we have no particular account; but there is little doubt that they will be found to correspond with the one we are now describing.

The abject and degraded character which the Count de Buffon, with equal eloquence and absurdity, has drawn of the whole tribe of woodpeckers, belongs not to the elegant and sprightly bird now before us. How far it is applicable to any of them will be examined hereafter. He is not "constrained to drag out an insipid existence in boring the bark and hard fibres of trees to extract his prey," for he frequently finds in the loose mouldering ruins of an old stump (the capital of a nation of pismires) more than is sufficient for the wants of a whole week. *He* cannot be said to "lead a mean and gloomy life, without an intermission of labour," who usually feasts by the first peep of dawn, and spends the early and sweetest hours of morning on the highest peaks of the tallest trees, calling on his mate or companions; or pursuing and gambolling with them round the larger limbs and body of the tree for hours together; for such are really his habits. Can it be said, that "necessity never grants an interval of sound repose"

* Picus cafer, Turton's Linn.

to that bird who, while other tribes are exposed to all the peltings of the midnight storm, lodges dry and secure in a snug chamber of his own constructing? or that "the narrow circumference of a tree circumscribes *his* dull round of life," who, as seasons and inclination inspire, roams from the frigid to the torrid zone, feasting on the abundance of various regions? Or is it a proof that "his appetite is never softened by delicacy of taste," because he so often varies his bill of fare, occasionally preferring to animal food the rich milkiness of young Indian corn, and the wholesome and nourishing berries of the wild cherry, sour gum, and red cedar? Let the reader turn to the faithful representation of him given in the plate, and say whether his looks be "sad and melancholy." It is truly ridiculous and astonishing that such absurdities should escape the lips or pen of one so able to do justice to the respective merits of every species; but Buffon had too often a favourite theory to prop up, that led him insensibly astray; and so, forsooth, the whole family of woodpeckers must look sad, sour, and be miserable, to satisfy the caprice of a whimsical philosopher, who takes it into his head that they are, and ought to be so!

But the Count is not the only European who has misrepresented and traduced this beautiful bird. One has given him brown legs;* another a yellow neck;† a third has declared him a cuckoo;‡ and, in an English translation of Linnæus's "System of Nature," lately published, he is characterised as follows: "Body, striated with black and gray; cheeks, red; chin, black; never climbs on trees;" § which is just as correct as if, in describing the human species, we should say,—Skin, striped with black and green; cheeks, blue; chin, orange; never walks on foot, &c. The pages of natural history should resemble a faithful mirror, in which mankind may recognise the true images of the living originals; instead of which, we find

* See Encyc. Brit. art. Picus. † Latham. ‡ Klein.

§ P. griseo nigroque transversim striatus——truncos arborum non scandit.—Ind. Orn., vol. i. p. 242.

this department of them too often like the hazy and rough medium of wretched window-glass, through whose crooked protuberances everything appears so strangely distorted, that one scarcely knows their most intimate neighbours and acquaintances.

The gold-winged woodpecker has the back and wings above of a dark umber, transversely marked with equidistant streaks of black; upper part of the head an iron-gray: cheeks and parts surrounding the eyes, a fine cinnamon colour: from the lower mandible a strip of black, an inch in length, passes down each side of the throat, and a lunated spot, of a vivid blood red, covers the hind head, its two points reaching within half an inch of each eye; the sides of the neck, below this, incline to a bluish gray; throat and chin, a very light cinnamon or fawn colour; the breast is ornamented with a broad crescent of deep black; the belly and vent, white, tinged with yellow, and scattered with innumerable round spots of black, every feather having a distinct central spot, those on the thighs and vent being heart-shaped and largest; the lower or inner side of the wing and tail, shafts of all the larger feathers, and indeed of almost every feather, are of a beautiful golden yellow; that on the shafts of the primaries being very distinguishable, even when the wings are shut; the rump is white, and remarkably prominent; the tail-coverts white, and curiously serrated with black; upper side of the tail, and the tip below, black, edged with light loose filaments of a cream colour, the two exterior feathers, serrated with whitish; shafts, black towards the tips, the two middle ones, nearly wholly so; bill, an inch and a half long, of a dusky horn colour, somewhat bent, ridged only on the top, tapering, but not to a point, that being a little wedge-formed; legs and feet, light blue; iris of the eye, hazel; length, twelve inches; extent, twenty. The female differs from the male chiefly in the greater obscurity of the fine colours, and in wanting the black mustaches on each side of the throat. This description, as well as the drawing, was taken from a very beautiful and perfect specimen.

Though this species, generally speaking, is migratory, yet they often remain with us in Pennsylvania during the whole winter. They also inhabit the continent of North America, from Hudson's Bay to Georgia; and have been found by voyagers on the northwest coast of America. They arrive at Hudson's Bay in April, and leave it in September. Mr Hearne, however, informs us, that "the gold-winged woodpecker is almost the only species of woodpecker that winters near Hudson's Bay." The natives there call it *Ou-thee-quan-nor-ow,* from the golden colour of the shafts and lower side of the wings. It has numerous provincial appellations in the different States of the Union, such as "High-hole," from the situation of its nest, and "Hittock," "Yucker," "Piut," "Flicker," by which last it is usually known in Pennsylvania. These names have probably originated from a fancied resemblance of its notes to the sound of the words; for one of its most common cries consists of two notes, or syllables, frequently repeated, which, by the help of the hearer's imagination, may easily be made to resemble any or all of them.

BLACK-THROATED BUNTING. (*Emberiza Americana.*)

PLATE III.—FIG. 2.

Calandra pratensis, the May-bird, *Bartram*, p. 291.—*Peale's Museum*, No. 5952. —*Arct. Zool.* 228.—Emberiza Americana, *Ind. Orn.* p. 44.

EMBERIZA AMERICANA.—LINNÆUS.*

Fringilla Americana, *Bonap. Synop.* 107.

OF this bird I have but little to say. They arrive in Pennsylvania from the south about the middle of May; abound

* America has no birds perfectly typical with the *Emberizæ* of Europe; the group appears to assume two forms, under modifications, that of *E. miliaria*, with the bill of considerable strength, and that of the weaker make, of *E. schæniculus*. To the former will be allied our present species; under the latter will rank the small *F. socialis, melodia, and palustris,*

in the neighbourhood of Philadelphia, and seem to prefer level fields covered with rye-grass, timothy, or clover, where they build their nest, fixing it in the ground, and forming it of fine dried grass. The female lays five white eggs, sprinkled with specks and lines of black. Like most part of their genus, they are nowise celebrated for musical powers. Their whole song consists of five notes, or, more properly, of two notes; the first repeated twice, and slowly, the second thrice, and rapidly, resembling *chip, chip, che che ché.* Of this ditty, such as it is, they are by no means parsimonious, for, from their first arrival for the space of two or three months, every level field of grain or grass is perpetually serenaded with *chip, chip, che che ché.* In their shape and manners they very much resemble the yellowhammer of Britain (*E. citrinella*); like them, they are fond of mounting to the top of some half-grown tree, and there chirruping for half-an-hour at a time. In travelling through different parts of New York and Pennsylvania in spring and summer, wherever I came to level fields of deep grass, I have constantly heard these birds around me. In August they become mute; and soon after, that is, towards the beginning of September, leave us altogether.

The black-throated bunting is six inches and a half in length; the upper part of the head is of a dusky greenish yellow; neck, dark ash; breast, inside shoulders of the wing, line over the eye, and at the lower angle of the bill, yellow; chin, and space between the bill and eye, white; throat,

&c.; the form is further represented in North America by *Plectrophanes* and *Pipilo*, and may be said to run into the Finches by means of the latter, and Mr Swainson's genus, *Zonotrichia.* The principal variations are the want, or smallness, of the palatial knob, and the wideness of the upper mandible, which exceeds that of the lower, while the reverse is the case in the true birds. Vieillot, I believe, proposed *Passerina* for some birds, but included many that were not so nearly allied, and Bonaparte has proposed *Spiza* to receive them, and to stand as a sub-genus of *Fringilla.* We think the form, colouring, and markings, joined with their song and habit, associates them much closer to *Emberiza,* and as such have at present retained them.—Ed.

covered with a broad, oblong, somewhat heart-shaped patch of black, bordered on each side with white; back, rump, and tail, ferruginous, the first streaked with black; wings, deep dusky, edged with a light clay colour; lesser coverts and whole shoulder of the wing, bright bay; belly and vent, dull white; bill, light blue, dusky above, strong and powerful for breaking seeds; legs and feet, brown; iris of the eye, hazel. The female differs from the male in having little or no black on the breast, nor streak of yellow over the eye; beneath the eye she has a dusky streak, running in the direction of the jaw. In all those I opened, the stomach was filled with various seeds, gravel, eggs of insects, and sometimes a slimy kind of earth or clay.

This bird has been figured by Latham, Pennant, and several others. The former speaks of a bird which he thinks is either the same, or nearly resembling it, that resides in summer in the country about Hudson's Bay, and is often seen associating in flights with the geese;* this habit, however, makes me suspect that it must be a different species; for, while with us here, the black-throated bunting is never gregarious, but is almost always seen singly, or in pairs, or, at most, the individuals of one family together.

BLUE BIRD. (*Sylvia sialis.*)

PLATE III.—Fig. 3.

Le rouge gorge bleu, *Buffon*, v. 212. *Pl. enl.* 390.—Blue Warbler, *Lath.* ii. 446.—*Catesb.* i. 47.—Motacilla sialis, *Linn. Syst.* 336.—*Bartram*, p. 291.—*Peale's Museum*, No. 7188.

SIALIA WILSONII.—Swainson.†

The Blue Redbreast, *Edw.* pl. 24.—Saxicola sialis, *Bonap. Synop.* p. 89.—Erythaca (Sialia) Wilsonii, *North. Zool.* ii. p. 210.

The pleasing manners and sociable disposition of this little bird entitle him to particular notice. As one of the first

* Latham, Synopsis, Supplement, p. 158.

† This beautiful species, interesting both as regards its domestic economy, and the intimate link which it fills up in the natural system, has

messengers of spring, bringing the charming tidings to our very doors, he bears his own recommendation always along with him, and meets with a hearty welcome from everybody.

Though generally accounted a bird of passage, yet, so early as the middle of February, if the weather be open, he usually makes his appearance about his old haunts, the barn, orchard, and fence posts. Storms and deep snows sometimes succeeding, he disappears for a time; but about the middle of March is again seen, accompanied by his mate, visiting the box in the

been dedicated, by Mr Swainson, to our author. It remained a solitary individual, until the discovery of a Mexican species by that gentleman, described under the title of *S. Mexicana;* and the return of the last overland Arctic expedition brought forward a third, confirming the views that were before held regarding it. According to these, it will range among the *Saxicolinæ*, whence it had been previously removed from *Sylvia* by Vieillot and Bonaparte, and it will hold the place, in North and South America, of the robin of Europe, and the stonechats of that country and Africa; while, in New Holland, the *Muscicapa multicolor*, now bearing the generic title of *Petroica*, with some allied species, will represent it. The old species ranges extensively over North America, and the northern parts of the south continent, extending also to some of the islands: the newly-discovered one appears confined to a more northern latitude. It has been described in the second volume of the "Northern Zoology," under the name of *S. Arctica*, and I now add the information contained in that valuable work:—

"Colour of the dorsal aspect, ultramarine blue; the webs of the tertiaries, and the tips of the inner margins of the quill and tail-feathers, dull umber brown; the base of the plumage, blackish gray. *Under surface*—the cheeks, throat, breast, and insides of the wings, greenish blue, bordering on the abdomen to grayish blue; vent-feathers, and under tail-coverts, white; tail beneath, and inside of the quill-feathers, olive brown, with a strong tinge of blue; bill and feet, pitch black; form, in general, that of *S. Wilsonii*, but the bill is considerably narrower at the base, and proportionably larger, straighter, and less notched, and bent at the tip of the upper mandible; its breadth is equal to its depth; wings, three quarters of an inch shorter than the tail; the second quill-feather is the longest; the first and third are equal, and about a line shorter; the tenth is an inch and a half shorter than the second; tail, forked, or deeply emarginated, the central feathers being more than half an inch shorter than the exterior ones; legs and feet, similarly formed with those of *S. Wilsonii;* length, seven inches nine lines."—ED.

garden, or the hole in the old apple tree, the cradle of some generations of his ancestors. "When he first begins his amours," says a curious and correct observer, "it is pleasing to behold his courtship, his solicitude to please and to secure the favour of his beloved female. He uses the tenderest expressions, sits close by her, caresses and sings to her his most endearing warblings. When seated together, if he espies an insect delicious to her taste, he takes it up, flies with it to her, spreads his wing over her, and puts it in her mouth." * If a rival makes his appearance,—for they are ardent in their loves,—he quits her in a moment, attacks and pursues the intruder as he shifts from place to place, in tones that bespeak the jealousy of his affection, conducts him with many reproofs, beyond the extremities of his territory, and returns to warble out his transports of triumph beside his beloved mate. The preliminaries being thus settled, and the spot fixed on, they begin to clean out the old nest and the rubbish of the former year, and to prepare for the reception of their future offspring. Soon after this, another sociable little pilgrim (*Motacilla domestica*, house wren) also arrives from the south, and, finding such a snug berth preoccupied, shows his spite, by watching a convenient opportunity, and, in the absence of the owner, popping in and pulling out sticks; but takes special care to make off as fast as possible.

The female lays five, and sometimes six eggs, of a pale blue colour; and raises two, and sometimes three broods in a season; the male taking the youngest under his particular care while the female is again sitting. Their principal food are insects, particularly large beetles, and others of the coleopterous kinds that lurk among old, dead, and decaying trees. Spiders are also a favourite repast with them. In the fall, they occasionally regale themselves on the berries of the sour gum, and, as winter approaches, on those of the red cedar, and on the fruit of a rough hairy vine that runs up and cleaves fast to the trunks of trees. Ripe persimmons is another of

* Letter from Mr William Bartram to the author.

their favourite dishes, and many other fruits and seeds which I have found in their stomachs at that season, which, being no botanist, I am unable to particularise. They are frequently pestered with a species of tape-worm, some of which I have taken from their intestines of an extraordinary size, and, in some cases, in great numbers. Most other birds are also plagued with these vermin, but the blue bird seems more subject to them than any I know, except the woodcock. An account of the different species of vermin, many of which, I doubt not, are nondescripts, that infest the plumage and intestines of our birds, would of itself form an interesting publication; but, as this belongs more properly to the entomologist, I shall only, in the course of this work, take notice of some of the most remarkable; and occasionally represent them on the same plate with those birds upon which they are usually found.

The usual spring and summer song of the blue bird is a soft, agreeable, and oft-repeated warble, uttered with open quivering wings, and is extremely pleasing. In his motions and general character, he has great resemblance to the robin redbreast of Britain; and, had he the brown olive of that bird, instead of his own blue, could scarcely be distinguished from him. Like him, he is known to almost every child; and shows as much confidence in man by associating with him in summer, as the other by his familiarity in winter. He is also of a mild and peaceful disposition, seldom fighting or quarrelling with other birds. His society is courted by the inhabitants of the country, and few farmers neglect to provide for him, in some suitable place, a snug little summer-house, ready fitted and rent free. For this he more than sufficiently repays them by the cheerfulness of his song, and the multitude of injurious insects which he daily destroys. Towards fall, that is, in the month of October, his song changes to a single plaintive note, as he passes over the yellow many-coloured woods; and its melancholy air recalls to our minds the approaching decay of the face of nature. Even after the trees are stript of their leaves, he still lingers over his native fields, as if loth to leave them. About the middle or end of November, few or none of them are seen; but, with every return of mild and open weather, we hear his plaintive note amidst the fields, or in the air, seeming to deplore the devastations of winter. Indeed, he appears scarcely ever totally to forsake us; but to follow fair weather through all its journeyings till the return of spring.

Such are the mild and pleasing manners of the blue bird, and so universally is he esteemed, that I have often regretted that no pastoral muse has yet arisen in this Western woody world to do justice to his name, and endear him to us still more by the tenderness of verse, as has been done to his representative in Britain, the robin redbreast. A small acknowledgment of this kind I have to offer, which the reader, I hope, will excuse as a tribute to rural innocence.

When winter's cold tempests and snows are no more,
  Green meadows and brown furrowed fields re-appearing,
The fishermen hauling their shad to the shore,
  And cloud-cleaving geese to the lakes are a-steering;
When first the lone butterfly flits on the wing,
  When red glow the maples, so fresh and so pleasing,
O then comes the blue bird, the herald of spring!
  And hails with his warblings the charms of the season.

Then loud-piping frogs make the marshes to ring;
  Then warm glows the sunshine, and fine is the weather;
The blue woodland flowers just beginning to spring,
  And spicewood and sassafras budding together:
O then to your gardens, ye housewives, repair,
  Your walks border up, sow and plant at your leisure;
The blue bird will chant from his box such an air,
  That all your hard toils will seem truly a pleasure!

He flits through the orchard, he visits each tree,
  The red-flowering peach, and the apple's sweet blossoms;
He snaps up destroyers wherever they be,
  And seizes the caitiffs that lurk in their bosoms;
He drags the vile grub from the corn it devours,
  The worms from their webs, where they riot and welter;
His song and his services freely are ours,
  And all that he asks is—in summer a shelter.

The ploughman is pleased when he gleans in his train,
Now searching the furrows—now mounting to cheer him;
The gard'ner delights in his sweet, simple strain,
And leans on his spade to survey and to hear him;
The slow ling'ring schoolboys forget they'll be chid,
While gazing intent as he warbles before them,
In mantle of sky-blue, and bosom so red,
That each little loiterer seems to adore him.

When all the gay scenes of the summer are o'er,
And autumn slow enters so silent and sallow,
And millions of warblers, that charmed us before,
Have fled in the train of the sun-seeking swallow;
The blue bird, forsaken, yet true to his home,
Still lingers, and looks for a milder to-morrow,
Till forced by the horrors of winter to roam,
He sings his adieu in a lone note of sorrow.

While spring's lovely season, serene, dewy, warm,
The green face of earth, and the pure blue of heaven,
Or love's native music have influence to charm,
Or sympathy's glow to our feelings are given,
Still dear to each bosom the blue bird shall be;
His voice, like the thrillings of hope, is a treasure;
For, through bleakest storms, if a calm he but see,
He comes to remind us of sunshine and pleasure!

The blue bird, in summer and fall, is fond of frequenting open pasture fields; and there perching on the stalks of the great mullein, to look out for passing insects. A whole family of them are often seen thus situated, as if receiving lessons of dexterity from their more expert parents, who can espy a beetle crawling among the grass at a considerable distance; and, after feeding on it, instantly resume their former position.* But whoever informed Dr Latham that "this bird is never

* The very habits of our European *Saxicolæ* are here described; they invariably seek the summit of some elevation, a hillock, a stone, bush, or some of the taller wild plants, and if occasionally on a tree, the topmost branch is always preferred; there they perch, uttering their monotonous call, which increases in anxiety and frequency as we approach the nest, or the young before they are able to fly; or they alight at intervals, run for some distance, and again remount to a fresh station. When not annoyed, they retain the same elevated situations, looking

seen on trees, though it makes its nest in the holes of them!"* might as well have said, that the Americans are never seen in the streets, though they build their houses by the sides of them. For what is there in the construction of the feet and claws of this bird to prevent it from perching? Or what sight more common to an inhabitant of this country than the blue bird perched on the top of a peach or apple tree, or among the branches of those reverend broad-armed chestnut trees that stand alone in the middle of our fields, bleached by the rains and blasts of ages?

The blue bird is six inches and three quarters in length, the wings remarkably full and broad; the whole upper parts are of a rich sky-blue, with purple reflections; the bill and legs are black; inside of the mouth and soles of the feet, yellow, resembling the colour of a ripe persimmon; the shafts of all the wing and tail-feathers are black; throat, neck, breast, and sides, partially under the wings, chestnut; wings, dusky black at the tips; belly and vent, white; sometimes the secondaries are exteriorly light brown, but the bird has in that case not arrived at his full colour. The female is easily distinguished by the duller cast of the back, the plumage of which is skirted with light brown, and by the red on the breast being much fainter, and not descending nearly so low as in the male; the secondaries are also more dusky. This species is found over the whole United States; in the Bahama Islands, where many of them winter; as also in Mexico, Brazil, and Guiana.

Mr Edwards mentions, that the specimen of this bird which he was favoured with was sent from the Bermudas; and, as

out for food, taking the insects seldom on the wing, but generally by a sudden spring, or leap down, and returning immediately with the prey in their bill, where it is retained for a few minutes, while they repeat their uniform note. The young, as soon as they are able to fly, have the same manners with their parents, and at the season when these are first on the wing, some extensive commons have appeared almost entirely in motion with our common species.—ED.

* Synopsis, vol. ii. p. 446, 40.

these islands abound with the cedar, it is highly probable that many of those birds pass from our continent thence, at the commencement of winter, to enjoy the mildness of that climate as well as their favourite food.

As the blue bird is so regularly seen in winter, after the continuance of a few days of mild and open weather, it has given rise to various conjectures as to the place of his retreat. Some supposing it to be in close, sheltered thickets, lying to the sun; others the neighbourhood of the sea, where the air is supposed to be more temperate, and where the matters thrown up by the waves furnish him with a constant and plentiful supply of food. Others trace him to the dark recesses of hollow trees, and subterraneous caverns, where they suppose he dozes away the winter, making, like Robinson Crusoe, occasional reconnoitering excursions from his castle, whenever the weather happens to be favourable. But amidst the snows and severities of winter, I have sought for him in vain in the most favourable sheltered situations of the middle States; and not only in the neighbourhood of the sea, but on both sides of the mountains.* I have never, indeed, explored the depths of caverns in search of him, because I would as soon expect to meet with tulips and butterflies there as blue birds; but, among hundreds of woodmen, who have cut down trees of all sorts and at all seasons, I have never heard one instance of these birds being found so immured in winter; while, in the whole of the middle and eastern States, the same general observation seems to prevail, that the blue bird always makes his appearance in winter after a few days of mild and open weather. On the other hand, I have myself found them numerous in the woods of North and South Carolina, in the depth of winter; and I have also been assured by different gentlemen of respectability, who have resided in the islands of Jamaica, Cuba, and the Bahamas and Bermudas, that this

* I speak of the species here generally. Solitary individuals are found, particularly among our cedar trees, sometimes in the very depth of winter.

very bird is common there in winter. We also find, from the works of Hernandez, Piso, and others, that it is well known in Mexico, Guiana, and Brazil; and, if so, the place of its winter retreat is easily ascertained, without having recourse to all the trumpery of holes and caverns, torpidity, hybernation, and such ridiculous improbabilities.

Nothing is more common in Pennsylvania than to see large flocks of these birds, in spring and fall, passing at considerable heights in the air; from the south in the former, and from the north in the latter season. I have seen, in the month of October, about an hour after sunrise, ten or fifteen of them descend from a great height, and settle on the top of a tall detached tree, appearing, from their silence and sedateness, to be strangers, and fatigued. After a pause of a few minutes, they began to dress and arrange their plumage, and continued so employed for ten or fifteen minutes more; then, on a few warning notes being given, perhaps by the leader of the party, the whole remounted to a vast height, steering in a direct line for the southwest. In passing along the chain of the Bahamas towards the West Indies, no great difficulty can occur, from the frequency of these islands; nor even to the Bermudas, which are said to be six hundred miles from the nearest part of the continent. This may seem an extraordinary flight for so small a bird; but it is, nevertheless, a fact that it is performed. If we suppose the blue bird in this case to fly only at the rate of a mile per minute, which is less than I have actually ascertained him to do over land, ten or eleven hours would be sufficient to accomplish the journey; besides the chances he would have of resting-places by the way, from the number of vessels that generally navigate those seas. In like manner, two days at most, allowing for numerous stages for rest, would conduct him from the remotest regions of Mexico to any part of the Atlantic States. When the natural history of that part of the continent and its adjacent isles is better known, and the periods at which its birds of passage arrive and depart are truly ascertained, I have no doubt but these suppositions will be fully corroborated.

ORCHARD ORIOLE. (*Oriolus mutatus.*)

PLATE IV.

Peale's Museum, No. 1508.—Bastard Baltimore, *Catesby*, i. 49.—Le Baltimore Batard, *Buffon*, iii. 233. *Pl. enl.* 506.—Oriolus spurius, *Gmelin*, *Syst.* i. p. 389.—*Lath. Syn.* ii. p. 433, 20. p. 437, 24.—*Bartram*, p. 290.

ICTERUS SPURIUS.—BONAPARTE.

Icterus spurius, *Bonap. Synop.* p. 51.—The Orchard Oriole, *Aud.* i. 221, pl. xlii.

THERE are no circumstances, relating to birds, which tend so much to render their history obscure and perplexing, as the various changes of colour which many of them undergo. These changes are in some cases periodical; in others progressive; and are frequently so extraordinary, that, unless the naturalist has resided for years in the country which the birds inhabit, and has examined them at almost every season, he is extremely liable to be mistaken and imposed on by their novel appearance. Numerous instances of this kind might be cited from the pages of European writers, in which the same bird has been described two, three, and even four different times, by the same person, and each time as a different kind. The species we are now about to examine is a remarkable example of this; and as it has never, to my knowledge, been either accurately figured or described, I have devoted one plate to the elucidation of its history.

The Count de Buffon, in introducing what he supposed to be the male of this bird, but which appears evidently to have been the female of the baltimore oriole, makes the following observations, which I give in the words of his translator:—"This bird is so called (spurious baltimore), because the colours of its plumage are not so lively as in the preceding (*Baltimore O.*) In fact, when we compare these birds, and find an exact correspondence in everything except the colours, and not even in the distribution of these, but only in the different tints they assume, we cannot hesitate to infer that the spurious baltimore is a variety of a more generous race,

Oriolus Spurius, Orchard Oriole. 1. Female. 2. and 3. Males of the second and third years. 4. Male in complete plumage. a. Egg of the Orchard Oriole. b. Egg of the Baltimore Oriole.

degenerated by the influence of climate, or some other accidental cause."

How the influence of climate could affect one portion of a species and not the other, when both reside in the same climate, and feed nearly on the same food, or what accidental cause could produce a difference so striking, and also so regular, as exists between the two, are, I confess, matters beyond my comprehension. But if it be recollected that the bird which the Count was thus philosophising upon was nothing more than the female Baltimore oriole, which exactly corresponds to the description of his male bastard baltimore, the difficulties at once vanish, and with them the whole superstructure of theory founded on this mistake. Dr Latham, also, while he confesses the great confusion and uncertainty that prevail between the true and bastard baltimore and their females, considers it highly probable that the whole will be found to belong to one and the same species, in their different changes of colour. In this conjecture, however, the worthy naturalist has likewise been mistaken; and I shall endeavour to point out the fact, as well as the source of this mistake.

And here I cannot but take notice of the name which naturalists have bestowed on this bird, and which is certainly remarkable. Specific names, to be perfect, ought to express some peculiarity common to no other of the genus, and should, at least, be consistent with truth; but, in the case now before us, the name has no one merit of the former, nor even that of the latter to recommend it, and ought henceforth to be rejected as highly improper, and calculated, like that of *goatsucker*, and many others equally ridiculous, to perpetuate that error from which it originated. The word *bastard*, among men, has its determinate meaning; but when applied to a whole species of birds, perfectly distinct from any other, originally deriving their peculiarities of form, manners, colour, &c., from the common source of all created beings, and perpetuating them, by the usual laws of generation, as unmixed and independent as any other, is, to call it by no worse name, a gross absurdity. Should the reader be displeased at this, I beg leave to remind him, that, as the faithful historian of our feathered tribes, I must be allowed the liberty of vindicating them from every misrepresentation whatever, whether originating in ignorance or prejudice; and of allotting to each respective species, as far as I can distinguish, that rank and place in the great order of nature to which it is entitled.

To convince the foreigner (for Americans have no doubt on the subject) that the present is a distinct species from the baltimore, it might be sufficient to refer to the figure of the latter in Plate I., and to fig. 4., Plate IV., of this work. I will, however, add, that I conclude this bird to be specifically different from the baltimore, from the following circumstances: its size—it is less, and more slender; its colours, which are different, and *very differently disposed;* the form of its bill, which is sharper pointed and more bent; the form of its tail, which is not *even*, but *wedged;* its notes, which are neither so full nor so mellow, and uttered with much more rapidity; its mode of building, and the materials it uses, both of which are different; and lastly, the shape and colour of the eggs of each (see figs. *a* and *b*), which are evidently unlike. If all these circumstances—and I could enumerate a great many more—be not sufficient to designate this as a distinct species, by what criterion, I would ask, are we to discriminate between a *variety* and an *original* species, or to assure ourselves that the great horned owl is not, in fact, a *bastard* goose, or the carrion crow a mere *variety* of the humming bird?

These mistakes have been occasioned by several causes; principally by the changes of colour to which the birds are subject, and the distance of Europeans from the country they inhabit. Catesby, it is true, while here, described and figured the baltimore, and perhaps was the first who published figures of either species; but he entirely omitted saying anything of the female, and, instead of the male and female of the present species, as he thought, he has only figured the male in two of his different dresses ; and succeeding compilers have followed and repeated the same error. Another cause may be assigned, viz., the extreme shyness of the female orchard oriole, represented at fig. 1. This bird has hitherto escaped the notice of European naturalists, or has been mistaken for another species, or perhaps for a young bird of the first season, which it almost exactly resembles. In none of the numerous works on ornithology has it ever before appeared in its proper character ; though the male has been known to Europeans for more than a century, and has usually been figured in one of his dresses as male, and in another as female ; these varying according to the fluctuating opinions of different writers. It is amusing to see how gentlemen have groped in the dark in pairing these two species of orioles, of which the following examples may be given :—

Buffon's and Latham's baltimore oriole.	*Male*—Male baltimore. *Female*—Male orchard oriole, fig. 4.
Spurious baltimore of ditto.	*Male*—Female baltimore. *Female*—Male orchard oriole, fig. 2.
Pennant's baltimore oriole.	*Male*—Male baltimore. *Female*—Young male baltimore.
Spurious oriole of ditto.	*Male*—Male orchard oriole, fig. 4. *Female*—Ditto ditto, fig. 2.
Catesby's baltimore oriole.	*Male*—Male baltimore. *Female*—Not mentioned.
Spurious baltimore of ditto.	*Male*—Male orchard oriole, fig. 2. *Female*—Ditto ditto, fig. 4.

Among all these authors, Catesby is doubtless the most inexcusable, having lived for several years in America, where he had an opportunity of being more correct : yet, when it is considered that the female of this bird is so much shyer than the male ; that it is seldom seen ; and that, while the males are flying around and bewailing an approach to their nest, the females keep aloof, watching every movement of the enemy in restless but silent anxiety ; it is less to be wondered at, I say, that two birds of the same kind, but different in plumage, making their appearance together at such times, should be taken for male and female of the same nest, without doubt or examination, as, from that strong sympathy for each other's distress which prevails so universally among them at this season, it is difficult sometimes to distinguish between the sufferer and the sympathising neighbour.

The female of the orchard oriole, fig. 1, is six inches and a half in length, and eleven inches in extent; the colour above is a yellow olive, inclining to a brownish tint on the back; the wings are dusky brown, lesser wing-coverts tipt with yellowish white, greater coverts and secondaries exteriorly edged with the same, primaries slightly so; tail, rounded at the extremity, the two exterior feathers three quarters of an inch shorter than the middle ones; whole lower parts, yellow; bill and legs, light blue; the former bent a little, very sharp pointed, and black towards the extremity; iris of the eye, hazel; pupil, black. The young male of the first season corresponds nearly with the above description. But in the succeeding spring he makes his appearance with a large patch of black marking the front, lores, and throat, as represented in fig. 2. In this stage, too, the black sometimes makes its appearance on the two middle feathers of the tail; and slight stains of reddish are seen commencing on the sides and belly. The rest of the plumage as in the female: this continuing nearly the same, on the same bird, during the remainder of the season. At the same time, other individuals are found, as represented by fig. 3, which are at least birds of the third summer. These are mottled with black and olive on the upper parts of the back, and with reddish bay and yellow on the belly, sides, and vent, scattered in the most irregular manner, not alike in any two individuals; and, generally, the two middle feathers of the tail are black, and the others centred with the same colour. This bird is now evidently approaching to its perfect plumage, as represented in fig. 4, where the black spreads over the whole head, neck, upper part of the back, breast, wings, and tail; the reddish bay, or bright chestnut, occupying the lower part of the breast, the belly, vent, rump, tail-coverts, and three lower rows of the lesser wing-coverts. The black on the head is deep and velvety; that of the wings inclining to brown; the greater wing-coverts are tipt with white. In the same orchard, and at the same time, males in each of these states of plumage may be found, united to their respective plain-coloured mates.

In all these, the manners, mode of building, food, and notes, are, generally speaking, the same, differing no more than those of any other individuals belonging to one common species. The female appears always nearly the same.

I have said that these birds construct their nests very differently from the baltimores. They are so particularly fond of frequenting orchards, that scarcely one orchard in summer is without them. They usually suspend their nest from the twigs of the apple tree; and often from the extremities of the outward branches. It is formed exteriorly of a particular species of long, tough, and flexible grass, knit or sewed through and through in a thousand directions, as if actually done with a needle. An old lady of my acquaintance, to whom I was one day showing this curious fabrication, after admiring its texture for some time, asked me, in a tone between joke and earnest, whether I did not think it possible to learn these birds to darn stockings? This nest is hemispherical, three inches deep by four in breadth; the concavity scarcely two inches deep by two in diameter. I had the curiosity to detach one of the fibres, or stalks of dried grass, from the nest, and found it to measure thirteen inches in length, and in that distance was thirty-four times hooked through and returned, winding round and round the nest! The inside is usually composed of wool, or the light downy appendages attached to the seeds of the *Platanus occidentalis*, or button-wood, which form a very soft and commodious bed. Here and there the outward work is extended to an adjoining twig, round which it is strongly twisted, to give more stability to the whole, and prevent it from being overset by the wind.

When they choose the long pendant branches of the weeping willow to build in, as they frequently do, the nest, though formed of the same materials, is made much deeper, and of slighter texture. The circumference is marked out by a number of these pensile twigs, that descend on each side like ribs, supporting the whole; their thick foliage, at the same time, completely concealing the nest from view. The depth in this case is increased to four or five inches, and the whole is made much slighter. These long pendant branches, being sometimes twelve and even fifteen feet in length, have a large sweep in the wind, and render the first of these precautions necessary, to prevent the eggs or young from being thrown out; and the close shelter afforded by the remarkable thickness of the foliage is, no doubt, the cause of the latter. Two of these nests, such as I have here described, are now lying before me, and exhibit not only art in the construction, but judgment in adapting their fabrication so judiciously to their particular situations. If the actions of birds proceeded, as some would have us believe, from the mere impulses of that thing called *instinct*, individuals of the same species would uniformly build their nest in the same manner, wherever they might happen to fix it; but it is evident from those just mentioned, and a thousand such circumstances, that they reason *à priori*, from cause to consequence; providently managing with a constant eye to future necessity and convenience.

The eggs, one of which is represented on the same plate (fig. *a*), are usually four, of a very pale bluish tint, with a few small specks of brown and spots of dark purple. An egg of the baltimore oriole is exhibited beside it (fig. *b*); both of these were minutely copied from nature, and are sufficient of themselves to determine, beyond all possibility of doubt, the identity of the two species. I may add, that Mr Charles W. Peale, proprietor of the Museum in Philadelphia, who, as a practical naturalist, stands deservedly first in the first rank of American connoisseurs, and who has done more for the promotion of that sublime science than all our speculative theorists together, has expressed to me his perfect conviction of the changes which these birds pass through, having himself examined them both in spring and towards the latter part of summer, and having at the present time in his possession thirty or forty individuals of this species, in almost every gradation of change.

The orchard oriole, though partly a dependent on the industry of the farmer, is no sneaking pilferer, but an open and truly beneficent friend. To all those countless multitudes of destructive bugs and caterpillars that infest the fruit trees in spring and summer, preying on the leaves, blossoms, and embryo of the fruit, he is a deadly enemy; devouring them whenever he can find them, and destroying, on an average, some hundreds of them every day, without offering the slightest injury to the fruit, however much it may stand in his way. I have witnessed instances where the entrance to his nest was more than half closed up by a cluster of apples, which he could have easily demolished in half a minute; but, as if holding the property of his patron sacred, or considering it as a natural bulwark to his own, he slid out and in with the greatest gentleness and caution. I am not sufficiently conversant in entomology to particularise the different species of insects on which he feeds, but I have good reason for believing that they are almost altogether such as commit the greatest depredations on the fruits of the orchard; and, as he visits us at a time when his services are of the greatest value, and, like a faithful guardian, takes up his station where the enemy is most to be expected, he ought to be held in respectful esteem, and protected by every considerate husbandman. Nor is the gaiety of his song one of his least recommendations. Being an exceedingly active, sprightly, and restless bird, he is on the ground —on the trees—flying and carolling in his hurried manner, in almost one and the same instant. His notes are shrill and lively, but uttered with such rapidity, and seeming confusion, that the ear is unable to follow them distinctly. Between these, he has a single note, which is agreeable and interesting. Wherever he is protected, he shows his confidence and gratitude by his numbers and familiarity. In the Botanic

Gardens of my worthy and scientific friends, the Messrs Bartrams of Kingsess, which present an epitome of almost everything that is rare, useful, and beautiful in the vegetable kingdom of this western continent, and where the murderous gun scarce ever intrudes, the orchard oriole revels without restraint through thickets of aromatic flowers and blossoms, and, heedless of the busy gardener that labours below, hangs his nest in perfect security on the branches over his head.

The female sits fourteen days; the young remain in the nest ten days afterwards, before they venture abroad, which is generally about the middle of June. Nests of this species, with eggs, are sometimes found so late as the 20th of July, which must either belong to birds that have lost their first nest, or it is probable that many of them raise two broods in the same season, though I am not positive of the fact.

The orchard orioles arrive in Pennsylvania rather later than the baltimores, commonly about the first week in May, and extend as far as the province of Maine. They are also more numerous towards the mountains than the latter species. In traversing the country near the Blue Ridge, in the month of August, I have seen at least five of this species for one of the baltimore. Early in September, they take their departure for the south; their term of residence here being little more than four months. Previous to their departure, the young birds become gregarious, and frequent the rich extensive meadows of the Schuylkill, below Philadelphia, in flocks of from thirty to forty, or upwards. They are easily raised from the nest, and soon become agreeable domestics. One which I reared and kept through the winter, whistled with great clearness and vivacity at two months old. It had an odd manner of moving its head and neck, slowly and regularly, and in various directions, when intent on observing anything without stirring its body. This motion was as slow and regular as that of a snake. When at night a candle was brought into the room, it became restless, and evidently dissatisfied, fluttering about the cage, as if seeking to get out; but when the cage was

Drawn from Nature by A. Wilson. *Engraved by W. H. Lizars.*

1. Great American Shrike or Butcher Bird. 2. Pine Grossbeak. 3. Ruby crown'd Wren. 4. Shore Lark.

5.

placed on the same table with the candle, it seemed extremely well pleased, fed and drank, drest, shook, and arranged its plumage, sat as close to the light as possible, and sometimes chanted a few broken, irregular notes in that situation, as I sat writing or reading beside it. I also kept a young female of the same nest during the greatest part of winter, but could not observe, in that time, any change in its plumage.*

GREAT AMERICAN SHRIKE, OR BUTCHER BIRD.
(*Lanius excubitor.*)†

PLATE V.—FIG. 1.

La pie grische-grise, *Buffon*, i. 296. *Pl. enl.* 445.—*Peale's Museum*, No. 664.—White Whisky John, *Phil. Trans.* lxii. 386.—*Arct. Zool.* ii. No. 127.

LANIUS BOREALIS.—VIEILLOT.

Lanius borealis, *Vieill.*—*North. Zool.* ii. 3.

THE form and countenance of this bird bespeak him full of courage and energy ; and his true character does not belie his

* This bird is interesting, as showing the remarkable change of colour which takes place in the group, and which, in many instances, has been the occasion of a multiplication of species. It will rank with the baltimore bird in the *Icterus* of Brisson, and they will form the only individuals belonging to the northern continent of America. According to Audubon, the flesh of the orchard oriole is esteemed by the Creoles of Louisiana, and at the season when the broods have collected, and feed most upon insects in the moist meadows, they are procured for the table in considerable abundance.—ED.

† Wilson has marked this species with a note of doubt, showing the accuracy of his observation where he had such slender means of making out species ; a mistake also into which C. L. Bonaparte, with greater opportunities, has also fallen. Vieillot seems to have been the first to distinguish it, and Mr Swainson has satisfactorily pointed out the difference in the "Northern Zoology." *Lanius excubitor* is not found at all in America, and this species seems to fill up its want; the chief differences are in the size, *Lanius borealis* being larger. The female is of a browner shade; with more gray underneath ; the former a distribution of colour in the females unknown among those bearing similar shades; in habits they in every way agree.—ED.

appearance, for he possesses these qualities in a very eminent degree. He is represented on the plate rather less than his true size, but in just proportion, and with a fidelity that will enable the European naturalist to determine whether this be really the same with the great cinereous shrike (*Lanius excubitor*, Linn.) of the eastern continent or not; though the progressive variableness of the plumage, passing, according to age, and sometimes to climate, from ferruginous to pale ash, and even to a bluish white, renders it impossible that this should be an exact representation of every individual.

This species is by no means numerous in the lower parts of Pennsylvania, though most so during the months of November, December, and March. Soon after this, it retires to the north, and to the higher inland parts of the country to breed. It frequents the deepest forests; builds a large and compact nest in the upright fork of a small tree, composed outwardly of dry grass and whitish moss, and warmly lined within with feathers. The female lays six eggs of a pale cinereous colour, thickly marked at the greater end with spots and streaks of rufous. She sits fifteen days. The young are produced early in June, sometimes towards the latter end of May; and during the greater part of the first season are of a brown ferruginous colour on the back.

When we compare the beak of this species with his legs and claws, they appear to belong to two very different orders of birds; the former approaching, in its conformation, to that of the Accipitrine; the latter to those of the pies; and, indeed, in his food and manners he is assimilated to both. For though man has arranged and subdivided this numerous class of animals into separate tribes and families, yet nature has united these to each other by such nice gradations, and so intimately, that it is hardly possible to determine where one tribe ends, or the succeeding commences. We therefore find several eminent naturalists classing this genus of birds with the Accipitrine, others with the pies. Like the former, he preys occasionally on other birds; and, like the latter, on insects, particularly grasshoppers, which I believe to be his principal food; having at almost all times, even in winter, found them in his stomach. In the month of December, and while the country was deeply covered with snow, I shot one of these birds near the head waters of the Mohawk River, in the State of New York, the stomach of which was entirely filled with large black spiders. He was of a much purer white above than any I have since met with, though evidently of the same species with the present; and I think it probable that the males become lighter coloured as they advance in age, till the minute transverse lines of brown on the lower parts almost disappear.

In his manners he has more resemblance to the pies than to birds of prey, particularly in the habit of carrying off his surplus food, as if to hoard it for future exigencies; with this difference, that crows, jays, magpies, &c., conceal theirs at random, in holes and crevices, where, perhaps, it is forgotten, or never again found; while the butcher bird sticks his on thorns and bushes, where it shrivels in the sun, and soon becomes equally useless to the hoarder. Both retain the same habits in a state of confinement, whatever the food may be that is presented to them.

This habit of the shrike, of seizing and impaling grasshoppers and other insects on thorns, has given rise to an opinion that he places their carcases there by way of baits, to allure small birds to them, while he himself lies in ambush to surprise and destroy them. In this, however, they appear to allow him a greater portion of reason and contrivance than he seems entitled to, or than other circumstances will altogether warrant; for we find that he not only serves grasshoppers in this manner, but even small birds themselves, as those have assured me who have kept them in cages in this country, and amused themselves with their manœuvres. If so, we might as well suppose the farmer to be inviting crows to his corn when he hangs up their carcases around it, as the butcher bird to be decoying small birds by a display of the dead bodies of their comrades!

In the *Transactions of the American Philosophical Society*, vol. iv. p. 124, the reader may find a long letter on this subject from Mr John Heckewelder of Bethlehem to Dr Barton, the substance of which is as follows:—That on the 17th of December 1795, he (Mr Heckewelder) went to visit a young orchard which had been planted a few weeks before, and was surprised to observe on every one of the trees one, and on some two or three grasshoppers, stuck down on the sharp thorny branches; that, on inquiring of his tenant the reason of this, he informed him that they were stuck there by a small bird of prey, called by the Germans, *neuntödter* (ninekiller), which caught and stuck nine grasshoppers a day; and he supposed, that, as the bird itself never fed on grasshoppers, it must do it for pleasure. Mr Heckewelder now recollected that one of these ninekillers had, many years before, taken a favourite bird of his out of his cage at the window, since which, he had paid particular attention to it; and being perfectly satisfied that it lived entirely on mice and small birds, and, moreover, observing the grasshoppers on the trees all fixed in natural positions, as if alive, he began to conjecture that this was done to decoy such small birds as feed on these insects to the spot, that he might have an opportunity of devouring them. "If it were true," says he, "that this little hawk had stuck them up for himself, how long would he be in feeding on one or two hundred grasshoppers? But if it be intended to seduce the smaller birds to feed on these insects, in order to have an opportunity of catching them, that number, or even one-half, or less, may be a good bait all winter," &c.

This is, indeed, a very pretty fanciful theory, and would entitle our bird to the epithet *fowler*, perhaps with more propriety than *lanius*, or *butcher*; but, notwithstanding the attention which Mr Heckewelder professes to have paid to this bird, he appears not only to have been ignorant that grasshoppers were, in fact, the favourite food of this ninekiller, but never once to have considered that grasshoppers would be but a very insignificant and tasteless bait for our winter birds, which are chiefly those of the finch kind, that feed almost exclusively on hard seeds and gravel; and among whom five hundred grasshoppers might be stuck up on trees and bushes, and remain there untouched by any of them for ever. Besides, where is his necessity of having recourse to such refined stratagems, when he can, at any time, seize upon small birds by mere force of flight? I have seen him, in an open field, dart after one of our small sparrows with the rapidity of an arrow, and kill it almost instantly. Mr William Bartram long ago informed me, that one of these shrikes had the temerity to pursue a snow bird (*F. Hudsonia*) into an open cage, which stood in the garden, and, before they could arrive to its assistance, had already strangled and scalped it, though he lost his liberty by the exploit. In short, I am of opinion, that his resolution and activity are amply sufficient to enable him to procure these small birds whenever he wants them, which, I believe, is never but when hard pressed by necessity, and a deficiency of his favourite insects; and that the crow or the blue jay may, with the same probability, be supposed to be laying baits for mice and flying squirrels when they are hoarding their Indian-corn, as he for birds while thus disposing of the exuberance of his favourite food. Both the former and the latter retain the same habits in a state of confinement; the one filling every seam and chink of his cage with grain, crumbs of bread, &c., and the other sticking up, not only insects, but flesh, and the bodies of such birds as are thrown in to him, on nails or sharpened sticks fixed up for the purpose. Nor, say others, is this practice of the shrike difficult to be accounted for. Nature has given to this bird a strong, sharp, and powerful beak, a broad head, and great strength in the muscles of his neck; but his legs, feet, and claws are by no means proportionably strong, and are unequal to the task of grasping and tearing his prey, like those of the owl and falcon kind. He, therefore, wisely avails himself of the powers of the former, both in strangling his prey, and in tearing it to pieces while feeding.

The character of the butcher bird is entitled to no common degree of respect. His activity is visible in all his motions; his courage and intrepidity beyond every other bird of his size (one of his own tribe only excepted, *L. tyrannus*, or king bird); and in affection for his young, he is surpassed by no other. He associates with them in the latter part of summer, the whole family hunting in company. He attacks the largest hawk or eagle in their defence, with a resolution truly astonishing, so that all of them respect him, and, on every occasion, decline the contest. As the snows of winter approach, he descends from the mountainous forests, and from the regions of the north, to the most cultivated parts of the country, hovering about our hedgerows, orchards, and meadows, and disappears again early in April.

The great American shrike is ten inches in length, and thirteen in extent; the upper part of the head, neck, and back, is pale cinereous; sides of the head, nearly white, crossed with a bar of black that passes from the nostril, through the eye, to the middle of the neck; the whole under parts, in some specimens, are nearly white, in others more dusky, and thickly marked with minute transverse curving lines of light brown; the wings are black, tipt with white, with a single spot of white on the primaries, just below their coverts; the scapulars, or long downy feathers that fall over the upper part of the wing, are pure white; the rump and tail-coverts, a very fine gray or light ash; the tail is cuneiform, consisting of twelve feathers, the two middle ones wholly black, the others tipt more and more with white to the exterior ones, which are nearly all white; the legs, feet, and claws are black; the beak straight; thick, of a light blue colour, the upper mandible furnished with a sharp process, bending down greatly at the point, where it is black, and beset at the base with a number of long black hairs or bristles; the nostrils are also thickly covered with recumbent hairs; the iris of the eye is a light hazel; pupil, black. The figure on the plate will give a perfect idea of the bird. The female is easily distinguished by being ferruginous on the back and head; and having the band of black extending only behind the eye, and of a dirty brown or burnt colour; the under parts are also something rufous, and the curving lines more strongly marked; she is rather less than the male, which is different from birds of prey in general, the females of which are usually the larger of the two.

In the "Arctic Zoology," we are told that this species is frequent in Russia, but does not extend to Siberia; yet one was taken within Behring's Straits, on the Asiatic side, in lat. 66°; and the species probably extends over the whole continent of North America from the Western Ocean. Mr Bell, while on his travels through Russia, had one of these birds given him, which he kept in a room, having fixed up a sharpened stick for him in the wall; and on turning small birds loose in the room, the butcher bird instantly caught them by the throat in such a manner as soon to suffocate them; and then stuck them on the stick, pulling them on with bill and claws; and so served as many as were turned loose, one after another, on the same stick.*

PINE GROSBEAK. (*Loxia enucleator.*)

PLATE V.—FIG. 2.

Loxia enucleator, *Linn. Syst.* i. p. 299, 3.—Le dur bec, ou gros bec de Canada, *Buffon*, iii. p. 457. *Pl. enl.* 135, 1.—*Edw.* 123, 124.—*Lath. Syn.* iii. p. 111, 5.—*Peale's Museum*, No. 5652.

CORYTHUS ENUCLEATOR.—CUVIER.†

Loxia enucleator, *Penn. Arct. Zool.* ii. p. 348.—Corythus enucleator, *Cuv. Regn. Anim.* i. p. 391.—*Fleem. Br. Zool.* p. 76.—Bouvreuil dur bec, Pyrrhula enucleator, *Temm.* i. 333.—Pine Grosbeak, Pyrrhula enucleator, *Selby Orn. Ill.* i. 256, pl. 53.—Pyrrhula enucleator, *Bonap. Syn.* 114.

THIS is perhaps one of the gayest plumaged land birds that frequent the inhospitable regions of the north, whence they

* Edwards, vii. 231.

† This interesting species seems nowhere of common occurrence; it is very seldom seen in collections; and boxes of skins, either from dif-

are driven, as if with reluctance, by the rigours of winter, to visit Canada and some of the northern and middle States, returning to Hudson's Bay so early as April. The specimen from which our drawing was taken was shot on a cedar tree, a few miles to the north of Philadelphia, in the month of December; and a faithful resemblance of the original, as it then appeared, is exhibited in the plate. A few days afterwards, another bird of the same species was killed not far from Gray's Ferry, four miles south from Philadelphia, which proved to be a female. In this part of the State of Pennsylvania, they are rare birds, and seldom seen. As they do not, to my knowledge, breed in any part of this State, I am unable, from personal observation, to speak of their manners or musical talents.

ferent parts of Europe or America, can seldom rank the pine grosbeak among their number. The testimony of all travellers in America, who have attended to nature, correspond in their accounts, and one of the latest, Mr Audubon, has mentioned it to me as of extreme scarcity. In this country they seem to be of equal rarity, though they are generally placed in our list of British birds without any remark. Pennant observes (Arct. Zool. ii. 348), that he has seen them in the forests of Invercauld; and Mr Selby says (Br. Orn. 257), that, from the testimony of the gamekeepers, whom he had an opportunity of speaking with in the Highlands, they may be ranked only as occasional visitants. I am aware, however, of no instance of their being killed in this country. Pennant infers, from those which he saw in the month of August, that they breed here. "Such a conclusion," Mr Selby justly remarks, "ought scarcely to be inferred from this fact, as a sufficient interval of time had elapsed for these individuals to have emigrated from Norway, or other northern countries, to Scotland, after incubation, as they are known to breed as early as May in their natural haunts." I have been unable to find any trace whatever of their ever breeding in this country; most of the migrating species breed very early, and those that change their station for the sake of finding a breeding-place, commence the office of building, &c., immediately on their arrival, a necessary circumstance to enable the young to perform their migration before the change of season. Cuvier has formed his genus *Corythus* of this individual, which still remains the only one that has yet been placed in it; but I am of opinion that the crimson-necked bullfinch (*Pyrrhula frontalis*, Say) should stand very near, or with it. Their alliance to the true bullfinches is very great, and Mr Swainson's genus, *Crithagra*, may form another near ally.—Ed.

Mr Pennant says, they sing on their first arrival in the country round Hudson's Bay, but soon become silent; make their nest on trees, at a small height from the ground, with sticks, and line it with feathers. The female lays four white eggs, which are hatched in June. Forster observes, that they visit Hudson's Bay only in May, on their way to the north; and are not observed to return in the autumn; and that their food consists of birchwillow buds, and others of the same nature.*

The pine grosbeak measures nine inches in length, and fourteen inches in extent; the head, neck, breast, and rump are of a rich crimson, palest on the breast; the feathers on the middle of the back are centred with arrow-shaped spots of black, and skirted with crimson, which gives the plumage a considerable flush of red there; those on the shoulders are of a deep slate colour, partially skirted with red, and light ash. The greater wing-coverts and next superior row are broadly tipt with white, and slightly tinged with reddish; wings and tail, black, edged with light brown; tail, considerably forked; lower part of the belly, ash colour; vent-feathers, skirted with white, and streaked with black; legs, glossy black; bill, a brownish horn colour, very thick, short, and hooked at the point; the upper mandible overhanging the lower considerably, approaching in its form to that of the parrot; base of the bill, covered with recumbent hairs of a dark brown colour. The whole plumage, near the roots, as in most other birds, is of a deep bluish ash colour. The female was half an inch shorter, and answered nearly to the above description; only, those parts that in the male were crimson, were in her of a dirty yellowish colour. The female, according to Forster, referred to above, has those parts which in the male are red, more of an orange tint; and he censures Edwards for having represented the female of too bright a red. It is possible that my specimen of the female might have been a bird of the first season, not come to its full colours. Those figured by Mr Edwards † were both brought from Hudson's Bay, and appear

* Philosophical Transactions, lxii. 402.

† Edw. iii. 124.

to be the same with the one now before us, though his colouring of the female differs materially from his description.

If this, as Mr Pennant asserts, be the same species with that of the eastern continent, it would seem to inhabit almost the whole extent of the arctic regions. It is found in the north of Scotland, where Pennant suspects it breeds. It inhabits Europe as far north as Drontheim; is common in all the pine forests of Asia, in Siberia, and the north of Russia; is taken in autumn about Petersburg, and brought to market in great numbers. It returns to Lapland in spring; is found in Newfoundland, and on the western coast of North America.*

Were I to reason from analogy, I would say, that, from the great resemblance of this bird to the purple finch (*Fringilla purpurea*), it does not attain its full plumage until the second summer; and is subject to considerable change of colour in moulting, which may have occasioned all the differences we find concerning it in different authors. But this is actually ascertained to be the case; for Mr Edwards saw two of these birds alive in London, in cages; the person in whose custody they were, said they came from Norway; that they had moulted their feathers, and were not afterwards so beautiful as they were at first. One of them, he says, was coloured very much like the green finch (*L. chloris*). The purple finch, though much smaller, has the rump, head, back, and breast, nearly of the same colour as the pine grosbeak, feeds in the same manner, on the same food, and is also subject to like changes of colour.

Since writing the above, I have kept one of these pine grosbeaks, a male, for more than half a year. In the month of August those parts of the plumage which were red became of a greenish yellow, and continue so still. In May and June its song, though not so loud as some birds of its size, was extremely clear, mellow, and sweet. It would warble out this for a whole morning together, and acquired several of the notes of a red bird (*L. cardinalis*), that hung near it. It is ex-

* Pennant.

ceedingly tame and familiar, and when it wants food or water, utters a continual melancholy and anxious note. It was caught in winter near the North River, thirty or forty miles above New York.

RUBY-CROWNED WREN. (*Sylvia calendula.*)

PLATE V.—Fig. 3.

Le Roitlet rubis, *Buff.* v. 373.—*Edw.* 254.—*Lath. Syn.* ii. 511.—*Arct. Zool.* 320.—Regulus cristatus alter vertice rubini coloris, *Bartram*, p. 292.—*Peale's Museum*, No. 7244.

REGULUS CALENDULUS.—Stephens.*

Regulus calendulus, *Steph. Cont. Sh. Zool.* vol. x. p. 760.—*Bonap. Synop.* 91.

This little bird visits us early in the spring, from the south, and is generally first found among the maple blossoms about the beginning of April. These failing, it has recourse to those of the peach, apple, and other fruit trees, partly for the tops of the sweet and slender stamina of the flowers, and partly for the winged insects that hover among them. In the middle of summer, I have rarely met with these birds in Pennsylvania; and as they penetrate as far north as the country round Hudson's Bay, and also breed there, it accounts for their late arrival here, in fall. They then associate with the different species of titmouse and the golden-crested wren; and are particularly numerous in the month of October and beginning of November, in orchards, among the decaying leaves of the apple trees, that at that season are infested with great numbers of small black-winged insects, among which they make great havoc. I have often regretted the painful necessity one is under of taking away the lives of such inoffensive, useful little creatures, merely to obtain a more perfect knowledge of the species; for they appear so busy, so active, and unsuspecting, as to continue searching about the same twig, even after their

* See note to *Regulus cristatus.*

companions have been shot down beside them. They are more remarkably so in autumn, which may be owing to the great number of young and inexperienced birds which are then among them ; and frequently at this season, I have stood under the tree, motionless, to observe them, while they gleaned among the low branches sometimes within a foot or two of my head. They are extremely adroit in catching their prey ; have only at times a feeble chirp ; visit the tops of the tallest trees, as well as the lowest bushes ; and continue generally for a considerable time among the branches of the same tree, darting about from place to place ; appearing, when on the top of a high maple, no bigger than humble bees.

The ruby-crowned wren is four inches long, and six in extent ; the upper parts of the head, neck, and back, are of a fine greenish olive, with a considerable tinge of yellow ; wings and tail, dusky purplish brown, exteriorly edged with yellow olive ; secondaries, and first row of wing-coverts, edged and tipt with white, with a spot of deep purplish brown across the secondaries, just below their coverts ; the hind head is ornamented with an oblong lateral spot of vermilion, usually almost hid by the other plumage ; round the eye, a ring of yellowish white ; whole under parts, of the same tint ; legs, dark brown ; feet and claws, yellow ; bill, slender, straight, not notched, furnished with a few black hairs at the base ; inside of the mouth, orange. The female differs very little in its plumage from the male, the colours being less lively, and the bird somewhat less. Notwithstanding my utmost endeavours, I have never been able to discover their nest ; though, from the circumstance of having found them sometimes here in summer, I am persuaded that they occasionally breed in Pennsylvania ; but I know several birds, no larger than this, that usually build on the extremities of the tallest trees in the woods ; which I have discovered from their beginning before the leaves are out. Many others, no doubt, choose similar situations ; and should they delay building until the woods are thickened with leaves, it is no easy matter to discover them. In fall, they are so extremely fat, as almost to dissolve between the fingers as you open them; owing to the great abundance of their favourite insects at that time.

SHORE LARK. (*Alauda alpestris.*)

PLATE V.—FIG. 4.

Alauda alpestris, *Linn. Syst.* 289.—*Lath. Synop.* ii. 385.—*Peale's Museum,* No. 5190.—Alauda campestris, gutture flavo, *Bartram,* p. 290.—L'Alouette de Virginia, *Buff.* v. 55.—*Catesb.* i. 32.

ALAUDA ALPESTRIS.—LINNÆUS.

Alauda alpestris alouette à Hause col noir, *Temm.* i. 279.—*Bonap. Synop.* 102.—*Vieill. Gal. des Ois.* pl. 155, p. 256.—Alauda cornuta, *Swain. Synop., Birds of Mexico, Phil. Mag. & Ann.* 1827, p. 434.—*North. Zool.* ii. p. 245.

THIS is the most beautiful of its genus, at least in this part of the world. It is one of our winter birds of passage, arriving from the north in the fall; usually staying with us the whole winter, frequenting sandy plains and open downs, and is numerous in the Southern States, as far as Georgia, during that season. They fly high, in loose scattered flocks; and at these times have a single cry, almost exactly like the sky lark of Britain. They are very numerous in many tracts of New Jersey; and are frequently brought to Philadelphia market. They are then generally very fat, and are considered excellent eating. Their food seems principally to consist of small round compressed black seeds, buckwheat, oats, &c., with a large proportion of gravel. On the flat commons, within the boundaries of the city of Philadelphia, flocks of them are regularly seen during the whole winter. In the stomach of these I have found, in numerous instances, quantities of the eggs or larvæ of certain insects mixed with a kind of slimy earth. About the middle of March they generally disappear, on their route to the north.* Forster informs us, that they visit the

* In winter, says Pennant, they retire to the southern provinces in great flights; but it is only by severe weather that they reach Virginia

environs of Albany Fort in the beginning of May; but go farther north to breed; that they feed on grass seeds and buds of the sprig birch, and run into small holes, keeping close to the ground, from whence the natives call them *Chi-chup-pi-sue.** This same species appears also to be found in Poland, Russia, and Siberia, in winter, from whence they also retire farther north on the approach of spring, except in the north-east parts, and near the high mountains.†

The length of this bird is seven inches, the extent twelve inches; the forehead, throat, sides of the neck, and line over the eye are of a delicate straw, or Naples yellow, elegantly relieved by a bar of black, that passes from the nostril to the eye, below which it falls, rounding, to the depth of three quarters of an inch; the yellow on the forehead and over the eye is bounded within, for its whole length, with black, which covers part of the crown; the breast is ornamented with a broad fan-shaped patch of black: this, as well as all the other spots of black, are marked with minute curves of yellow points; back of the neck, and towards the shoulders, a light drab tinged with lake; lesser wing-coverts, bright cinnamon; greater wing-coverts, the same, interiorly dusky, and tipt with whitish; back and wings, drab-coloured, tinged with reddish, each feather of the former having a streak of dusky black down its centre; primaries, deep dusky, tipt and edged with whitish; exterior feathers, most so; secondaries, broadly edged with light drab, and scolloped at the tips; tail, forked,

and Carolina. They frequent sandhills on the seashore, and feed on the seaside oats, or *Uniola paniculata.* They have a single note, like the sky lark in winter.—Temminck mentions them as birds of passage in Germany, and that they breed also in Asia. One or two specimens have lately been killed in England, so that their geographic range is pretty considerable. The *Alauda calandra* of Linnæus is introduced into the "Northern Zoology" as an inhabitant of the Fur countries, on the authority of a specimen in the British Museum, and will stand as the second lark found in that country.—Ed.

* Philosophical Transactions, vol. lxii. p. 398.

† Arctic Zoology.

black; the two middle feathers, which by some have been mistaken for the coverts, are reddish drab, centred with brownish black; the two outer ones on each side, exteriorly edged with white; breast, of a dusky vinous tinge, and marked with spots or streaks of the same; the belly and vent, white; sides, streaked with bay; bill short (Latham, in mistake, says seven inches*), of a dusky blue colour; tongue, truncate and bifid; legs and claws, black; hind heel, very long, and almost straight; iris of the eye, hazel. One glance at the figure on the plate will give a better idea than the whole of this minute description, which, however, has been rendered necessary by the errors of others. The female has little or no black on the crown; and the yellow on the front is narrow, and of a dirty tinge.

There is a singular appearance in this bird, which I have never seen taken notice of by former writers, viz., certain long black feathers, which extend, by equal distances beyond each other, above the eyebrow; these are longer, more pointed, and of a different texture from the rest around them; and the bird possesses the power of erecting them, so as to appear as if horned, like some of the owl tribe. Having kept one of these birds alive for some time, I was much amused at this odd appearance, and think it might furnish a very suitable specific appellation, viz., *Alauda cornuta*, or horned lark. These horns become scarcely perceivable after the bird is dead. The head is slightly crested.

Shore lark and sky lark are names by which this species is usually known in different parts of the Union. They are said to sing well, mounting in the air, in the manner of the song lark of Europe; but this is only in those countries where they breed. I have never heard of their nests being found within the territory of the United States.

* Synopsis, vol. ii. p. 385.

MARYLAND YELLOW-THROAT. (*Sylvia Marilandica.*)

PLATE VI.—FIG. 1.

Turdus trichas, *Linn. Syst.* i. 293.—*Edw.* 237.—Yellow-breasted Warbler, *Arct. Zool.* ii. No. 283. *Id.* 284.—Le Figuier aux joues noires, *Buff.* v. 292.—La Fauvette à poitrine jaune de la Louisiane, *Buff.* v. 162. *Pl. enl.* 709, fig. 2.—*Lath. Syn.* iv. 433, 32.—*Peale's Museum*, No. 6902.

TRICHUS PERSONATUS.—SWAINSON.*

Trichas personatus, *Swain. Zool. Journ.* No. 10, p. 167.—The Yellow-breasted Warbler, or Maryland Yellow-throat, *Aud.* i. pl. 23, p. 121.

THIS is one of the humble inhabitants of briers, brambles, alder bushes, and such shrubbery as grow most luxuriantly in low, watery situations; and might with propriety be denominated *Humility*, its business or ambition seldom leading it

* Mr Swainson has formed from this species his genus *Trichas*, and bestowed upon it the new and appropriate name of *personatus*, or *masked*. *Marilandica* of Brisson and Wilson could scarcely be retained, *Trichas* of Linnæus having the priority. The latter is now converted into a generic term; and as the species does not seem entirely confined to Maryland, another and more appropriate than either will perhaps make less confusion than the attempts to restore some old one. Mr Swainson makes the following remarks upon the genus:—"This form is intimately connected with *Synalaxis*, and two or three other groups peculiar to Africa and Australia. Feebleness of flight and strength of foot separate these birds from the typical genera; while the strength and curvature of the hind claw forbid us to associate them with the true *Motacillæ*."

The female is figured on Plate XVIII. of this volume, where it is mentioned as one of the birds whose nest the cow bunting selects to deposit her eggs in. "The nest," according to Mr Audubon, "is placed on the ground, and partly sunk in it: it is now and then covered over in the form of an oven, from which circumstance children name this warbler the *oven-bird*. It is composed externally of withered leaves and grass, and is lined with hair. The eggs are from four to six, of a white colour, spreckled with light brown, and are deposited about the middle of May. Sometimes two broods are reared in a season. I have never observed the egg of the cow bunting in the nests of the second brood."

The male birds do not attain their full plumage until the second spring.—ED.

Drawn from Nature by A. Wilson

Engraved by W. H. Lizars

1. Maryland Yellow-throat. 2. Yellow-breasted Chat. 3 Summer Red Bird. 4. Female. 5. Indigo Bird. 6. American Redstart

6.

higher than the tops of the underwood. Insects and their larvæ are its usual food. It dives into the deepest of the thicket, rambles among the roots, searches round the stems, examines both sides of the leaf, raising itself on its legs, so as to peep into every crevice; amusing itself at times with a very simple, and not disagreeable, song or twitter, *whĭtititee, whĭtititee, whĭtititee;* pausing for half a minute or so, and then repeating its notes as before. It inhabits the whole United States from Maine to Florida, and also Louisiana; and is particularly numerous in the low, swampy thickets of Maryland, Pennsylvania, and New Jersey. It is by no means shy; but seems deliberate and unsuspicious, as if the places it frequented, or its own diminutiveness, were its sufficient security. It often visits the fields of growing rye, wheat, barley, &c., and no doubt performs the part of a friend to the farmer, in ridding the stalks of vermin, that might otherwise lay waste his fields. It seldom approaches the farmhouse or city, but lives in obscurity and peace, amidst its favourite thickets. It arrives in Pennslyvania about the middle or last week of April, and begins to build its nest about the middle of May: this is fixed on the ground, among the dried leaves, in the very depth of a thicket of briers, sometimes arched over, and a small hole left for entrance; the materials are dry leaves and fine grass, lined with coarse hair; the eggs are five, white, or semi-transparent, marked with specks of reddish brown. The young leave the nest about the 22d of June; and a second brood is often raised in the same season. Early in September they leave us, returning to the south.

This pretty little species is four inches and three quarters long, and six inches and a quarter in extent; back, wings, and tail, green olive, which also covers the upper part of the neck, but approaches to cinereous on the crown; the eyes are inserted in a band of black, which passes from the front, on both sides, reaching half way down the neck; this is bounded above by another band of white, deepening into light blue;

throat, breast, and vent, brilliant yellow; belly, a fainter tinge of the same colour; inside coverts of the wings, also yellow; tips and inner vanes of the wings, dusky brown; tail, cuneiform, dusky, edged with olive green; bill, black, straight, slender, of the true *Motacilla* form, though the bird itself was considered as a species of thrush by Linnæus; but very properly removed to the genus *Motacilla* by Gmelin; legs, flesh coloured; iris of the eye, dark hazel. The female wants the black band through the eye, has the bill brown, and the throat of a much paler yellow. This last, I have good reason to suspect, has been described by Europeans as a separate species; and that from Louisiana, referred to in the synonyms, appears evidently the same as the former, the chief difference, according to Buffon, being in its wedged tail, which is likewise the true form of our own species; so that this error corrected will abridge the European nomenclature of two species. Many more examples of this kind will occur in the course of our description.

YELLOW-BREASTED CHAT.* (*Pipra polyglotta.*)

PLATE VI.—Fig. 2.

Muscicapa viridis, *Gmel. Syst.* i. 936.—Le Merle vert de la Caroline, *Buffon*, iii. 396.—Chattering Flycatcher, *Arct. Zool.* ii. No. 266.—*Lath. Synop.* iii. 350, 48. —Garrulus Australis, *Bartram*, 290.—*Peale's Museum*, No. 6661.

ICTERIA VIRIDIS.—Bonaparte.

Icteria dumicola, *Vieill. Gal. des Ois.* pl. 85, p. 119.—Icteria viridis, *Bonap. Synop.* p. 69.

This is a very singular bird. In its voice and manners, and the habit it has of keeping concealed, while shifting and

* The Prince of Musignano remarks, when speaking of this bird, in his excellent "Observations on the Nomenclature of Wilson's Ornithology," "It is not a little remarkable that Wilson should have introduced this genus in his 'Ornithology.' The bird he placed in it has certainly no relation to the Manakins, nor has any one of that genus been found within the United States. This bird has been placed by authors in half

vociferating around you, it differs from most other birds with which I am acquainted, and has considerable claims to originality of character. It arrives in Pennslyvania about the first week in May, and returns to the south again as soon as its young are able for the journey, which is usually about the middle of August; its term of residence here being scarcely four months. The males generally arrive several days before the females, a circumstance common with many other of our birds of passage.

When he has once taken up his residence in a favourite situation, which is almost always in close thickets of hazel, brambles, vines, and thick underwood, he becomes very jealous of his possessions, and seems offended at the least intrusion; scolding every passenger as soon as they come within view, in a great variety of odd and uncouth monosyllables, which it is difficult to describe, but which may be readily imitated, so as to deceive the bird himself, and draw him after you for half a quarter of a mile at a time, as I have sometimes amused myself in doing, and frequently without once seeing him. On these occasions, his responses are constant and rapid, strongly expressive of anger and anxiety; and while the bird itself remains unseen, the voice shifts from place to place, among the bushes, as if it proceeded from a spirit. First is heard a repetition of short notes, resembling the whistling of the wings of a duck or teal, beginning loud and rapid, and falling lower and slower, till they end in detached notes; then a succession of others, something like the barking of young puppies, is followed by a variety of hollow, guttural sounds, each eight or ten times repeated, more like those proceeding from the throat of a quadruped than that of a bird; which are succeeded by others not unlike the mewing of a cat, but considerably hoarser. All

a dozen different genera. It was arranged in *Muscicapa* by Gmelin, Latham, and Pennant; in *Turdus*, by Brisson and Buffon; in *Ampelis*, by Sparrman; and in *Tanagra*, by Desmarest. I was at first inclined to consider it as a *Vireo;* but, after having dwelt more upon the characters and habits of this remarkable species, I have concluded to adopt *Icteria* as an independent genus, agreeably to *Vieillot*."—ED.

these are uttered with great vehemence, in such different keys, and with such peculiar modulations of voice, as sometimes to seem at a considerable distance, and instantly as if just beside you, now on this hand, now on that; so that, from these manœuvres of ventriloquism, you are utterly at a loss to ascertain from what particular spot or quarter they proceed. If the weather be mild and serene, with clear moonlight, he continues gabbling in the same strange dialect, with very little intermission, during the whole night, as if disputing with his own echoes; but probably with a design of inviting the passing females to his retreat; for, when the season is farther advanced, they are seldom heard during the night.

About the middle of May they begin to build. Their nest is usually fixed in the upper part of a bramble bush, in an almost impenetrable thicket; sometimes in a thick vine or small cedar; seldom more than four or five feet from the ground. It is composed outwardly of dry leaves; within these are laid thin strips of the bark of grape-vines, and the inside is lined with fibrous roots of plants, and fine dry grass. The female lays four eggs, slightly flesh-coloured, and speckled all over with spots of brown or dull red. The young are hatched in twelve days; and make their first excursion from the nest about the second week in June. A friend of mine, an amateur in canary birds, placed one of the chat's eggs under a hen canary, who brought it out; but it died on the second day; though she was so solicitous to feed and preserve it, that her own eggs, which required two days more sitting, were lost through her attention to this.

While the female of the chat is sitting, the cries of the male are still more loud and incessant. When once aware that you have seen him, he is less solicitous to conceal himself; and will sometimes mount up into the air, almost perpendicularly, to the height of thirty or forty feet, with his legs hanging; descending as he rose, by repeated jerks, as if highly irritated, or, as is vulgarly said, "dancing mad." All this noise and gesticulation we must attribute to his extreme affection for his mate and young; and when we consider the great distance which in all probability he comes, the few young produced at a time, and that seldom more than once in the season, we can see the wisdom of Providence very manifestly in the ardency of his passions.

Mr Catesby seems to have first figured the yellow-breasted chat; and the singularity of its manners has not escaped him. After repeated attempts to shoot one of them, he found himself completely baffled: and was obliged, as he himself informs us, to employ an Indian for that purpose, who did not succeed without exercising all his ingenuity. Catesby also observed its dancing manœuvres, and supposed that it always flew with its legs extended; but it is only in these paroxysms of rage and anxiety that this is done, as I have particularly observed.

The food of these birds consists chiefly of large black beetles, and other coleopterous insects; I have also found whortleberries frequently in their stomach, in great quantities, as well as several other sorts of berries.* They are very numerous in the neighbourhood of Philadelphia, particularly on the borders of rivulets, and other watery situations, in hedges, thickets, &c., but are seldom seen in the forest, even where there is underwood. Catesby indeed asserts, that they are only found on the banks of large rivers, two or three hundred miles from the sea; but, though this may be the case in South Carolina, yet in Maryland and New Jersey, and also in New York, I have met with these birds within two hours' walk of the sea, and in some places within less than a mile of the shore. I have not been able to trace him to any of the West India islands; though they certainly retire to Mexico, Guiana, and Brazil, having myself seen skins of these birds in the possession of a French gentleman, which were brought from the two latter countries.

By recurring to the synonyms at the beginning of this article, it will be perceived how much European naturalists have differed in classing this bird. That the judicious Mr

* Vieillot mentions the fruit of the *Solanum Carolinense* as a particular favourite of this bird.—Ed.

Pennant, Gmelin, and even Dr Latham, however, should have arranged it with the flycatchers, is certainly very extraordinary; as neither in the particular structure of its bill, tongue, feet, nor in its food or manners, has it any affinity whatever to that genus. Some other ornithologists have removed it to the tanagers; but the bill of the chat, when compared with that of the summer red bird in the same plate, bespeaks it at once to be of a different tribe. Besides, the tanagers seldom lay more than two or three eggs; the chat usually four: the former build on trees; the latter in low thickets. In short, though this bird will not exactly correspond with any known genus, yet the form of its bill, its food, and many of its habits, would almost justify us in classing it with the genus *Pipra* (Manakin), to which family it seems most nearly related.

The yellow-breasted chat is seven inches long, and nine inches in extent; the whole upper parts are of a rich and deep olive green, except the tips of the wings, and interior vanes of the wing and tail-feathers, which are dusky brown; the whole throat and breast is of a most brilliant yellow, which also lines the inside of the wings, and spreads on the sides immediately below; the belly and vent are white; the front, slate coloured, or dull cinereous; lores, black; from the nostril, a line of white extends to the upper part of the eye, which it nearly encircles; another spot of white is placed at the base of the lower mandible, the bill is strong, slightly curved, sharply ridged on the top, compressed, overhanging a little at the tip, not notched, pointed, and altogether black; the tongue is tapering, more fleshy than those of the *Muscicapa* tribe, and a little lacerated at the tip; the nostril is oval, and half covered with an arching membrane; legs and feet, light blue, hind claw rather the strongest, the two exterior toes united to the second joint.

The female may be distinguished from the male by the black and white adjoining the eye being less intense or pure than in the male; and in having the inside of the mouth of a dirty flesh colour, which, in the male, is black; in other respects, their plumage is nearly alike.

SUMMER RED BIRD. (*Tanagra æstiva.*)

PLATE VI.—FIGS. 3 AND 4.

Tanagra Mississippensis, *Lath. Ind. Orn.* i. 421, 5.—Mexican Tanager, *Lath. Synop.* iii. 219, 5 B.—Tanagra variegata, *Ind. Orn.* i. 421, 6.—Tanagra æstiva, *Ind. Orn.* i. 422, 7.—Muscicapa rubra, *Linn. Syst.* i. 326, 8.—*Buff.* vi. 252. *Pl. enl.* 741.—*Catesby, Car.* i. 56.—Merula flammula, Sandhill Red Bird, *Bartram*, 299.—*Peale's Museum*, No. 6134.

PYRANGA ÆSTIVA.—VIEILLOT.

Subgenus Pyranga,* Tanagra estiva, *Bonap. Synop.* p. 105.

THE change of colour which this bird is subject to during the first year, and the imperfect figure first given of it by Catesby, have deceived the European naturalists so much, that four different species have been formed out of this one, as appears by the above synonyms, all of which are referable to the present species, the summer red bird. As the female differs so much in colour from the male, it has been thought proper to represent them both; the female having never, to my knowledge, appeared in any former publication; and all the figures of the other that I have seen being little better than caricatures, from which a foreigner can form no just conception of the original.

The male of the summer red bird (fig. 3) is wholly of a rich vermilion colour, most brilliant on the lower parts, except the inner vanes and tips of the wings, which are of a dusky brown;

* *Pyranga* has been used by Vieillot to designate a group among the *Tanagers*, having the bill of considerable strength, and furnished on the upper mandible with an obtuse tooth,—a structure which has been taken by Desmarest to denote his *Tanagras coluriens*, or shrike-like tanagers. They are also the *Tanagras cardinal* of Cuvier. Bonaparte, again, retains Vieillot's group, but only as a subgenus to *Tanagra.*

It is composed of nine or ten species, three only being found in North America. They are generally of rich, sometimes gaudy plumage, and require more than one year to arrive at maturity. They live in pairs, and feed on insects, berries, or soft seeds.—ED.

the bill is disproportionably large, and inflated, the upper mandible furnished with a process, and the whole bill of a yellowish horn colour; the legs and feet are light blue, inclining to purple; the eye, large, the iris of a light hazel colour; the length of the whole bird, seven inches and a quarter; and between the tips of the expanded wings, twelve inches. The female (fig. 4) differs little in size from the male; but is, above, of a brownish yellow olive, lightest over the eye; throat, breast, and whole lower part of the body, of a dull orange yellow; tips and interior vanes of the wings, brown; bill, legs, and eye, as in the male. The nest is built in the woods, on the horizontal branch of a half-grown tree, often an evergreen, at the height of ten or twelve feet from the ground; composed, outwardly, of broken stalks of dry flax, and lined with fine grass; the female lays three light blue eggs; the young are produced about the middle of June; and I suspect that the same pair raise no more than one brood in a season, for I have never found their nests but in May or June. Towards the middle of August, they take their departure for the south, their residence here being scarcely four months. The young are, at first, of a green olive above, nearly the same colour as the female below, and do not acquire their full tints till the succeeding spring or summer.

The change, however, commences the first season before their departure. In the month of August, the young males are distinguished from the females by their motley garb; the yellow plumage below, as well as the olive green above, first becoming stained with spots of a buff colour, which gradually brighten into red; these being irregularly scattered over the whole body, except the wings and tail, particularly the former, which I have often found to contain four or five green quills in the succeeding June. The first of these birds I ever shot was green winged; and conceiving it at that time to be a nondescript, I made a drawing of it with care; and on turning to it at this moment, I find the whole of the primaries, and two of the secondaries, yellowish green, the rest of the plumage a full red. This was about the middle of May. In the month of August, of the same year, being in the woods with the gun, I perceived a bird of very singular plumage, and having never before met with such an oddity, instantly gave chase to it. It appeared to me, at a small distance, to be sprinkled all over with red, green, and yellow. After a great deal of difficulty—for the bird had taken notice of my eagerness, and had become extremely shy—I succeeded in bringing it down; and found it to be a young bird of the same species with the one I had killed in the preceding May, but less advanced to its fixed colours; the wings entirely of a greenish yellow, and the rest of the plumage spotted, in the most irregular manner, with red, yellow, brown, and greenish. This is the *variegated* tanager, referred to in the synonyms prefixed to this article. Having, since that time, seen them in all their stages of colour, during their residence here, I have the more satisfaction in assuring the reader that the whole four species mentioned by Dr Latham are one and the same. The two figures in our plate represent the male and female in their complete plumage, and of their exact size.

The food of these birds consists of various kinds of bugs and large black beetles. In several instances, I have found the stomach entirely filled with the broken remains of humble bees. During the season of whortleberries, they seem to subsist almost entirely on these berries; but, in the early part of the season, on insects of the above description. In Pennsylvania, they are a rare species, having myself sometimes passed a whole summer without seeing one of them; while in New Jersey, even within half a mile of the shore, opposite the city of Philadelphia, they may generally be found during the season.

The note of the male is a strong and sonorous whistle, resembling a loose trill or shake on the notes of a fife, frequently repeated; that of the female is rather a kind of chattering, approaching nearly to the rapid pronunciation of *chicky-tucky-tuck, chicky-tucky-tuck,* when she sees any person approaching the neighbourhood of her nest. She is, however, rarely seen, and usually mute, and scarcely to be distinguished from the colour of the foliage at a distance; while the loquacity and brilliant red of the male make him very conspicuous; and when seen among the green leaves, particularly if the light falls strongly on his plumage, he has a most beautiful and elegant appearance. It is worthy of remark, that the females of almost all our splendid feathered birds are drest in plain and often obscure colours, as if Providence meant to favour their personal concealment, and, consequently, that of their nest and young, from the depredations of birds of prey; while, among the latter, such as eagles, owls, hawks, &c., which are under no such apprehension, the females are uniformly covered with richer coloured plumage than the males.

The summer red bird delights in a flat sandy country covered with wood, and interspersed with pine trees; and is consequently more numerous towards the shores of the Atlantic than in the interior. In both Carolinas, and in Georgia and Florida, they are in great plenty. In Mexico some of them are probably resident, or at least winter there, as many other of our summer visitants are known to do. In the northern States they are very rare; and I do not know that they have been found either in Upper or Lower Canada. Du Pratz, in his "History of Louisiana," has related some particulars of this bird, which have been repeated by almost every subsequent writer on the subject, viz., that "it inhabits the woods on the Mississippi, and collects against winter a vast magazine of maize, which it carefully conceals with dry leaves, leaving only a small hole for entrance; and is so jealous of it, as never to quit its neighbourhood, except to drink." It is probable, though I cannot corroborate the fact, that individuals of this species may winter near the Mississippi; but that, in a climate so moderate, and where such an exuberance of fruits, seeds, and berries is to be found, even during winter, this, or any other bird, should take so much pains in hoarding a vast quantity of Indian-corn, and attach itself so closely to it, is rather apocryphal. The same writer, vol. ii. p. 24, relates similar particulars of the cardinal grosbeak (*Loxia cardinalis*), which, though it winters in Pennsylvania, where the climate is much more severe, and where the length and rigours of that season would require a far larger magazine, and be a threefold greater stimulus to hoarding, yet has no such habit here. Besides, I have never found a single grain of Indian-corn in the stomach of the summer red bird, though I have examined many individuals of both sexes. On the whole, I consider this account of Du Pratz's in much the same light with that of his countryman, Charlevoix, who gravely informs us, that the owls of Canada lay up a store of live mice for winter; the legs of which they first break, to prevent them from running away, and then feed them carefully, and fatten them, till wanted for use.*

Its manners—though neither its bill nor tongue—partake very much of those of the flycatcher; for I have frequently observed both male and female, a little before sunset, in parts of the forest clear of underwood, darting after winged insects, and continuing thus engaged till it was almost dusk.

INDIGO BIRD. (*Fringilla cyanea.*)

PLATE VI.—FIG. 5.

Tanagra cyanea, *Linn. Syst.* i. 315.—Le Ministre, *Buff.* iv. 86.—Indigo Bunting, *Arct. Zool.* ii. No. 235.—*Lath. Synop.* iii. 205, 63.—Blue Linnet, *Edw.* 273.—*Peale's Museum*, No. 6002.—Linaria cyanea, *Bart.* p. 290.

FRINGILLA CYANEA.—WILSON.

Fringilla cyanea, *Bonap. Synop.*† p. 107.

THIS is another of those rich plumaged tribes that visit us in spring from the regions of the south. It arrives in Pennsyl-

* Travels in Canada, vol. i. p. 239, Lond. 1761, 8vo.

† By a letter from my friend Mr Swainson, I am informed that the Prince of Musignano intends to form a genus of this bird; I have there-

vania on the second week in May, and disappears about the middle of September. It is numerous in all the settled parts of the middle and eastern States; in the Carolinas and Georgia it is also abundant. Though Catesby says that it is only found at a great distance from the sea, yet round the city of New York, and in many places along the shores of New Jersey, I have met with them in plenty. I may also add, on the authority of Mr William Bartram, that "they inhabit the continent and sea-coast islands, from Mexico to Nova Scotia, from the sea-coast west beyond the Apalachian and Cherokee mountains." * They are also known in Mexico, where they probably winter. Its favourite haunts, while with us, are about gardens, fields of deep clover, the borders of woods, and roadsides, where it is frequently seen perched on the fences. In its manners it is extremely active and neat, and a vigorous and pretty good songster. It mounts to the highest tops of a large tree, and chants for half an hour at a time. Its song is not one continued strain, but a repetition of short notes, commencing loud and rapid, and falling, by almost imperceptible gradations, for six or eight seconds, till they seem hardly articulate, as if the little minstrel were quite exhausted; and, after a pause of half a minute, or less, commences again as before. Some of our birds sing only in spring, and then chiefly in the morning, being comparatively mute during the heat of noon; but the indigo bird chants with as much animation under the meridian sun, in the month of July, as in the month of May; and continues his song occasionally to the middle or end of August. His usual note, when alarmed by an approach to his nest, is a sharp *chip*, like that of striking two hard pebbles smartly together.

Notwithstanding the beauty of his plumage, the vivacity with which he sings, and the ease with which he can be reared

fore provisionally added its present name, not wishing to interfere where I am acquainted with the intentions of another. It appears to range with the *Tanagrinæ*.—Ed.

* Travels, p. 299.

and kept, the indigo bird is seldom seen domesticated. The few I have met with were taken in trap cages; and such of any species rarely sing equal to those which have been reared by hand from the nest. There is one singularity which, as it cannot be well represented in the figure, may be mentioned here, viz., that, in some certain lights, his plumage appears of a rich sky-blue, and in others of a vivid verdigris green; so that the same bird, in passing from one place to another before your eyes, seems to undergo a total change of colour. When the angle of incidence of the rays of light reflected from his plumage is acute, the colour is green; when obtuse, blue. Such, I think, I have observed to be uniformly the case, without being optician enough to explain why it is so. From this, however, must be excepted the colour of the head, which, being of a very deep blue, is not affected by a change of position.

The nest of this bird is usually built in a low bush, among rank grass, grain, or clover, suspended by two twigs, one passing up each side; and is composed outwardly of flax, and lined with fine dry grass. I have also known it to build in the hollow of an apple tree. The eggs, generally five, are blue, with a blotch of purple at the great end.

The indigo bird is five inches long, and seven inches in extent; the whole body is of a rich sky-blue, deepening on the head to an ultramarine, with a tinge of purple; the blue on the body, tail, and wings, varies in particular lights to a light green, or verdigris colour, similar to that on the breast of a peacock; wings, black, edged with light blue, and becoming brownish towards the tips; lesser coverts, light blue; greater, black, broadly skirted with the same blue; tail, black, exteriorly edged with blue; bill, black above, whitish below, somewhat larger in proportion than finches of the same size usually are, but less than those of the genus *Emberiza*, with which Mr Pennant has classed it, though I think improperly, as the bird has much more of the form and manners of the genus *Fringilla*, where I must be permitted to place it; legs and feet, blackish brown. The female is of a light flaxen colour, with the wings dusky black, and the cheeks, breast, and whole lower parts, a clay colour, with streaks of a darker colour under the wings, and tinged in several places with bluish. Towards fall, the male, while moulting, becomes nearly of the colour of the female, and in one which I kept through the winter, the rich plumage did not return for more than two months ; though I doubt not, had the bird enjoyed his liberty and natural food under a warm sun, this brownness would have been of shorter duration. The usual food of this species is insects and various kinds of seeds.

AMERICAN REDSTART. (*Muscicapa ruticilla.*)

PLATE VI.—FIG. 6.

Muscicapa ruticilla, *Linn. Syst.* i. 236, 10.—*Gmel. Syst.* i. 935.—Motacilla flavicauda, *Gmel. Syst.* i. 997 (female).—Le gobe-mouche d'Amerique, *Briss. Orn.* ii. 383, 14. *Pl. enl.* 566, fig. 1, 2.—Small American Redstart, *Edw.* 80. *Id.* 257 (female).—Yellow-tailed Warbler, *Arct. Zool.* ii. No. 301. *Id.* ii. No. 282.—*Lath. Syn.* iv. 427, 18.—*Arct. Zool.* ii. No. 301 (female).—*Peale's Museum,* No. 6658.

SETOPHAGA RUTICILLA.—SWAINSON.*

Muscicapa ruticilla, *Bonap. Synop.* p. 68.—Setophaga ruticilla, *North. Zool.* ii. 223.—Setophaga, *Swain. N. Groups, Zool. Journ.* Sept. 1827, p. 360.

THOUGH this bird has been classed by several of our most respectable ornithologists among the warblers, yet in no species are the characteristics of the genus *Muscicapa* more decisively marked ; and, in fact, it is one of the most expert flycatchers of its tribe. It is almost perpetually in motion ; and will pursue a retreating party of flies from the tops of the tallest trees, in an almost perpendicular, but zigzag direction,

* This bird forms the type of *Setophaga,* Swainson ; a genus formed of a few species belonging entirely to the New World, and intimately connected with the fan-tailed flycatchers of Australia, the *Rhippiduræ* of Vigors and Horsfield.

The young bird is figured on Plate XLV. Fig. 2. of Vol. II.—ED.

to the ground, while the clicking of its bill is distinctly heard; and I doubt not but it often secures ten or twelve of these in a descent of three or four seconds. It then alights on an adjoining branch, traverses it lengthwise for a few moments, flirting its expanded tail from side to side, and suddenly shoots off, in a direction quite unexpected, after fresh game, which it can discover at a great distance. Its notes, or twitter, though animated and sprightly, are not deserving the name of song; sometimes they are *weése, weése, weése,* repeated every quarter of a minute, as it skips among the branches; at other times this twitter varies to several other chants, which I can instantly distinguish in the woods, but cannot find words to imitate. The interior of the forest, the borders of swamps and meadows, deep glens covered with wood, and wherever flying insects abound, there this little bird is sure to be seen. It makes its appearance in Pennsylvania, from the south, late in April, and leaves us again about the beginning of September. It is very generally found over the whole United States; and has been taken at sea, in the fall, on its way to St Domingo,* and other of the West India islands, where it winters, along with many more of our summer visitants. It is also found in Jamaica, where it remains all winter.†

The name redstart, evidently derived from the German *rothsterts* (red tail), has been given this bird from its supposed resemblance to the redstart of Europe (*Motacilla phœnicurus*); but besides being decisively of a different genus, it is very different both in size and in the tints and disposition of the colours of its plumage. Buffon goes even so far as to question whether the differences between the two be more than what might be naturally expected from change of climate. This eternal reference of every animal of the New World to that of the Old, if adopted to the extent of this writer, with all the transmutations it is supposed to have produced, would leave us in doubt whether even the

* Edwards. † Sloane.

Ka-te-dids* of America were not originally nightingales of the Old World, degenerated by the inferiority of the food and climate of this upstart continent. We have in America many different species of birds that approach so near in resemblance to one another, as not to be distinguished but by the eye of a naturalist, and on a close comparison; these live in the same climate, feed on the same food, and are, I doubt not, the same now as they were five thousand years ago; and, ten thousand years hence, if the species then exist, will be found marked with the same nice discriminations as at present. It is therefore surprising that two different species, placed in different quarters of the world, should have certain near resemblances to one another, without being bastards, or degenerated descendants, the one of the other, when the whole chain of created beings seem united to each other by such amazing gradations, that bespeak not random chance and accidental degeneracy, but the magnificent design of an incomprehensibly wise and omnipotent Creator.

The American redstart builds frequently in low bushes, in the fork of a small sapling, or on the drooping branches of the elm, within a few feet of the ground; outwardly it is formed of flax, well wound together, and moistened with its saliva, interspersed here and there with pieces of lichen, and lined with a very soft downy substance. The female lays five white eggs, sprinkled with gray and specks of blackish. The male is extremely anxious for its preservation, and, on a person's approaching the place, will flirt about within a few feet, seeming greatly distressed.†

The length of this species is five inches; extent, six and a quarter; the general colour above is black, which covers the whole head and neck, and spreads on the upper part of the

* A species of *Gryllus*, well known for its lively chatter during the evenings and nights of September and October.

† Mr Audubon says, "The nest is slight, composed of lichens and dried fibres of rank weeds or grape vines, nicely lined with soft cotton materials."—P. 203.—Ed.

breast in a rounding form, where, as well as on the head and neck, it is glossed with steel blue; sides of the breast below this, black; the inside of the wings, and upper half of the wing-quills, are of a fine aurora colour; but the greater and lesser coverts of the wings, being black, conceal this, and the orange or aurora colour appears only as a broad transverse band across the wings; from thence to the tip, they are brownish; the four middle feathers of the tail are black, the other eight of the same aurora colour, and black towards the tips; belly and vent, white, slightly streaked with pale orange; legs, black; bill, of the true *Muscicapa* form, triangular at the base, beset with long bristles, and notched near the point; the female has not the rich aurora band across the wing; her back and crown are cinereous, inclining to olive; the white below is not so pure; lateral feathers of the tail and sides of the breast, greenish yellow; middle tail-feathers, dusky brown. The young males of a year old are almost exactly like the female, differing in these particulars, that they have a yellow band across the wings which the female has not, and the back is more tinged with brown; the lateral tail-feathers are also yellow; middle ones, brownish black; inside of the wings, yellow. On the third season they receive their complete colours; and as males of the second year, in nearly the dress of the female, are often seen in the woods, having the same notes as the full plumaged male, it has given occasion to some people to assert that the females sing as well as the males; and others have taken them for another species. The fact, however, is as I have stated it. This bird is too little known by people in general to have any provincial name.

CEDAR BIRD. (*Ampelis Americana.*)

PLATE VII.—FIG. 1.

Ampelis garrulus, *Linn. Syst.* i. 297, 1, β.—Bombycilla Carolinensis, *Brisson*, ii. 337, 1. *Id.* 8vo. i. 251.—Chatterer of Carolina, *Catesb.* i. 46.—*Arct. Zool.* ii. No. 207.—*Lath. Syn.* iii. 93, 1, A.—*Edw.* 242.—*Cook's Last Voyage*, ii. 518. —*Ellis's Voyage*, ii. 13.—*Peale's Museum*, No. 5608.

BOMBYCILLA AMERICANA.—SWAINSON.

Le jaseur du cèdre, Bombycilla cedorum, *Vieill. Gal. des Ois.* pl. cxviii. p. 186.—Bombycilla Carolinensis, *Bonap. Synop.* p. 59.—Bombycilla Americana, *North. Zool.* ii. p. 239.

THE figure of the cedar bird which accompanies this description was drawn from a very beautiful specimen, and exhibits the form of its crest when erected, which gives it so gay and elegant an appearance. At pleasure it can lower and contract this so closely to its head and neck as not to be observed. The plumage of these birds is of an exquisitely fine and silky texture, lying extremely smooth and glossy. Notwithstanding the name *chatterers* given to them, they are perhaps the most silent species we have, making only a feeble, lisping sound, chiefly as they rise or alight. They fly in compact bodies of from twenty to fifty; and usually alight so close together on the same tree, that one half are frequently shot down at a time. In the months of July and August, they collect together in flocks, and retire to the hilly parts of the State, the Blue Mountains, and other collateral ridges of the Alleghany, to enjoy the fruit of the *Vaccinium uliginosum*, whortleberries, which grow there in great abundance; whole mountains, for many miles, being almost entirely covered with them; and where, in the month of August, I have myself found the cedar birds numerous. In October they descend to the lower, cultivated parts of the country, to feed on the berries of the sour gum and red cedar, of which last they are immoderately fond; and thirty or forty may sometimes be seen fluttering among

Drawn from Nature by A. Wilson. Engraved by W. H. Lizars

1. Cedar Bird. 2. Red bellied Woodpecker. 3. Yellow-throated Flycatcher. 4. Purple Finch.

7.

the branches of one small cedar tree, plucking off the berries.* They are also found as far south as Mexico, as appears from the accounts of Fernandez, Seba,† and others. Fernandez saw them near Tetzeuco, and calls them *Coquantotl ;* says they delight to dwell in the mountainous parts of the country, and that their flesh and song are both indifferent.‡ Most of our epicures here are, however, of a different opinion as to their palatableness ; for, in the fall and beginning of summer, when they become very fat, they are in considerable esteem for the table ; and great numbers are brought to the market of Philadelphia, where they are sold from twelve to twenty-five cents per dozen. During the whole winter and spring they are occasionally seen ; and about the 25th of May appear in numerous parties, making great havoc among the early

* They appear all to be berry-eaters, at least during winter. Those of Europe have generally been observed to feed on the fruit of the mountain ash, and one or two killed near Carlisle, which I had an opportunity of examining, were literally crammed with hollyberries. "The appetite of the cedar bird," Audubon remarks, "is of so extraordinary a nature as to prompt it to devour every fruit or berry that comes in its way. In this manner they gorge themselves to such excess as sometimes to be unable to fly, and suffer themselves to be taken by the hand ; and I have seen some, which, though wounded and confined to a cage, have eaten apples until suffocation deprived them of life."—P. 227. "But they are also excellent flycatchers, spending much of their time in the pursuit of winged insects: this is not, however, managed with the vivacity or suddenness of true flycatchers, but with a kind of listlessness. They start from the branches, and give chase to the insects, ascending after them for a few yards, or move horizontally towards them, and as soon as the prey is secured, return to the spot, where they continue watching with slow motions of the head. This amusement is carried on during evening, and longer at the approach of autumn, when the berries become scarce. They become very fat during the season of fruits, and are then so tender and juicy as to be sought after by every epicure for the table, —a basketful of these birds is sometimes sent as a Christmas present." —P. 223.—Ed.

† The figure of this bird in Seba's voluminous work is too wretched for criticism ; it is there called "Oiseau Xomotl, d'Amerique, huppée." Seb. ii. p. 66, t. 65, fig. 5.

‡ Hist. Av. Nov. Hisp. 55.

cherries, selecting the best and ripest of the fruit. Nor are they easily intimidated by the presence of Mr Scarecrow; for I have seen a flock deliberately feasting on the fruit of a loaded cherry tree, while on the same tree one of these *guardian angels,* and a very formidable one too, stretched his stiffened arms, and displayed his dangling legs, with all the pomposity of authority. At this time of the season, most of our resident birds, and many of our summer visitants, are sitting or have young; while, even on the 1st of June, the eggs in the ovary of the female cedar bird are no larger than mustard seed; and it is generally the 8th or 10th of that month before they begin to build. These last are curious circumstances, which it is difficult to account for, unless by supposing that incubation is retarded by a scarcity of suitable food in spring, berries and other fruit being their usual fare. In May, before the cherries are ripe, they are lean, and little else is found in their stomachs than a few shrivelled cedar berries, the refuse of the former season, and a few fragments of beetles and other insects, which do not appear to be their common food; but in June, while cherries and strawberries abound, they become extremely fat; and, about the 10th or 12th of that month, disperse over the country in pairs to breed; sometimes fixing on the cedar, but generally choosing the orchard for that purpose. The nest is large for the size of the bird, fixed in the forked or horizontal branch of an apple tree, ten or twelve feet from the ground; outwardly, and at bottom, is laid a mass of coarse dry stalks of grass, and the inside is lined wholly with very fine stalks of the same material. The eggs are three or four, of a dingy bluish white, thick at the great end, tapering suddenly, and becoming very narrow at the other; marked with small roundish spots of black of various sizes and shades; and the great end is of a pale dull purple tinge, marked likewise with touches of various shades of purple and black. About the last week in June the young are hatched, and are at first fed on insects and their larvæ, but, as they advance in growth, on berries of various kinds.

These facts I have myself been an eyewitness to. The female, if disturbed, darts from the nest in silence to a considerable distance; no notes of wailing or lamentation are heard from either parent, nor are they even seen, notwithstanding you are in the tree examining the nest and young. These nests are less frequently found than many others, owing, not only to the comparatively few numbers of the birds, but to the remarkable muteness of the species. The season of love, which makes almost every other small bird musical, has no such effect on them, for they continue, at that interesting period, as silent as before.

This species is also found in Canada, where it is called *Recollet,* probably, as Dr Latham supposes, from the colour and appearance of its crest resembling the hood of an order of friars of that denomination. It has also been met with by several of our voyagers on the northwest coast of America, and appears to have an extensive range.

Almost all the ornithologists of Europe persist in considering this bird as a variety of the European chatterer (*A. garrulus*), with what justice or propriety a mere comparison of the two will determine.* The European species is very nearly twice the cubic bulk of ours; has the whole lower parts of a uniform dark vinous bay; the tips of the wings streaked with lateral bars of yellow; the nostrils, covered with bristles; † the feathers

* The small American species, figured by our author, was by many considered as only the American variety of that which was thought to belong to Europe and Asia alone. The fallacy of this opinion was decided by the researches of several ornithologists, and latterly confirmed by the discovery in America of the *B. garrulus* itself, the description of which will form part of Vol. III.

The genus *Bombycilla* of Brisson is generally adopted for these two birds, and will now also contain a third very beautiful and nearly allied species, discovered in Japan by the enterprising, but unfortunate, naturalist Seibold, and figured in the *Planches Coloriées* of M. Temminck, under the name of *B. phœnicoptera.* It may be remarked, that the last wants the waxlike appendages to the wings and tail, at least so they are represented in M. Temminck's plate; but our own species sometimes wants them also.—ED.

† Turton.

on the chin, loose and tufted; the wings, black; and the markings of white and black on the sides of the head different from the American, which is as follows:—Length, seven inches; extent, eleven inches; head, neck, breast, upper part of the back and wing-coverts, a dark fawn colour, darkest on the back, and brightest on the front; head, ornamented with a high-pointed, almost upright, crest; line from the nostril over the eye to the hind head, velvety black, bordered above with a fine line of white, and another line of white passes from the lower mandible; chin, black, gradually brightening into fawn colour, the feathers there lying extremely close; bill, black; upper mandible nearly triangular at the base, without bristles, short, rounding at the point, where it is deeply notched; the lower, scolloped at the tip, and turning up; tongue, as in the rest of the genus, broad, thin, cartilaginous, and lacerated at the end; belly, yellow; vent, white; wings, deep slate, except the two secondaries next the body, whose exterior vanes are of a fawn colour, and interior ones white, forming two whitish stripes there, which are very conspicuous; rump and tail-coverts, pale light blue; tail, the same, gradually deepening into black, and tipt for half an inch with rich yellow. Six or seven, and sometimes the whole nine, secondary feathers of the wings are ornamented at the tips with small red oblong appendages, resembling red sealing-wax; these appear to be a prolongation of the shafts, and to be intended for preserving the ends, and consequently the vanes, of the quills, from being broken and worn away by the almost continual fluttering of the bird among thick branches of the cedar. The feathers of those birds which are without these appendages are uniformly found ragged on the edges, but smooth and perfect in those on whom the marks are full and numerous. These singular marks have been usually considered as belonging to the male alone, from the circumstance, perhaps, of finding female birds without them. They are, however, common to both male and female. Six of the latter are now lying before me, each with large and numerous clusters of eggs, and having the waxen appendages in full perfection. The young birds do not receive them until the second fall, when, in moulting time, they may be seen fully formed, as the feather is developed from its sheath. I have once or twice found a solitary one on the extremity of one of the tail-feathers. The eye is of a dark blood colour; the legs and claws, black; the inside of the mouth, orange; gap, wide; and the gullet capable of such distension as often to contain twelve or fifteen cedar berries, and serving as a kind of craw to prepare them for digestion. No wonder, then, that this gluttonous bird, with such a mass of food almost continually in its throat, should want both the inclination and powers for vocal melody, which would seem to belong to those only of less gross and voracious habits. The chief difference in the plumage of the male and female consists in the dulness of the tints of the latter, the inferior appearance of the crest, and the narrowness of the yellow bar on the tip of the tail.

Though I do not flatter myself with being able to remove that prejudice from the minds of foreigners which has made them look on this bird also as a degenerate and not a distinct species from their own; yet they must allow that the change has been very great, very uniform, and universal all over North America, where I have never heard that the European species has been found; or, even if it were, this would only show more clearly the specific difference of the two, by proving that climate or food could never have produced these differences in either when both retain them, though confined to the same climate.

But it is not only in the colour of their plumage that these two birds differ, but in several important particulars in their manners and habits. The breeding place of the European species is absolutely unknown; supposed to be somewhere about the Polar Regions; from whence, in winter, they make different, and very irregular excursions to various parts of Europe; seldom advancing farther south than the north of

England, in lat. 54° N., and so irregularly, that many years sometimes elapse between their departure and reappearance, which, in more superstitious ages, has been supposed to portend some great national calamity. On the other hand, the American species inhabits the whole extensive range between Mexico and Canada, and perhaps much farther both northerly and southerly, building and rearing their young in all the intermediate regions, often in our gardens and orchards, within a few yards of our houses. Those of our fellow-citizens who have still any doubts, and wish to examine for themselves, may see beautiful specimens of both birds in the superb collection of Mr Charles W. Peale of Philadelphia, whose magnificent museum is indeed a national blessing, and will be a lasting honour to his memory.

In some parts of the country they are called crown birds ; in others, cherry birds, from their fondness for that fruit. They also feed on ripe persimmons, small winter grapes, bird cherries, and a great variety of other fruits and berries. The action of the stomach on these seeds and berries does not seem to injure their vegetative powers ; but rather to promote them, by imbedding them in a calcareous case, and they are thus transported to and planted in various and distant parts by these little birds. In other respects, however, their usefulness to the farmer may be questioned ; and in the general chorus of the feathered songsters they can scarcely be said to take a part. We must, therefore, rank them far below many more homely and minute warblers, their neighbours, whom Providence seems to have formed both as allies to protect the property of the husbandman from devouring insects, and as musicians to cheer him, while engaged in the labours of the field, with their innocent and delightful melody.

RED-BELLIED WOODPECKER. (*Picus Carolinus.*)

PLATE VII.—Fig. 2.

Picus Carolinus, *Linn. Syst.* i. 174, 10.—Pic varie de la Jamaique, *Buffon*, vii. 72. *Pl. enl.* 597.—Picus varius medius Jamaicensis, *Sloan. Jam.* 299, 15.—Jamaica Woodpecker, *Edw.* 244.—*Catesb.* i. 19, fig. 2.—*Arct. Zool.* ii. No. 161.—*Lath. Syn.* ii. 570, 17. *Id.* 571, 17 A. *Id.* β.—L'Epeiche rayé de la Louisiane, *Buff.* vii. 73. *Pl. enl.* 692.—*Peale's Museum*, No. 1944.

COLAPTES CAROLINUS.—Swainson.

Picus Carolinus, *Bonap. Synop.* p. 45.—Picus erythrauchen, *Wagl. Syst. Av.* No. 38.

This species possesses all the restless and noisy habits so characteristic of its tribe. It is more shy and less domestic than the red-headed one (*P. erythrocephalus*), or any of the other spotted woodpeckers. It is also more solitary. It prefers the largest, high-timbered woods, and tallest decayed trees of the forest; seldom appearing near the ground, on the fences, or in orchards, or open fields; yet where the trees have been deadened, and stand pretty thick, in fields of Indian-corn, as is common in new settlements, I have observed it to be very numerous, and have found its stomach sometimes completely filled with that grain.* Its voice is hoarser than any of the others; and its usual note, *chow*,

* This species will also range in the genus *Colaptes*, but will present a more aberrant form. In it we have the compressed and slightly bent shape of the bill, becoming stronger and more angular; we have the barred plumage of the upper parts, but that of the head is uniform, and only slightly elongated behind; and in the wings and tail the shafts of the quills lose their strength and beautiful colour. In Wilson's description of the habits, we also find them agreeing with the modifications of form. It prefers the more solitary recesses of lofty forests; and, though capable of turning and twisting, and possessing a great part of the activity of the nuthatch and titmice, it seldom appears about orchards or upon the ground; yet it occasionally visits the cornfields, and feeds on the grain; and, as remarked above, is "capable of subsisting on coarser and more various fare." These modifications of habit we shall always find in unison with the structure; and we cannot too much admire the wisdom that has thus mutually adapted them to the various offices they are destined to fill.—Ed.

has often reminded me of the barking of a little lapdog. It is a most expert climber, possessing extraordinary strength in the muscles of its feet and claws, and moves about the body and horizontal limbs of the trees with equal facility in all directions. It rattles like the rest of the tribe on the dead limbs, and with such violence as to be heard, in still weather, more than half a mile off, and listens to hear the insects it has alarmed. In the lower side of some lofty branch that makes a considerable angle with the horizon, the male and female, in conjunction, dig out a circular cavity for their nest, sometimes out of the solid wood, but more generally into a hollow limb, twelve or fifteen inches above where it becomes solid. This is usually performed early in April. The female lays five eggs of a pure white, or almost semi-transparent; and the young generally make their appearance towards the latter end of May or beginning of June, climbing up to the higher parts of the tree, being as yet unable to fly. In this situation they are fed for several days, and often become the prey of the hawks. From seeing the old ones continuing their caresses after this period, I believe that they often, and perhaps always, produce two broods in a season. During the greatest part of the summer, the young have the ridge of the neck and head of a dull brownish ash; and a male of the third year has received his complete colours.

The red-bellied woodpecker is ten inches in length and seventeen in extent; the bill is nearly an inch and a half in length, wedged at the point, but not quite so much grooved as some others, strong, and of a bluish black colour; the nostrils are placed in one of these grooves, and covered with curving tufts of light brown hairs, ending in black points; the feathers on the front stand more erect than usual, and are of a dull yellowish red; from thence, along the whole upper part of the head and neck, down the back, and spreading round to the shoulders, is of the most brilliant golden glossy red; the whole cheeks, line over the eye, and under side of the neck, are a pale buff colour, which, on the breast and belly, deepens into a yellowish ash, stained on the belly with a blood red; the vent and thigh feathers are dull white, marked down their centres with heart-formed and long arrow-pointed spots of black. The back is black, crossed with transverse curving lines of white; the wings are also black; the lesser wing-coverts circularly tipt, and the whole primaries and secondaries beautifully crossed with bars of white, and also tipt with the same; the rump is white, interspersed with touches of black; the tail-coverts, white near their extremities; the tail consists of ten feathers, the two middle ones black, their interior webs or vanes white, crossed with diagonal spots of black; these, when the edges of the two feathers just touch, coincide and form heart-shaped spots; a narrow sword-shaped line of white runs up the exterior side of the shafts of the same feathers; the next four feathers on each side are black; the outer edges of the exterior ones, barred with black and white, which, on the lower side, seems to cross the whole vane, as in the figure; the extremities of the whole tail, except the outer feather, are black, sometimes touched with yellowish or cream colour; the legs and feet are of a bluish green, and the iris of the eye red. The tongue, or os hyöides, passes up over the hind head, and is attached, by a very elastic retractile membrane, to the base of the right nostril; the extremity of the tongue is long, horny, very pointed, and thickly edged with barbs; the other part of the tongue is worm-shaped. In several specimens, I found the stomach nearly filled with pieces of a species of fungus that grows on decayed wood,* and, in all, with great numbers of insects, seeds, gravel, &c. The female differs from the male in having the crown, for an inch, of a fine ash, and the black not so intense; the front is reddish, as in the male, and the whole hind head, down

* Most probably swallowed with the insects which infest and are nourished in the various *Boleti polypori*, &c., but forming no part of their real food.—Ed.

to the back, likewise of the same rich red as his. In the bird from which this latter description was taken I found a large cluster of minute eggs, to the number of fifty or upwards, in the beginning of the month of March.

This species inhabits a large extent of country, in all of which it seems to be resident, or nearly so. I found them abundant in Upper Canada, and in the northern parts of the State of New York, in the month of November; they also inhabit the whole Atlantic States as far as Georgia, and the southern extremity of Florida, as well as the interior parts of the United States, as far west as Chilicothe, in the State of Ohio, and, according to Buffon, Louisiana. They are said to be the only woodpeckers found in Jamaica; though I question whether this be correct; and to be extremely fond of the capsicum or Indian pepper.* They are certainly much hardier birds, and capable of subsisting on coarser and more various fare, and of sustaining a greater degree of cold, than several other of our woodpeckers. They are active and vigorous; and, being almost continually in search of insects that injure our forest trees, do not seem to deserve the injurious epithets that almost all writers have given them. It is true they frequently perforate the timber in pursuit of these vermin; but this is almost always in dead and decaying parts of the tree, which are the nests and nurseries of millions of destructive insects. Considering matters in this light, I do not think their services overpaid by all the ears of Indian-corn they consume; and would protect them, within my own premises, as being more useful than injurious.

* Sloane.

YELLOW-THROATED FLYCATCHER. (*Muscicapa sylvicola.*)

PLATE VII.—Fig. 3.

Peale's Museum, No. 6827.

VIREO FLAVIFRONS.*—Vieillot.

Vireo flavifrons, *Bonap. Synop.* p. 70.

The summer species is found chiefly in the woods, hunting among the high branches, and has an indolent and plaintive note, which it repeats, with some little variation, every ten or twelve seconds, like *preeò preeà*, &c. It is often heard in company with the red-eyed flycatcher (*Muscicapa olivacea*) or whip-tom-kelly of Jamaica; the loud energetic notes of the latter, mingling with the soft, languid warble of the former, pro-

* *Vireo* is a genus originally formed by Vieillot to contain an American group of birds, since the formation of which several additions have been made by Bonaparte and Swainson of species which were not at first contemplated as belonging to it.

The group is peculiar to both continents of America,—they inhabit woods, feed on insects and berries, and in their manner have considerable alliance to the warblers and flycatchers. By Mr Swainson they are placed among the *Ampelidæ*, or berry-eaters, but with a mark of uncertainty whether they should stand here or at the extremity of some other family. The Arctic Expedition has added a new species much allied to *V. olivaceus.* Mr Swainson has dedicated it to the venerable naturalist Bartram, the intimate friend of Wilson, and mentions that, on comparing seventeen species, *Vireo Bartramii* was much smaller, the colours rather brighter, the wings considerably shorter and more rounded, and the first quill always shorter than the fifth,—that *V. olivaceus* is confined to North America, while *V. Bartramii* extends to Brazil. The species of the Arctic Expedition were procured by Mr David Douglas on the banks of the Columbia. Mr Swainson also met with the species in the Brazils; and, from specimens sent to us by that gentleman, I have no hesitation in considering them distinct, and of at once recognising the differences he has pointed out.

Mr Audubon has figured another species, which will rank as an addition to this genus, and, if proved new, will stand as *Vireo Vigorsii;* he has only met with a single individual in Pennsylvania, and enters into no description of its history, or distinctions from other allied birds.—Ed.

ducing an agreeable effect, particularly during the burning heat of noon, when almost every other songster but these two is silent. Those who loiter through the shades of our magnificent forests at that hour, will easily recognise both species. It arrives from the south early in May; and returns again with its young about the middle of September. Its nest, which is sometimes fixed on the upper side of a limb, sometimes on a horizontal branch among the twigs, generally on a tree, is composed outwardly of thin strips of the bark of grape vines, moss, lichens, &c., and lined with fine fibres of such-like substances; the eggs, usually four, are white, thinly dotted with black, chiefly near the great end. Winged insects are its principal food.

Whether this species has been described before or not, I must leave to the sagacity of the reader who has the opportunity of examining European works of this kind to discover.* I have met with no description in Pennant, Buffon, or Latham, that will properly apply to this bird, which may perhaps be owing to the imperfection of the account, rather than ignorance of the species, which is by no means rare.

The yellow-throated flycatcher is five inches and a half long, and nine inches from tip to tip of the expanded wings; the upper part of the head, sides of the neck, and the back, are of a fine yellow olive; throat, breast, and line over the eye, which it nearly encircles, a delicate lemon yellow, which, in a lighter tinge, lines the wings; belly and vent, pure silky white; lesser wing-coverts, lower part of the back and rump, ash; wings, deep brown, almost black, crossed with two white bars; primaries edged with light ash, secondaries with white; tail a little forked, of the same brownish black with the wings, the three exterior feathers edged on each vane with white; legs and claws, light blue; the two exterior toes united to the middle one, as far as the second joint; bill, broad at the base, with three or four slight bristles, the upper mandible overhanging the lower at the point, near which it is deeply notched;

* See orange-throated warbler, Latham, Syn. ii. 481, 103.

tongue, thin, broad, tapering near the end, and bifid ; the eye is of a dark hazel, and the whole bill of a dusky light blue. The female differs very little in colour from the male; the yellow on the breast and round the eye is duller, and the white on the wings less pure.

PURPLE FINCH. (*Fringilla purpurea.*)

PLATE VII.—Fig. 4.

Fringilla purpurea, *Gmel. Syst.* i. 923.—Bouvreuil violet de la Caroline, *Buff.* iv. 395.—Purple Finch, *Arct. Zool.* ii. No. 258.—*Catesb.* i. 41.—*Lath. Synop.* iii. 275, 39.—Crimson-headed Finch, *Arct. Zool.* ii. No. 257.—*Lath. Synop.* iii. 275, 39.—*Gmel. Syst.* i. 864.—Fringilla rosea Pallas, iii. 699, 26.—Hemp Bird, *Bartram*, 291.—Fringilla purpurea, *Id.* 291.—*Peale's Museum*, No. 6504.

ERYTHROSPIZA PURPUREA.—Bonaparte.

Fringilla purpurea, *Bonap. Synop.* p. 114.—Purple Finch, *Aud.* i. p. 24. Pl. iv. —Fringilla purpurea, Crested Purple Finch, *North. Zool.* ii. p. 264.—Erythrospiza purpurea, *Osserv. di C. L. Bonap. Sulla Sec. Ed. del. Cuv. Reg. Anim.* p. 80.

This is a winter bird of passage, coming to us in large flocks from the north in September and October; great numbers remaining with us in Pennslyvania during the whole winter, feeding on the seeds of the poplar, button-wood, juniper, cedar, and on those of many rank weeds that flourish in rich bottoms and along the margin of creeks. When the season is very severe, they proceed to the south, as far at least as Georgia, returning north early in April. They now frequent the elm trees, feeding on the slender but sweet covering of the flowers; and as soon as the cherries put out their blossoms, feed almost exclusively on the stamina of the flowers; afterwards the apple blossoms are attacked in the same manner; and their depredations on these continue till they disappear, which is usually about the 10th or middle of May. I have been told, that they sometimes breed in the northern parts of New York, but have never met with their nests. About the middle of September, I have found these birds numerous on Long Island, and round Newark in New Jersey. They fly at a considerable height in the air, and their note is a single *chink*, like that of the rice bird. They possess great boldness and spirit, and, when caught, bite violently, and hang by the bill from your hand, striking with great fury; but they are soon reconciled to confinement, and in a day or two are quite at home. I have kept a pair of these birds upwards of nine months to observe their manners. One was caught in a trap, the other was winged with the gun; both are now as familiar as if brought up from the nest by the hand, and seem to prefer hemp-seed and cherry blossoms to all other kinds of food. Both male and female, though not crested, are almost constantly in the habit of erecting the feathers of the crown; they appear to be of a tyrannical and domineering disposition, for they nearly killed an indigo bird, and two or three others that were occasionally placed with them, driving them into a corner of the cage, standing on them, and tearing out their feathers, striking them on the head, munching their wings, &c., till I was obliged to interfere; and even if called to, the aggressor would only turn up a malicious eye to me for a moment, and renew his outrage as before. They are a hardy vigorous bird. In the month of October, about the time of their first arrival, I shot a male, rich in plumage, and plump in flesh, but which wanted one leg, that had been taken off a little above the knee; the wound had healed so completely, and was covered with so thick a skin, that it seemed as though it had been so for years. Whether this mutilation was occasioned by a shot, or in party quarrels of its own, I could not determine; but our invalid seemed to have used his stump either in hopping or resting, for it had all the appearance of having been brought in frequent contact with bodies harder than itself.

This bird is a striking example of the truth of what I have frequently repeated in this work, that in many instances the same bird has been more than once described by the same person as a different species; for it is a fact which time will establish, that the crimson-headed finch of Pennant and Latham, the purple finch of the same and other naturalists, the hemp bird of Bartram, and the *Fringilla rosea* of Pallas, are one and the same, viz., the purple finch, the subject of the present article.*

The purple finch is six inches in length and nine in extent; head, neck, back, breast, rump, and tail-coverts, dark crimson, deepest on the head and chin, and lightest on the lower part of the breast; the back is streaked with dusky; the wings and tail are also dusky black, edged with reddish; the latter a good deal forked; round the base of the bill, the recumbent feathers are of a light clay or cream colour; belly and vent, white; sides under the wings, streaked with dull reddish; legs, a dirty purplish flesh colour; bill, short, strong, conical, and of a dusky horn colour; iris, dark hazel; the feathers

* The present figure is that of an adult male, and that sex in the winter state is again figured and described in the second volume. Bonaparte has shown that Wilson is wrong in making the *F. rosea* of Pallas and the *Loxia erythrina* of Gmelin the same with his bird. Mr Swainson remarks, "We are almost persuaded that there are two distinct species of these purple finches, which not only Wilson, but all the modern ornithologists of America, have confounded under the same name." We may reasonably conclude, then, that another allied species may yet be discovered, and that perhaps Wilsón was wrong regarding birds which he took for the *F. rosea*.

F. purpurea and *Pyrrhula frontalis* of Say and Bonaparte will rank as a subgenus in *Pyrrhula*, and, from the description of their habits, approach very near to both the crossbills and pine grosbeaks.

By the attention of the Prince of Musignano, I have received his review of Cuvier's *Règne Animal*, and am now enabled to state from it the opinion of that ornithologist regarding the station of these birds. He agrees in the subordinate rank of the group, and its alliance to the finches, bullfinches, and *Coccothraustes* or hawkfinch, and proposes the subgeneric name of *Erythrospiza*, which I have provisionally adopted, having *Fringilla purpurea* of Wilson as typical, and containing *Pyrrhula frontalis*, Say and Bonap.; *P. githaginea*, Temm. Pl. Col.; *Loxia Siberica*, Falck.; *L. rosea*, Pall.; *L. erythrina*, Pall.; *P. synoica*, Temm. Pl. Col.; and *L. rubicilla*, Lath. According to the list of species which he has mentioned, and which we have no present opportunity of comparing with the true type, the group will have a very extensive distribution over America, Europe, Asia, and Africa.—Ed.

covering the ears are more dusky red than the other parts of the head. This is the male when arrived at its full colours. The female is nearly of the same size, of a brown olive or flaxen colour, streaked with dusky black; the head, seamed with lateral lines of whitish; above and below the hind part of the ear-feathers are two streaks of white; the breast is whitish, streaked with a light flax colour; tail and wings, as in the male, only both edged with dull brown, instead of red; belly and vent, white. This is also the colour of the young during the first, and to at least the end of the second season, when the males begin to become lighter yellowish, which gradually brightens to crimson; the female always retains nearly the same appearance. The young male bird of the first year may be distinguished from the female by the tail of the former being edged with olive green, that of the latter with brown. A male of one of these birds, which I kept for some time, changed in the month of October from red to greenish yellow, but died before it recovered its former colour.

BROWN CREEPER. (*Certhia familiaris.*)

PLATE VIII.—Fig. 1.

Little Brown Variegated Creeper, *Bartram*, 289.—*Peale's Museum*, No. 2434.

CERTHIA FAMILIARIS.—Linnæus.

Certhia familiaris, *Linn. Syst. Nat.* i. 469.—*Bonap. Synop.* p. 95.—The Creeper, *Bewick*, *Brit. Birds*, i. p. 148.—Le Grimpereau, *Temm. Man.* i. p. 410.—Common Creeper, *Selby Ill.* plate 39, vol. i. p. 116.

This bird agrees so nearly with the common European creeper (*Certhia familiaris*), that I have little doubt of their being one and the same species.* I have examined, at different

* I have compared numerous British specimens with skins from North America, and can find no differences that will entitle a separation of species. In this country they are very abundant, more so apparently in winter, so that we either receive a great accession from the more northern parts of Europe, or the colder season and diminished supply of

Drawn from Nature by A. Wilson

Engraved by W. H. Lizars

1. Brown Creeper. 2. Golden-crested Wren. 3. House Wren. 4. Black-capt Titmouse. 5. Crested Titmouse. 6. Winter Wren.

8.

times, great numbers of these birds, and have endeavoured to make a correct drawing of the male, that Europeans and others may judge for themselves; and the excellent artist to whom the plate was entrusted has done his part so well in the engraving, as to render the figure a perfect resemblance of the living original.

The brown creeper is an extremely active and restless little bird. In winter it associates with the small spotted woodpecker, nuthatch, titmouse, &c.; and often follows in their rear, gleaning up those insects which their more powerful bills had alarmed and exposed; for its own slender incurvated bill seems unequal to the task of penetrating into even the decayed wood; though it may into holes, and behind scales of the bark. Of the titmouse, there are generally present the individuals of a whole family, and seldom more than one or two of the others. As the party advances through the woods from tree to tree, our little gleaner seems to observe a good

food draws them from their woody solitudes nearer to the habitations of man. It is often said to be rare, an opinion no doubt arising from the difficulty of seeing it, and from its solitary and unassuming manners. A short quotation from a late author will best explain our meaning, and confirm the account of its manners, so correctly described above. "A retired inhabitant of the woods and groves, and not in any way conspicuous for voice or plumage, it passes its days with us, creating scarcely any notice or attention. Its small size, and the manner in which it procures its food, both tend to secret him from sight. In these pursuits its actions are more like those of a mouse than of a bird, darting like a great moth from tree to tree, uttering a faint trilling sound as it fixes on their boles, running round them in a spiral direction, when, with repeated wriggles, having gained the summit, it darts to the base of another, and commences again."

The present species will form the type and only individual yet discovered of the genus *Certhia*. The other birds described by our author as *Certhiæ* will all rank elsewhere; and the groups now known under the titles *Cinyris, Nectarinia*, &c., which were formerly included, making it of great extent, and certainly of very varied forms, will also with propriety hold their separate stations. The solitary type ranges in Europe, according to Pennant, as far north as Russia and Siberia, and Sandmore in Sweden. In North America it will extend nearly over the whole continent.—Ed.

deal of regularity in his proceedings; for I have almost always observed that he alights on the body near the root of the tree, and directs his course, with great nimbleness, upwards, to the higher branches, sometimes spirally, often in a direct line, moving rapidly and uniformly along, with his tail bent to the tree, and not in the hopping manner of the woodpecker, whom he far surpasses in dexterity of climbing, running along the lower side of the horizontal branches with surprising ease. If any person be near when he alights, he is sure to keep the opposite side of the tree, moving round as he moves, so as to prevent him from getting more than a transient glimpse of him. The best method of outwitting him, if you are alone, is, as soon as he alights, and disappears behind the trunk, take your stand behind an adjoining one, and keep a sharp look out twenty or thirty feet up the body of the tree he is upon,—for he generally mounts very regularly to a considerable height, examining the whole way as he advances. In a minute or two, hearing all still, he will make his appearance on one side or other of the tree, and give you an opportunity of observing him.

These birds are distributed over the whole United States; but are most numerous in the western and northern States, and particularly so in the depth of the forests, and in tracts of large-timbered woods, where they usually breed, visiting the thicker settled parts of the country in fall and winter. They are more abundant in the flat woods of the lower district of New Jersey than in Pennsylvania, and are frequently found among the pines. Though their customary food appears to consist of those insects of the coleopterous class, yet I have frequently found in their stomachs the seeds of the pine tree, and fragments of a species of fungus that vegetates in old wood, with generally a large proportion of gravel. There seems to be scarcely any difference between the colours and markings of the male and female. In the month of March, I opened eleven of these birds, among whom were several females, as appeared by the clusters of minute eggs with which their ovaries were filled, and also several well-marked males; and, on the most careful comparison of their plumage, I could find little or no difference; the colours, indeed, were rather more vivid and intense in some than in others; but sometimes this superiority belonged to a male, sometimes to a female, and appeared to be entirely owing to difference in age. I found, however, a remarkable and very striking difference in their sizes; some were considerably larger, and had the bill at least one-third longer and stronger than the others, and these I uniformly found to be males. I also received two of these birds from the country bordering on the Cayuga Lake, in New York State, from a person who killed them from the tree in which they had their nest. The male of this pair had the bill of the same extraordinary size with several others I had examined before; the plumage in every respect the same. Other males, indeed, were found at the same time, of the usual size. Whether this be only an accidental variety, or whether the male, when full-grown, be naturally so much larger than the female (as is the case with many birds), and takes several years in arriving at his full size, I cannot positively determine, though I think the latter most probable.

The brown creeper builds his nest in the hollow trunk or branch of a tree, where the tree has been shivered, or a limb broken off, or where squirrels or woodpeckers have wrought out an entrance, for nature has not provided him with the means of excavating one for himself. I have known the female begin to lay by the 17th of April. The eggs are usually seven, of a dull cinereous, marked with small dots of reddish yellow, and streaks of dark brown. The young come forth with great caution, creeping about long before they venture on wing. From the early season at which they begin to build, I have no doubts of their raising two broods during summer, as I have seen the old ones entering holes late in July.

The length of this bird is five inches, and nearly seven from the extremity of one wing to that of the other; the upper part of the head is of a deep brownish black; the back brown, and both streaked with white, the plumage of the latter being of a loose texture, with its filaments not adhering; the white is in the centre of every feather, and is skirted with brown; lower part of the back, rump, and tail-coverts, rusty brown, the last minutely tipt with whitish; the tail is as long as the body, of a light drab colour, with the inner web dusky, and consists of twelve quills, each sloping off and tapering to a point in the manner of the woodpeckers, but proportionably weaker in the shafts. In many specimens the tail was very slightly marked with transverse undulating waves of dusky, scarce observable; the two middle feathers the longest, the others on each side shortening, by one-sixth of an inch, to the outer one. The wing consists of nineteen feathers, the first an inch long, the fourth and fifth the longest, of a deep brownish black, and crossed about its middle with a curving band of rufous white, a quarter of an inch in breadth, marking ten of the quills; below this the quills are exteriorly edged, to within a little of their tips, with rufous white, and tipt with white; the three secondaries next the body are dusky white on their inner webs, tipt on the exterior margin with white, and, above that, alternately streaked laterally with black and dull white; the greater and lesser wing-coverts are exteriorly tipt with white; the upper part of the exterior edges of the former, rufous white; the line over the eye, and whole lower parts, are white, a little brownish towards the vent, but, on the chin and throat, pure, silky, and glistening; the white curves inwards about the middle of the neck. The bill is half an inch long, slender, compressed sidewise, bending downwards, tapering to a point, dusky above, and white below; the nostrils are oblong, half covered with a convex membrane, and without hairs or small feathers; the inside of the mouth is reddish; the tongue tapering gradually to a point, and horny towards the tip; the eye is dark hazel; the legs and feet, a dirty clay colour; the toes, placed three before and one behind, the two outer ones connected with the middle one to the first joint;

the claws rather paler, large, almost semicircular, and extremely sharp pointed ; the hind claw the largest. The figure in the plate represents a male of the usual size in its exact proportions, and but for the satisfaction of foreigners, might have rendered the whole of this prolix description unnecessary.

GOLDEN-CRESTED WREN. (*Sylvia regulus.*)

PLATE VIII.—Fig. 2.

Motacilla regulus, *Linn. Syst.* i. 338, 48.—*Lath. Syn.* iv. 508, 145.—*Edw.* 254.—*Peale's Museum*, No. 7246.

*REGULUS REGULOIDES.**—Jardine.

Regulus cristatus, *Bonap. Synop.* p. 91.—Female Golden-crowned Gold-crest, *Cont. of N. A. Orn.* i. pl. 2, p. 22.—Sylvia reguloides, *Sw.* MSS.

This diminutive species is a frequent associate of the one last described, and seems to be almost a citizen of the world at large, having been found not only in North and South America,

* The gold-crests, the common wrens, with an immense and varied host of species, were associated together in the genus *Sylvia*, until ornithologists began to look, not to the external characters in a limited view only, but in connection with the habits and affinities which invariably connect species together. Then many divisions were formed, and among these subordinate groups, *Regulus* of Ray was proposed for this small but beautiful tribe. It was used by Stephens, the continuator of Shaw's "Zoology," and by Bonaparte in his "Synopsis of North American Birds," and the first volume of his elegant continuation of Wilson. Mr Swainson makes this genus the typical form of the whole *Sylvianæ*, but designates it on that account under the title *Sylvia*. I have retained the old name of *Regulus*, on account of its former use by Ray, also from its having been adopted to this form by Stephens and Bonaparte, and lastly, as liable to create less confusion than the bringing forward of an old name (though denoting the typical affinity of the typical group) which has been applied to so many different forms in the same family.

Wilson was in error regarding the species here figured and the common gold-crest of Europe being identical, and Bonaparte has fallen into the same mistake when figuring the female. *Regulus cristatus* is exclusively European. *Regulus reguloides* appears yet exclusively North American. Upon comparing the two species minutely together, I find the following variations :—Length of *R. reguloides* three inches seven-the West Indies, and Europe, but even in Africa and India. The specimen from Europe, in Mr Peale's collection, appears to be in nothing specifically different from the American; and

eighths; of *R. cristatus*, from three inches and a half to three inches six-eighths. In *R. cristatus*, the bill is longer and more dilated at the base, and the under parts of the body are more tinged with olive; in *R. reguloides*, the orange part of the crest is much broader, and the black surrounding it, with the bar in front, broader and more distinct; the white streak above the eye is also better marked, and the nape of the neck has a pale ash-gray tinge, nearly wanting entirely in the British species.*

This very hardy and active tribe, with one exception, inhabits the temperate and northern climates, reaching even to the boundaries of the Arctic Circle. They are migratory in the more northern countries; and though some species are able to brave our severest winters, others are no doubt obliged, by want of food and a lower degree of cold, to quit the rigours of northern latitudes. The species of our author performs migrations northward to breed; and in Great Britain, at the commencement of winter, we have a regular accession to the numbers of our own gold-crest. If we examine their size, strength, and powers of flight, we must view the extent of their journeys with astonishment; they are indeed often so much exhausted, on their first arrival, as to be easily taken, and many sometimes even perish with the fatigue. A remarkable instance of a large migration is related by Mr Selby as occurring on the coast of Northumberland in 1822, when the sandhills and links were perfectly covered with them.

"On the 24th and 25th of October 1822, after a very severe gale, with thick fog, from the northeast (but veering, towards its conclusion, to the east and south of east), thousands of these birds were seen to arrive upon the seashore and sandbanks of the Northumbrian coast; many of them so fatigued by the length of their flight, or perhaps by the unfavourable shift of wind, as to be unable to rise again from the ground, and great numbers were in consequence caught or destroyed. This flight must have been immense in quantity, as its extent was traced through the whole length of the coasts of Northumberland and Durham. There appears little doubt of this having been a migration from the more northern provinces of Europe (probably furnished by the pine forests of Norway, Sweden, &c.), from the circumstance of its arrival being simultaneous with that of large flights of the woodcock, fieldfare, and redwing. Although I had never before witnessed the actual arrival of the

* There is a curious structure in the covering of the nostrils in most birds; where there is any addition to the horny substance, it is composed either of fine bristles or hairs, or of narrow feathers closely spread together. In the gold-crests it consists of a single plumelet on each side, the webs diverging widely.

the very accurate description given of this bird by the Count de Buffon agrees in every respect with ours. Here, as in Europe, it is a bird of passage, making its first appearance in Pennsylvania early in April, among the blossoms of the maple, often accompanied by the ruby-crowned wren, which, except in the markings of the head, it very much resembles. It is very frequent among evergreens, such as the pine, spruce,

gold-crested regulus, I had long felt convinced, from the great and sudden increase of the species during the autumnal and hyemal months, that our indigenous birds must be augmented by a body of strangers making these shores their winter's resort.

"A more extraordinary circumstance in the economy of this bird took place during the same winter (Memoirs of Wernerian Society, vol. v. p. 397), viz., the total disappearance of the whole tribe, *natives* as well as strangers, throughout Scotland and the north of England. This happened towards the conclusion of the month of January 1823, and a few days previous to the long-continued snowstorm so severely felt through the northern counties of England, and along the eastern parts of Scotland. The range and point of this migration are unascertained, but it must probably have been a distant one, from the fact of not a single pair having returned to breed, or pass the succeeding summer, in the situations they had been known always to frequent. Nor was one of the species to be seen till the following October, or about the usual time, as I have above stated, for our receiving an annual accession of strangers to our own indigenous birds."

They are chiefly if not entirely insectivorous, and very nimble and agile in search after their prey. They build their nests with great art; that of this country has it usually suspended near the extremity of a branch, and the outside beautifully covered with different mosses, generally similar to those growing upon the tree on which they build. In colours and the distribution of them, they closely agree, and all possess the beautiful golden crown, the well-known and admired mark of their common name. Our own island possesses only one, and though strong hopes have lately been raised of finding the second European species, *R. ignicapillus*, our endeavours have hitherto been unsuccessful. But I do not yet despair; they are so closely allied that a very near inspection is necessary to determine the individuals.

Mr Audubon has described and figured a bird under the name of *R. Cuvierii*, which may prove an addition to this genus. Only a single specimen was procured in Pennsylvania, and the species will rest on Mr Audubon's plate alone, until some others are obtained. The centre of the crest is described and represented of a rich vermilion.—ED.

cedar, juniper, &c., and in the fall is generally found in company with the two species of titmouse, brown creeper, and small spotted woodpecker. It is an active, unsuspicious, and diligent little creature, climbing and hanging, occasionally, among the branches, and sometimes even on the body of the tree, in search of the larvæ of insects attached to the leaves and stems, and various kinds of small flies, which it frequently seizes on wing. As it retires still farther north to breed, it is seldom seen in Pennsylvania from May to October, but is then numerous in orchards, feeding among the leaves of the apple trees, which, at that season, are infested with vast numbers of small black-winged insects. Its chirp is feeble, not much louder than that of a mouse; though, where it breeds, the male is said to have a variety of sprightly notes. It builds its nest frequently on the branches of an evergreen, covers it entirely round, leaving a small hole on one side for entrance, forming it outwardly of moss and lichens, and lining it warmly with down. The female lays six or eight eggs, pure white, with a few minute specks of dull red. Dr Latham, on whose authority this is given, observes, "It seems to frequent the oak trees in preference to all others. I have more than once seen a brood of these in a large oak, in the middle of a lawn, the whole little family of which, as soon as able, were in perpetual motion, and gave great pleasure to many who viewed them. The nest of one of these has also been made in a garden on a fir tree; it was composed of moss, the opening on one side, in shape roundish; it was lined with a downy substance, fixed with small filaments. It is said to sing very melodiously, very like the common wren, but weaker."* In Pennsylvania, they continue with us from October to December, and sometimes to January.

The golden-crested wren is four inches long, and six inches and a half in extent; back, a fine yellow olive; hind head and sides of the neck, inclining to ash; a line of white passes round the frontlet, extending over and beyond the eye on each side;

* Synopsis, ii. 509.

above this, another line or strip of deep black passes in the same manner, extending farther behind; between these two strips of black lies a bed of glossy golden yellow, which, being parted a little, exposes another of a bright flame colour, extending over the whole upper part of the head; when the little warbler flits among the branches in pursuit of insects, he opens and shuts this golden ornament with great adroitness, which produces a striking and elegant effect; lores, marked with circular points of black; below the eye is a rounding spot of dull white; from the upper mandible to the bottom of the ear-feathers runs a line of black, accompanied by another of white, from the lower mandible; breast, light cream colour; sides under the wings, and vent, the same; wings, dusky, edged exteriorly with yellow olive; greater wing-coverts, tipt with white, immediately below which a spot of black extends over several of the secondaries; tail, pretty long, forked, dusky, exterior vanes broadly edged with yellow olive; legs, brown, feet and claws, yellow; bill, black, slender, straight, evidently of the *Muscicapa* form, the upper mandible being notched at the point, and furnished at the base with bristles, that reach half way to its point; but what seems singular and peculiar to this little bird, the nostril on each side is covered by a single feather, that much resembles the antennæ of some butterflies, and is half the length of the bill. Buffon has taken notice of the same in the European. Inside of the mouth, a reddish orange; claws, extremely sharp, the hind one the longest. In the female, the tints and markings are nearly the same, only the crown or crest is pale yellow. These birds are numerous in Pennsylvania in the month of October, frequenting bushes that overhang streams of water, alders, briers, and particularly apple trees, where they are eminently useful in destroying great numbers of insects, and are at that season extremely fat.

HOUSE WREN. (*Sylvia domestica.*)

PLATE VIII.—FIG. 3.

Motacilla domestica (Regulus rufus), *Bartram*, 291.—*Peale's Museum*, No. 7283.

TROGLODYTES ŒDON.—VIEILLOT.

Troglodytes œdon, *Bonap. Synop.* p. 93, and note p. 439.—*North. Zool.* ii. p. 316.—The House Wren, *Aud.* pl. 83. *Orn. Biog.* i. 427.

THIS well-known and familiar bird arrives in Pennsylvania about the middle of April; and about the 8th or 10th of May, begins to build its nest, sometimes in the wooden cornice under the eaves, or in a hollow cherry tree, but most commonly in small boxes fixed on the top of a pole in or near the garden, to which he is extremely partial, for the great number of caterpillars and other larvæ with which it constantly supplies him. If all these conveniencies are wanting, he will even put up with an old hat, nailed on the weather boards, with a small hole for entrance; and, if even this be denied him, he will find some hole, corner, or crevice about the house, barn, or stable, rather than abandon the dwellings of man. In the month of June, a mower hung up his coat under a shed near a barn; two or three days elapsed before he had occasion to put it on again; thrusting his arm up the sleeve, he found it completely filled with some rubbish, as he expressed it, and, on extracting the whole mass, found it to be the nest of a wren completely finished, and lined with a large quantity of feathers. In his retreat, he was followed by the little forlorn proprietors, who scolded him with great vehemence for thus ruining the whole economy of their household affairs. The twigs with which the outward parts of the nest are constructed are short and crooked, that they may the better hook in with one another, and the hole or entrance is so much shut up, to prevent the intrusion of snakes or cats, that it appears almost impossible the body of the bird could be admitted; within this is a layer of fine dried stalks of grass, and lastly feathers. The eggs are six or seven, and sometimes nine, of a red purplish flesh colour, innumerable fine grains of that tint being thickly sprinkled over the whole egg. They generally raise two broods in a season; the first about the beginning of June, the second in July.*

* The wrens figured on this plate, and indeed all those of this northern continent, seem to be great favourites with the country people, to which distinction their utility in gardens in destroying caterpillars and noxious insects, their sprightly social manner, with their clean and neat appearance, fully entitle them. They form the genus *Troglodytes* of moderns, are limited in numbers, but distributed over Europe, America, and Africa; their habits are nearly alike, and the colours of the plumage are so similar, that some species are with difficulty distinguished from each other; and both those now figured have been confounded with that of this country, from which, however, the first differs, and the latter is still doubtful. The colours of the plumage are brown, with bars and crossings of darker shades, intermingled occasionally with spots and irregular blotches of yellowish white. They make very commodious nests, with a single entrance; all those with which we are acquainted are very prolific, breed more than once in the year, and lay at a time from twelve to sixteen eggs; they are always to be met with, but never in such profusion as their numerous broods would lead us to infer if all arrived at maturity. That of this country, though not so tame as to make use of a *ready-made* convenient breeding-place, is extremely familiar, and will build close by a window, or above a door, where there is a constant thoroughfare. It roosts, during the night, in holes of banks, ricks, or in the eaves of thatched houses, and generally seven or eight individuals will occupy one hole, flitting about, and disputing, as it were, which should enter first. These are beautiful provisions for their welfare, and the proportion of animal heat possessed necessarily by so small a bulk. Another curious particular in the economy of these little birds, is the many useless nests which are built, or, as they are sometimes called by boys, *cock nests*. These are never built so carefully, or in such private and recluse situations, as those intended for incubation, and are even sometimes left in an unfinished half-built state. I have never been able to satisfy myself whether they were the work of the male bird only, or of both conjointly; or to ascertain their use, whether really commenced with the view of breeding in them, or for roosting-places. The generally exposed situation in which they are placed, with the concealed spot chosen for those that have young, would argue against the former and the latter would, perhaps, require a greater reasoning power than most people would be willing to grant to this animal. They may, perhaps, be the first instinctive efforts of the young. Notwithstanding their

This little bird has a strong antipathy to cats; for, having frequent occasion to glean among the currant bushes, and other shrubbery in the garden, those lurking enemies of the feathered race often prove fatal to him. A box fitted up in the window of the room where I slept was taken possession of by a pair of wrens. Already the nest was built, and two eggs laid, when one day, the window being open, as well as the room door, the female wren, venturing too far into the room to reconnoitre, was sprung upon by grimalkin, who had planted herself there for the purpose; and, before relief could be given, was destroyed. Curious to see how the survivor would demean himself, I watched him carefully for several days. At first he sung with great vivacity for an hour or so, but, becoming uneasy, went off for half an hour; on his return, he chanted again as before, went to the top of the house, stable, and weeping willow, that she might hear him; but seeing no appearance of her, he returned once more, visited the nest,

small bulk, and tender-looking frame, they are very hardy, and brave the severest winters of this country; driven nearer to our houses from the necessity of food, they seem to rejoice in a hard clear frost, singing merrily on the top of some heap of brushwood, or sounding, in rapid succession, their note of alarm, when disturbed by any unwelcome visitor. A kitty-hunt, in a snowstorm, used to be a favourite amusement with boys; and many a tumble was got in the unseen ruggedness of the ground when in pursuit. At any time when annoyed, a hole, or thick heap of sticks, will form a refuge for this curious little bird, where it will either remain quiet until the danger is over, or, if there is any under way, will creep and run, escaping at another side; in like manner, it will duck and dive in the openings or hollows of the snow, and at the moment when capture seems inevitable, will escape at some distant opening, disappointing the hopes of the urchin who already anticipated possession.

We must here mention, in addition to the already described North American species, one figured by Mr Audubon, and dedicated to an artist, who will be long remembered by the British ornithologist, *Troglodytes Bewickii.* Mr Audubon has killed three specimens of it in Louisiana, and observes, "In shape, form, colour, and movements, it nearly resembles the great Carolina wren, and forms a kind of link between that bird and the house wren. It has not the quickness of motion, nor the liveliness of either of these birds."—Ed.

ventured cautiously into the window, gazed about with suspicious looks, his voice sinking to a low, melancholy note, as he stretched his little neck about in every direction. Returning to the box, he seemed for some minutes at a loss what to do, and soon after went off, as I thought, altogether, for I saw him no more that day. Towards the afternoon of the second day, he again made his appearance, accompanied with a new female, who seemed exceedingly timorous and shy, and who, after great hesitation, entered the box; at this moment the little widower or bridegroom seemed as if he would warble out his very life with ecstasy of joy. After remaining about half a minute in, they both flew off, but returned in a few minutes, and instantly began to carry out the eggs, feathers, and some of the sticks, supplying the place of the two latter with materials of the same sort; and ultimately succeeded in raising a brood of seven young, all of which escaped in safety.

The immense number of insects which this sociable little bird removes from the garden and fruit trees, ought to endear him to every cultivator, even if he had nothing else to recommend him; but his notes, loud, sprightly, tremulous, and repeated every few seconds with great animation, are extremely agreeable. In the heat of summer, families in the country often dine under the piazza, adjoining green canopies of mantling grape vines, gourds, &c., while overhead the thrilling vivacity of the wren, mingled with the warbling mimicry of the mocking bird, and the distant, softened sounds of numerous other songsters, that we shall hereafter introduce to the reader's acquaintance, form a soul-soothing and almost heavenly music, breathing peace, innocence, and rural repose. The European who judges of the song of this species by that of his own wren (*M. troglodytes*), will do injustice to the former, as in strength of tone and execution it is far superior, as well as the bird is in size, figure, and elegance of markings, to the European one. Its manners are also different; its sociability greater. It is no underground inhabitant; its nest is differently constructed, the number of its eggs fewer; it is also migratory; and has the tail and bill much longer. Its food is insects and caterpillars, and, while supplying the wants of its young, it destroys, on a moderate calculation, many hundreds a day, and greatly circumscribes the ravages of these vermin. It is a bold and insolent bird against those of the titmouse and woodpecker kind that venture to build within its jurisdiction; attacking them without hesitation, though twice its size, and generally forcing them to decamp. I have known him drive a pair of swallows from their newly formed nest, and take immediate possession of the premises, in which his female also laid her eggs, and reared her young. Even the blue bird, who claims an equal, and sort of hereditary right to the box in the garden, when attacked by this little impertinent, soon relinquishes the contest, the mild placidness of his disposition not being a match for the fiery impetuosity of his little antagonist. With those of his own species who settle and build near him, he has frequent squabbles; and when their respective females are sitting, each strains his whole powers of song to excel the other. When the young are hatched, the hurry and press of business leave no time for disputing, so true it is that idleness is the mother of mischief. These birds are not confined to the country; they are to be heard on the tops of the houses in the most central parts of our cities, singing with great energy. Scarce a house or cottage in the country is without at least a pair of them, and sometimes two; but unless where there is a large garden, orchard, and numerous outhouses, it is not often the case that more than one pair reside near the same spot, owing to their party disputes and jealousies. It has been said by a friend to this little bird, that "the esculent vegetables of a whole garden may, perhaps, be preserved from the depredations of different species of insects, by ten or fifteen pair of these small birds;"* and probably they might, were the combination practicable; but such a congregation of wrens about one garden is a

* Barton's Fragments, part i. p. 22.

phenomenon not to be expected but from a total change in the very nature and disposition of the species.

Having seen no accurate description of this bird in any European publication, I have confined my references to Mr Bartram and Mr Peale; but though Europeans are not ignorant of the existence of this bird, they have considered it, as usual, merely as a slight variation from the original stock (*M. troglodytes*), their own wren, in which they are, as usual, mistaken; the length and bent form of the bill, its notes, migratory habits, long tail, and red eggs, are sufficient specific differences.

The house wren inhabits the whole of the United States, in all of which it is migratory. It leaves Pennsylvania in September; I have sometimes, though rarely, seen it in the beginning of October. It is four inches and a half long, and five and three quarters in extent, the whole upper parts of a deep brown, transversely crossed with black, except the head and neck, which is plain; throat, breast, and cheeks, light clay colour; belly and vent, mottled with black, brown, and white; tail, long, cuneiform, crossed with black; legs and feet, light clay colour; bill, black, long, slightly curved, sharp pointed, and resembling that of the genus *Certhia* considerably; the whole plumage below the surface is bluish ash; that on the rump having large round spots of white, not perceivable unless separated with the hand. The female differs very little in plumage from the male.

BLACK-CAPT TITMOUSE. (*Parus atricapillus.*)

PLATE VIII.—Fig. 4.

Parus atricapillus, *Linn. Syst.* i. 341, 6.—*Gmel. Syst.* i. 1008.—La mésange à tête noire de Canada, *Buffon*, v. 408.—Canada Titmouse, *Arct. Zool.* i. No. 328.—*Lath. Syn.* iv. 542, 9.—*Peale's Museum*, No. 7380.

PARUS ATRICAPILLUS.—Linnæus.*

Parus atricapillus, *Bonap. Synop.* p. 100.—*North. Zool.* p. 226.

This is one of our resident birds, active, noisy, and restless; hardy beyond any of his size, braving the severest cold of our continent as far north as the country round Hudson's Bay, and always appearing most lively in the coldest weather. The males have a variety of very sprightly notes, which cannot, indeed, be called a song, but rather a lively, frequently repeated, and often varied twitter. They are most usually seen during the fall and winter, when they leave the depths of the woods, and approach nearer to the scenes of cultivation. At such seasons, they abound among evergreens, feeding on the seeds of the pine tree; they are also fond of sunflower seeds, and associate in parties of six, eight, or more, attended by the two species of nuthatch already described, the crested titmouse, brown creeper, and small spotted woodpecker; the whole forming a very nimble and restless company, whose food, manners, and dispositions are pretty much alike. About the middle of April they begin to build, choosing the deserted hole of a squirrel or woodpecker, and sometimes, with incredible labour, digging out one for themselves. The female lays six white eggs, marked with minute specks of red; the first brood appear about the beginning of June, and the second towards the end of July; the whole of the family continue to

* This is very closely allied to the *Parus palustris*, the marsh titmouse of Europe; but it is exclusively American, and ranges extensively to the north. The authors of the "Northern Zoology" mention them as one of the most common birds in the fur countries; a family inhabits almost every thicket.—Ed.

associate together during winter. They traverse the woods in regular progression, from tree to tree, tumbling, chattering, and hanging from the extremities of the branches, examining about the roots of the leaves, buds, and crevices of the bark for insects and their larvæ. They also frequently visit the orchards, particularly in fall, the sides of the barn and barn-yard, in the same pursuit, trees in such situations being generally much infested with insects. We, therefore, with pleasure, rank this little bird among the farmer's friends, and trust our rural citizens will always recognise him as such.

This species has a very extensive range; it has been found on the western coast of America, as far north as lat. 62°; it is common at Hudson's Bay, and most plentiful there during winter, as it then approaches the settlements in quest of food. Protected by a remarkably thick covering of long, soft, downy plumage, it braves the severest cold of those northern regions.

The black-capt titmouse is five inches and a half in length, and six and a half in extent; throat, and whole upper part of the head and ridge of the neck, black; between these lines a triangular patch of white, ending at the nostril; bill, black and short; tongue, truncate; rest of the upper parts, lead coloured or cinereous, slightly tinged with brown; wings, edged with white; breast, belly, and vent, yellowish white; legs, light blue; eyes, dark hazel. The male and female are nearly alike. The figure in the plate renders any further description unnecessary.

The upper parts of the head of the young are for some time of a dirty brownish tinge; and in this state they agree so exactly with the *Parus Hudsonicus*,* described by Latham, as to afford good grounds for suspecting them to be the same.

These birds sometimes fight violently with each other, and are known to attack young and sickly birds, that are incapable of resistance, always directing their blows against the skull.† Being in the woods one day, I followed a bird for some time,

* Hudson Bay titmouse, Synopsis, ii. 557.

† I have frequently heard this stated regarding the British titmice, particularly the greater, but I have never been able to trace it to any

the singularity of whose notes surprised me. Having shot him from off the top of a very tall tree, I found it to be the black-headed titmouse, with a long and deep indentation in the cranium, the skull having been evidently, at some former time, drove in and fractured, but was now perfectly healed. Whether or not the change of voice could be owing to this circumstance, I cannot pretend to decide.

CRESTED TITMOUSE. (*Parus bicolor.*)

PLATE VIII.—FIG. 5.

Parus bicolor, *Linn. Syst.* i. 544, 1.—La mésange huppée de la Caroline, *Buff.* v. 451.—Toupet Titmouse, *Arct. Zool.* i. No. 324.—*Lath. Syn.* iv. 544, 11.—*Peale's Museum*, No. 7364.

PARUS BICOLOR.—LINNÆUS.

Parus bicolor, *Bonap. Synop.* p. 100.—The Crested Titmouse, *Aud.* pl. 39. *Orn. Biog.* i. p. 198.

THIS is another associate of the preceding species; but more noisy, more musical, and more suspicious, though rather less active. It is, nevertheless, a sprightly bird, possessing a remarkable variety in the tones of its voice, at one time not much louder than the squeaking of a mouse, and a moment after whistling aloud, and clearly, as if calling a dog; and continuing this dog-call through the woods for half an hour at a time. Its high-pointed crest, or, as Pennant calls it, *toupet*, gives it a smart and not inelegant appearance. Its food corresponds with that of the foregoing; it possesses considerable strength in the muscles of its neck, and is almost perpetually digging into acorns, nuts, crevices, and rotten parts of the bark, after the larvæ of insects. It is also a constant resident here. When shot at and wounded, it fights with great spirit. When confined to a cage, it soon becomes familiar, and will subsist on hemp-seed, cherry stones, apple

authentic source ; it is perhaps exaggerated. Feeding on carrion, which they have also been represented to do, must in a wild state be from necessity. Mr Audubon asserts it as a fact, with regard to the *P. bicolor.* Mr Selby has seen *P. major* eat young birds.—ED.

seeds, and hickory nuts, broken and thrown into it. However, if the cage be made of willows, and the bird not much hurt, he will soon make his way through them. The great concavity of the lower side of the wings and tail of this genus of birds, is a strong characteristic, and well suited to their short irregular flight.

This species is also found over the whole United States; but is most numerous towards the north. It extends also to Hudson's Bay; and, according to Latham, is found in Denmark, and in the southern parts of Greenland, where it is called *Avingarsak.* If so, it probably inhabits the continent of North America from sea to sea.

The crested titmouse is six inches long, and seven inches and a half in extent. The whole upper parts, a dull cinereous or lead colour, except the front, which is black, tinged with reddish; whole lower parts, dirty white, except the sides under the wings, which are reddish orange; legs and feet, light blue; bill, black, short, and pretty strong; wing-feathers, relieved with dusky on their inner vanes; eye, dark hazel; lores, white; the head, elegantly ornamented with a high, pointed, almost upright crest; tail, a little forked, considerably concave below, and of the same colour above as the back; tips of the wings, dusky; tongue, very short, truncate, and ending in three or four sharp points. The female cannot be distinguished from the male by her plumage, unless in its being something duller, for both are equally marked with reddish orange on the sides under the wings, which some foreigners have made the distinguishing mark of the male alone.

The nest is built in a hollow tree, the cavity often dug by itself; the female begins to lay early in May; the eggs are usually six, pure white, with a few very small specks of red near the great end. The whole family, in the month of July, hunt together, the parents keeping up a continual chatter, as if haranguing and directing their inexperienced brood.*

* This beautiful and attractive race of birds, the genuine titmice, have a geographical distribution over the whole world,—South America

WINTER WREN. (*Sylvia troglodytes.*)

PLATE VIII.—FIG. 6.

Motacilla troglodytes? *Linn.*—*Peale's Museum*, No. 7284.

TROGLODYTES HYEMALIS?—VIEILLOT.

Troglodytes Europeus Leach, *Bonap. Synop.* p. 93.—Troglodytes hyemalis, *Vieill. Encyc. Méth.* ii. p. 470.—*North. Zool.* ii. p. 318.

THIS little stranger visits us from the north in the month of October, sometimes remaining with us all the winter, and is always observed, early in spring, on his route back to his breeding place. In size, colour, song, and manners, he

New Holland, and the islands in the South Pacific ocean excepted. In the latter countries, they seem represented by the genus *Pardalotus*, yet, however, very limited in numbers. They are more numerous in temperate, and even northern climates, than near the tropics; the greater numbers, both as to individuals and species, extend over Europe. In this country, when the want of foliage allows us to examine their manners, they form one of the most interesting of our winter visitants. I call them *visitants* only; for during summer they are occupied with the duties of incubation in retirement, amid the depths of the most solitary forests; and only at the commencement of winter, or during its rigours, become more domesticated, and flock in small parties, the amount of their broods, to our gardens and the vicinity of our houses; several species together, and generally in company with the gold-crested wrens. The activity of their motions in search of food, or in dispute with one another; the variety of their cries, from something very shrill and timid to loud and wild; their sometimes elegant, sometimes grotesque attitudes, contrasted by the difference of form; and the varied flights, from the short dart and jerk of the marsh and cole titmouse, or gold-crested wren, to the stringy successive line of the long-tailed one,—are objects which have, no doubt, called forth the notice of the ornithologist who has sometimes allowed himself to examine them in their natural abodes. The form of the different species is nearly alike, thick-set, stout, and short, the legs comparatively strong, the whole formed for active motion, and uniting strength for the removal of loose bark, moss, or even rotten wood, in search of their favourite food, insects; it, however, varies in two species of this country (one of which will form a separate subdivision), the long-tailed and the bearded titmice (*P. caudatus* and *biarmicus*), in the weaker frame and more lengthened shape of the tail; and it may be approaches nearer to the European wren (*M. troglodytes*) than any other species we have. During his residence here, he frequents the projecting banks of creeks, old roots, decayed

remarked, that both these make suspended nests, the one in woods, of a lengthened form and beautiful workmanship, generally hung near the extremity of a branch belonging to some thick silver, spruce, or Scotch fir ; the other balanced and waving among reeds, like some of the aquatic warblers ; while all the other species, and indeed all those abroad with whose nidification I am acquainted, choose some hollow tree or rent wall for their place of breeding. In a Brazilian species, figured by Temminck, the tail assumes a forked shape.

Insects are not their only food, though perhaps the most natural. When the season becomes too inclement for this supply, they become granivorous, and will plunder the farmyards, or eat grain and potatoes with the poultry and pigs. Some I have seen so domesticated (the common blue and greater titmice), as to come regularly during the storm to the windows for crumbs of bread. When confined, they become very docile, and will also eat pieces of flesh or fat. During winter, they roost in holes of trees or walls, eaves of thatched houses, or hay and corn ricks. When not in holes, they remain suspended with the back downwards or outwards. A common blue tomtit (and I have no doubt the same individual) has roosted for three years in the same spot, under one of the projecting capitals of a pillar, by the side of my own front door. The colours of the group are chaste and pleasing, as might have been expected from their distribution. There are, however, one or two exceptions in those figured by M. Temminck, from Africa. The general shades are black, gray, white, blue, and different tints of olive, sometimes reddish brown; and in these, when the brightest colours occur, the blue and yellow, they are so blended as not to be hard or offensive. Most of the species have some decided marks or colouring about the head, and the plumage is thick and downy and loose—a very necessary requisite to those which frequent the more northern latitudes.

Mr Audubon says, that this species sometimes forms a nest, by digging a hole for the purpose in the hardest wood with great industry and perseverance, although it is more frequently contented with the hole of the downy woodpecker, or some other small bird of that genus. We can hardly conceive that the crested titmouse, or indeed any of the race, had sufficient strength to dig its own nest. The bill, though very powerful when compared with the individual's bulk, is not formed on the principle of those which excavate for themselves. I lately received the nest of this species, taken from some hollow tree. The inside lining was almost entirely composed of the scales and cast-off exuvia of snakes.—ED.

logs, small bushes, and rushes near watery places; he even approaches the farmhouse, rambles about the wood pile, creeping among the interstices like a mouse. With tail erect, which is his constant habit, mounted on some projecting point or pinnacle, he sings with great animation. Even in the yards, gardens, and outhouses of the city, he appears familiar and quite at home. In short, he possesses almost all the habits of the European species. He is, however, migratory, which may be owing to the superior coldness of our continent. Never having met with the nest and eggs, I am unable to say how nearly they approximate to those of the former.

I can find no precise description of this bird, as an American species, in any European publication. Even some of our own naturalists seem to have confounded it with another very different bird, the marsh wren,* which arrives in Pennsylvania from the south in May, builds a globular or pitcher-shaped nest, which it suspends among the rushes and bushes by the river-side, lays five or six eggs of a dark fawn colour, and departs again in September. But the colours and markings of that bird are very unlike those of the winter wren, and its song altogether different. The circumstance of the one arriving from the north as the other returns to the south, and *vice versa,* with some general resemblance between the two, may have occasioned this mistake. They, however, not only breed in different regions, but belong to different genera, the marsh wren being decisively a species of *Certhia*, and the winter wren a true *Motacilla*. Indeed we have no less than five species of these birds in Pennsylvania, that, by a superficial observer, would be taken for one and the same; but between each of which nature has drawn strong, discriminating, and indelible lines of separation. These will be pointed out in their proper places.

If this bird, as some suppose, retires only to the upper regions of the country and mountainous forests to breed, as

* See Professor Barton's observations on this subject, under the article *Motacilla troglodytes?* Fragments, &c., p. 18; *ibid.* p. 12.

is the case with some others, it will account for his early and frequent residence along the Atlantic coast during the severest winters; though I rather suspect that he proceeds considerably to the northward; as the snow bird (*F. Hudsonia*), which arrives about the same time with the winter wren, does not even breed at Hudson's Bay, but passes that settlement in June, on his way to the northward; how much farther is unknown.

The length of the winter wren is three inches and a half; breadth, five inches; the upper parts are of a general dark brown, crossed with transverse touches of black, except the upper parts of the head and neck, which are plain; the black spots on the back terminate in minute points of dull white; the first row of wing-coverts is also marked with specks of white at the extremities of the black, and tipt minutely with black; the next row is tipt with points of white; the primaries are crossed with alternate rows of black and cream colour; inner vanes of all the quills, dusky, except the three secondaries next the body; tips of the wings, dusky; throat, line over the eye, sides of the neck, ear-feathers and breast, dirty white, with minute transverse touches of a drab or clay colour; sides under the wings, speckled with dark brown, black, and dirty white; belly and vent, thickly mottled with sooty black, deep brown, and pure white, in transverse touches; tail, very short, consisting of twelve feathers, the exterior one on each side a quarter of an inch shorter, the rest lengthening gradually to the middle ones; legs and feet, a light clay colour, and pretty stout; bill, straight, slender, half an inch long, and not notched at the point, of a dark brown or black above, and whitish below; nostril, oblong; eye, light hazel. The female wants the points of white on the wing-coverts. The food of this bird is derived from that great magazine of so many of the feathered race, insects and their larvæ, particularly such as inhabit watery places, roots of bushes, and piles of old timber.

It were much to be wished that the summer residence, nest,

and eggs of this bird, were precisely ascertained, which would enable us to determine whether it be, what I strongly suspect it is, the same species as the common domestic wren of Britain.*

RED-HEADED WOODPECKER. (*Picus erythrocephalus.*)

PLATE IX.—Fig. 1.

Picus erythrocephalus, *Linn. Syst.* i. 174, 7.—*Gmel. Syst.* i. 429.—Pic noir à domino rouge, *Buffon*, vii. 55. *Pl. enl.* 117.—*Catesby*, i. 20.—*Arct. Zool.* ii. No. 160. —*Lath. Syn.* ii. 561.—*Peale's Museum*, No. 1922.

MELANERPES ERYTHROCEPHALUS.—Swainson.†

Picus erythrocephalus, *Bonap. Synop.* p. 45.—*Wagl. Spec. Av. Picus*, No. 14.—The Red-headed Woodpecker, *Aud.* pl. 27, *Orn. Biog.* i. p. 141.—Melanerpes erythrocephalus, *North. Zool.* ii. p. 316.

There is perhaps no bird in North America more universally known than this. His tricoloured plumage, red, white, and

* There is a very great alliance between the British and American specimens; and all authors who have described this bird and that of Europe, have done so with uncertainty. Wilson evidently had a doubt, both from what he says, and from marking the species and his synonyms with a query. Vieillot had doubts, and Bonaparte goes a good deal on his authority, but points out no difference between the birds. Mr Swainson, in the "Northern Zoology," has described a bird as that of Vieillot's, killed on the shores of Lake Huron, and proves distinctly that the plumage and some of the relative proportions vary. It is likely that there are two American species concerned in this,—one northern, another extending to the south, and that one, perhaps, may be identical with that of Europe: one certainly seems distinct. I have retained *hyemalis* with a mark of doubt, it being impossible to determine those so closely allied without an examination of numerous species.—Ed.

† This will point out another of Mr Swainson's groups among the woodpeckers, equally distinct with *Colaptes*. The form is long and swallow-like; the bill more rounded than angular, the culmen quite round; the wings nearly as long as the tail. In their manners, they are extremely familiar; and during summer, feed almost entirely on the rich fruits and ripe grains of the country. The chaste and simple coloured *Picus bicolor*, from the *Minas Geraies*, I believe, will be another representative of this form.—Ed.

Drawn from Nature by A. Wilson

Engraved by W.H. Lizars

1. Red-headed Woodpecker. 2. Yellow-bellied W. 3. Hairy W. 4. Downy W.

9.

black, glossed with steel blue, is so striking and characteristic, and his predatory habits in the orchards and cornfields, added to his numbers, and fondness for hovering along the fences, so very notorious, that almost every child is acquainted with the red-headed woodpecker. In the immediate neighbourhood of our large cities, where the old timber is chiefly cut down, he is not so frequently found; and yet, at this present time (June 1808), I know of several of their nests within the boundaries of the city of Philadelphia. Two of these are in button-wood trees (*Platanus occidentalis*), and another in the decayed limb of an elm. The old ones, I observe, make their excursions regularly to the woods beyond the Schuylkill, about a mile distant, preserving great silence and circumspection in visiting their nests,—precautions not much attended to by them in the depth of the woods, because there the prying eye of man is less to be dreaded. Towards the mountains, particularly in the vicinity of creeks and rivers, these birds are extremely abundant, especially in the later end of summer. Wherever you travel in the interior at that season, you hear them screaming from the adjoining woods, rattling on the dead limbs of trees, or on the fences, where they are perpetually seen flitting from stake to stake, on the roadside, before you. Wherever there is a tree or trees of the wild cherry covered with ripe fruit, there you see them busy among the branches; and in passing orchards, you may easily know where to find the earliest, sweetest apples, by observing those trees on or near which the red-headed woodpecker is skulking; for he is so excellent a connoisseur in fruit, that wherever an apple or pear is found broached by him, it is sure to be among the ripest and best flavoured: when alarmed, he seizes a capital one by striking his open bill deep into it, and bears it off to the woods. When the Indian-corn is in its rich, succulent, milky state, he attacks it with great eagerness, opening a passage through the numerous folds of the husk, and feeding on it with voracity. The girdled, or deadened timber, so common among cornfields in the back settlements, are his favourite retreats, whence he sallies out to make his depredations. He is fond of the ripe berries of the sour gum, and pays pretty regular visits to the cherry trees when loaded with fruit. Towards fall he often approaches the barn or farmhouse, and raps on the shingles and weather boards: he is of a gay and frolicsome disposition; and half a dozen of the fraternity are frequently seen diving and vociferating around the high dead limbs of some large tree, pursuing and playing with each other, and amusing the passenger with their gambols. Their note or cry is shrill and lively, and so much resembles that of a species of tree-frog which frequents the same tree, that it is sometimes difficult to distinguish the one from the other.

Such are the *vicious* traits, if I may so speak, in the character of the red-headed woodpecker; and I doubt not but, from what has been said on this subject, that some readers would consider it meritorious to exterminate the whole tribe as a nuisance: and, in fact, the Legislatures of some of our provinces, in former times, offered premiums to the amount of twopence per head for their destruction.* But let us not condemn the species unheard: they exist—they must therefore be necessary.† If their merits and usefulness be found, on

* Kalm.

† The abundance of this species must be very great, and, from the depredations they commit, must be more felt. Mr Audubon says that a hundred have been shot in one day from a single cherry tree. In addition to their other bad habits, they carry off apples by thrusting in their bill as a spike, and thus supporting them. They also frequent pigeon-houses, and suck the eggs,—a habit not very common among this tribe; and, for the same purpose, enter the boxes prepared for the martins and blue birds. Another method of adding to their destruction, in Kentucky and the southern States, is in the following manner related by Audubon :—

"As soon as the red-heads have begun to visit a cherry or apple tree, a pole is placed along the trunk of the tree, passing up amongst the central branches, and extending six or seven feet above the highest twigs. The woodpeckers alight by preference on the pole, and whilst their body is close to it, a man, standing at the foot of the pole, gives it a twist below with the head of an axe, on the opposite side to that on which the examination, to preponderate against their vices, let us avail ourselves of the former, while we guard as well as we can against the latter.

Though this bird occasionally regales himself on fruit, yet his natural and most useful food is insects, particularly those numerous and destructive species that penetrate the bark and body of the tree to deposit their eggs and larvæ, the latter of which are well known to make immense havoc. That insects are his natural food is evident from the construction of his wedge-formed bill, the length, elasticity, and figure of his tongue, and the strength and position of his claws, as well as from his usual habits. In fact, insects form at least two-thirds of his subsistence; and his stomach is scarcely ever found without them. He searches for them with a dexterity and intelligence, I may safely say, more than human; he perceives, by the exterior appearance of the bark, where they lurk below; when he is dubious, he rattles vehemently on the outside with his bill, and his acute ear distinguishes the terrified vermin shrinking within to their inmost retreats, where his pointed and barbed tongue soon reaches them. The masses of bugs, caterpillars, and other larvæ, which I have taken from the stomachs of these birds, have often surprised me. These

woodpecker is, when, in consequence of the sudden vibration produced in the upper part, the bird is thrown off dead."

According to the same gentleman, many of the *red-heads* (a name by which they are universally known) remain in the southern districts of the United States during the whole winter. The greater number, however, pass to countries farther south. Their migration takes place during night, is commenced in the middle of September, and continues for a month or six weeks. They then fly high above the trees, far apart, like a disbanded army, propelling themselves by reiterated flaps of their wings at the end of each successive curve which they describe in their flight. The note which they emit at this time is different from the usual one, sharp, and easily heard from the ground, although the birds may be out of sight. At the dawn of day, the whole alight on the tops of the dead trees about the plantations, and remain in search of food until the approach of sunset, when they again, one after another, mount the air, and continue their journey.—Ed.

larvæ, it should be remembered, feed not only on the buds, leaves, and blossoms, but on the very vegetable life of the tree,—the alburnum, or newly forming bark and wood; the consequence is, that whole branches and whole trees decay under the silent ravages of these destructive vermin; witness the late destruction of many hundred acres of pine trees in the northeastern parts of South Carolina; * and the thousands of peach trees that yearly decay from the same cause. Will any one say, that taking half a dozen, or half a hundred, apples from a tree, is equally ruinous with cutting it down? or, that the services of a useful animal should not be rewarded with a small portion of that which it has contributed to preserve? We are told, in the benevolent language of the Scriptures, not to muzzle the mouth of the ox that treadeth out the corn; and why should not the same generous liberality be extended to this useful family of birds, which forms so powerful a phalanx against the inroads of many millions of destructive vermin?

The red-headed woodpecker is, properly speaking, a bird of passage; though, even in the eastern States, individuals are found during moderate winters, as well as in the States of New York and Pennsylvania; in Carolina, they are somewhat more numerous during that season, but not one-tenth of what are found in summer. They make their appearance in Pennsylvania about the 1st of May, and leave us about the middle of October. They inhabit from Canada to the Gulf of Mexico, and are also found on the western coast of North America. About the middle of May they begin to construct their nests, which, like the rest of the genus, they form in the body or large limbs of trees, taking in no materials, but smoothing it within to the proper shape and size. The female lays six eggs, of a pure white, and the young make their first

* In one place, on a tract of two thousand acres of pine land, on the Sampit River, near Georgetown, at least ninety trees in every hundred were destroyed by this pernicious insect,—a small, black-winged bug, resembling the weevil, but somewhat larger.

appearance about the 20th of June. During the first season, the head and neck of the young birds are blackish gray, which has occasioned some European writers to mistake them for females; the white on the wing is also spotted with black; but in the succeeding spring they receive their perfect plumage, and the male and female then differ only in the latter being rather smaller, and its colours not quite so vivid; both have the head and neck deep scarlet; the bill light blue, black towards the extremity, and strong; back, primaries, wing-coverts, and tail, black, glossed with steel blue; rump, lower part of the back, secondaries, and whole under parts from the breast downward, white; legs and feet, bluish green; claws, light blue; round the eye, a dusky narrow skin, bare of feathers; iris, dark hazel; total length, nine inches and a half; extent, seventeen inches. The figure on the plate was drawn and coloured from a very elegant living specimen.

Notwithstanding the care which this bird, in common with the rest of its genus, takes to place its young beyond the reach of enemies, within the hollows of trees, yet there is one deadly foe, against whose depredations neither the height of the tree nor the depth of the cavity is the least security. This is the black snake (*Coluber constrictor*), who frequently glides up the trunk of the tree, and, like a skulking savage, enters the woodpecker's peaceful apartment, devours the eggs or helpless young, in spite of the cries and flutterings of the parents; and, if the place be large enough, coils himself up in the spot they occupied, where he will sometimes remain for several days. The eager schoolboy, after hazarding his neck to reach the woodpecker's hole, at the triumphant moment when he thinks the nestlings his own, and strips his arm, launching it down into the cavity, and grasping what he conceives to be the callow young, starts with horror at the sight of a hideous snake, and almost drops from his giddy pinnacle, retreating down the tree with terror and precipitation. Several adventures of this kind have come to my knowledge; and one of them was attended with serious consequences, where both snake and boy fell to the ground; and a broken thigh, and long confinement, cured the adventurer completely of his ambition for robbing woodpeckers' nests.

YELLOW-BELLIED WOODPECKER. (*Picus varius.*)

PLATE IX.—FIG. 2.

Picus varius, *Linn. Syst.* i. 176, 20.—*Gmel. Syst.* i. 735.—Le pic varié de la Caroline, *Buff.* vii. 77. *Pl. enl.* 785.—Yellow-bellied Woodpecker, *Catesb.* i. 21.—*Arct. Zool.* ii. No. 166.—*Lath. Syn.* ii. 574, 20. *Id. Sup.* p. 109.—*Peale's Museum*, No. 2004.

DENDROCOPUS VARIUS.—SWAINSON.*

Picus varius, *Bonap. Synop.* p. 45.—*Wagl. Syst. Av. Picus*, No. 16.—Dendrocopus varius, *North. Zool.* ii. p. 309.

THIS beautiful species is one of our resident birds. It visits our orchards in the month of October in great numbers, is

* In this species and the two following, the little woodpecker of this country, and many others, we have the types of a subgenus (*Dendrocopus*, Koch) among the woodpeckers, which I have no hesitation in adopting, as containing a very marked group of black and white spotted birds, allied to confusion with each other. The genus is made use of for the first time in a British publication, the "Northern Zoology," by Mr Swainson, as the third subgenus of *Picus*. He thus remarks:—

"The third subgenus comprehends all the smaller black and white spotted woodpeckers of Europe and America. Some few occur in the mountainous parts of India; but, with these exceptions, the group, which is very extensive, seems to belong more particularly to temperate latitudes.

"It was met with by the Overland Expedition in flocks, on the banks of the Saskatchewan, in May. Its manners, at that period of the year, were strikingly contrasted with those of the resident woodpeckers; for, instead of flitting in a solitary way from tree to tree, and assiduously boring for insects, it flew about in crowded flocks in a restless manner, and kept up a continual chattering. Its geographical range is extensive; from the sixty-first parallel of latitude to Mexico."

Mr Swainson mentions having received a single specimen of a woodpecker from Georgia, closely allied to this, which he suspects to be undescribed; and, in the event of being correct, he proposes to dedicate it to Mr Audubon,—*Dendrocopus Audubonii*, Sw.—ED.

occasionally seen during the whole winter and spring, but seems to seek the depths of the forest to rear its young in; for during summer it is rarely seen among our settlements; and even in the intermediate woods, I have seldom met with it in that season. According to Brisson, it inhabits the continent from Cayenne to Virginia; and I may add, as far as to Hudson's Bay, where, according to Hutchins, they are called *Meksewè Paupastaow*; * they are also common in the States of Kentucky and Ohio, and have been seen in the neighbourhood of St Louis. They are reckoned by Georgi among the birds that frequent the Lake Baikal, in Asia; † but their existence there has not been satisfactorily ascertained.

The habits of this species are similar to those of the hairy and downy woodpeckers, with which it generally associates; and which are both represented on the same plate. The only nest of this bird which I have met with was in the body of an old pear tree, about ten or eleven feet from the ground. The hole was almost exactly circular, small for the size of the bird, so that it crept in and out with difficulty; but suddenly widened, descending by a small angle, and then running downward about fifteen inches. On the smooth solid wood lay four white eggs. This was about the 25th of May. Having no opportunity of visiting it afterwards, I cannot say whether it added any more eggs to the number; I rather think it did not, as it appeared at that time to be sitting.

The yellow-bellied woodpecker is eight inches and a half long, and in extent fifteen inches; whole crown, a rich and deep scarlet, bordered with black on each side, and behind forming a slight crest, which it frequently erects; ‡ from the nostrils, which are thickly covered with recumbent hairs, a narrow strip of white runs downward, curving round the breast; mixing with the yellowish white on the lower part of

* Latham. † Ibid.

‡ This circumstance seems to have been overlooked by naturalists.

the breast; throat, the same deep scarlet as the crown, bordered with black, proceeding from the lower mandible on each side, and spreading into a broad rounding patch on the breast; this black, in birds of the first and second year, is dusky gray, the feathers being only crossed with circular touches of black; a line of white, and below it another of black, proceed, the first from the upper part of the eye, the other from the posterior half of the eye, and both lose themselves on the neck and back; back, dusky yellow, sprinkled and elegantly waved with black; wings, black, with a large oblong spot of white; the primaries, tipt and spotted with white; the three secondaries next the body are also variegated with white; rump, white, bordered with black; belly, yellow; sides under the wings, more dusky yellow, marked with long arrow-heads of black; legs and feet, greenish blue; tail, black, consisting of ten feathers, the two outward feathers on each side tipt with white, the next totally black, the fourth edged on its inner vane half way down with white, the middle one white on its interior vane, and spotted with black; tongue, flat, horny for half an inch at the tip, pointed, and armed along its sides with reflected barbs; the other extremities of the tongue pass up behind the skull in a groove, and end near the right nostril; in birds of the first and second year they reach only to the crown; bill, an inch long, channelled, wedge-formed at the tip, and of a dusky horn colour. The female is marked nearly as the male, but wants the scarlet on the throat, which is whitish; she is also darker under the wings and on the sides of the breast. The young of the first season, of both sexes, in October have the crown sprinkled with black and deep scarlet; the scarlet on the throat may be also observed in the young males. The principal food of these birds is insects; and they seem particularly fond of frequenting orchards, boring the trunks of the apple trees in their eager search after them. On opening them, the liver appears very large, and of a dirty gamboge colour; the stomach strongly muscular, and generally filled with fragments of beetles and gravel. In the morning, they are extremely active in the orchards, and rather shyer than the rest of their associates. Their cry is also different, but, though it is easily distinguishable in the woods, cannot be described by words.

HAIRY WOODPECKER. (*Picus villosus.*)

PLATE IX.—FIG. 3.

Picus villosus, *Linn. Syst.* i. 175, 16.—Pic chevelu de Virginie, *Buffon*, vii. 7.—Pic varié male de Virginie, *Pl. enl.* 754.—Hairy Woodpecker, *Catesby*, i. 19, fig. 2.—*Arct. Zool.* ii. No. 164.—*Lath. Syn.* ii. 572, 18. *Id. Sup.* 108.—*Peale's Museum*, No. 1988.

DENDROCOPUS VILLOSUS.—SWAINSON.

Picus villosus, *Bonap. Synop.* p. 46.—*Wagl. Syst. Av. Picus*, 22.—Dendrocopus villosus, *North. Zool.* ii. p. 305.

THIS is another of our resident birds, and, like the former, a haunter of orchards, and borer of apple trees, an eager hunter of insects, their eggs and larvæ, in old stumps and old rails, in rotten branches and crevices of the bark; having all the characters of the woodpecker strongly marked. In the month of May he retires with his mate to the woods, and either seeks out a branch already hollow, or cuts out an opening for himself. In the former case, I have known his nest more than five feet distant from the mouth of the hole; and in the latter he digs first horizontally, if in the body of the tree, six or eight inches, and then downward, obtusely, for twice that distance; carrying up the chips with his bill, and scraping them out with his feet. They also not unfrequently choose the orchard for breeding in, and even an old stake of the fence, which they excavate for this purpose. The female lays five white eggs, and hatches in June. This species is more numerous than the last in Pennsylvania, and more domestic; frequently approaching the farmhouse and skirts of the town. In Philadelphia I have many times observed them examining old ragged trunks of the willow and poplar while people were passing immediately below. Their cry is strong, shrill, and tremulous; they have also a single note, or *chuck*, which they often repeat, in an eager manner, as they hop about, and dig into the crevices of the tree. They inhabit the continent from Hudson's Bay to Carolina and Georgia.

The hairy woodpecker is nine inches long and fifteen in extent; crown, black; line over and under the eye, white; the eye is placed in a black line, that widens as it descends to the back; hind head, scarlet, sometimes intermixed with black; nostrils, hid under remarkably thick, bushy, recumbent hairs or bristles; under the bill are certain long hairs thrown forward and upward, as represented in the figure; bill, a bluish horn colour, grooved, wedged at the end, straight, and about an inch and a quarter long; touches of black, proceeding from the lower mandible, end in a broad black strip that joins the black on the shoulder; back, black, divided by a broad lateral strip of white, the feathers composing which are loose and unwebbed, resembling hairs,—whence its name; rump and shoulders of the wing, black; wings, black, tipped and spotted with white, three rows of spots being visible on the secondaries, and five on the primaries; greater wing-coverts, also spotted with white; tail, as in the others, cuneiform, consisting of ten strong shafted and pointed feathers, the four middle ones black, the next partially white, the two exterior ones white, tinged at the tip with a brownish burnt colour; tail-coverts, black; whole lower side, pure white; legs, feet, and claws, light blue, the latter remarkably large and strong; inside of the mouth, flesh coloured; tongue, pointed, beset with barbs, and capable of being protruded more than an inch and a half; the os hyöides, in this species, passes on each side of the neck, ascends the skull, passes down towards the nostril, and is wound round the bone of the right eye, which projects considerably more than the left for its accommodation. The great mass of hairs that cover the nostril appears to be designed as a protection to the front of the head when the bird is engaged in digging holes into the wood. The membrane which encloses the brain, in this, as in all the other species of woodpeckers, is also of extraordinary strength, no doubt to prevent any bad effects from violent concussion while the bird is employed in digging for food. The female wants the red on the hind head; and the white below is tinged with brownish. The manner of flight of these birds has been already described under a former species, as consisting of alternate risings and sinkings. The hairy woodpeckers generally utter a loud tremulous scream as they set off, and when they alight. They are hard to kill; and, like the red-headed woodpecker, hang by the claws, even of a single foot, as long as a spark of life remains, before they drop.

This species is common at Hudson's Bay, and has lately been found in England.* Dr Latham examined a pair which were shot near Halifax, in Yorkshire; and, on comparing the male with one brought from North America, could perceive no difference, but in a slight interruption of the red that marked the hind head of the former; a circumstance which I have frequently observed in our own. The two females corresponded exactly.

* This, I believe, is a mistake; and although this bird is beginning to creep into our fauna in the rank of an occasional visitant, I can find no authentic trace of the hairy woodpecker being ever killed in Great Britain. It is a bird belonging to a northern climate; and although it closely resembles a native species, it can never be mistaken, with any ordinary examination or comparison. The halifax in Yorkshire will turn out in reality the halifax of the New World.—Ed.

DOWNY WOODPECKER. (*Picus pubescens.*)

PLATE IX.—FIG. 4.

Picus pubescens, *Linn. Syst.* i. 175, 15.—*Gmel. Syst.* i. 435.—Petit pic varié de Virginie, *Buffon*, vii. 76.—Smallest Woodpecker, *Catesby*, i. 21.—*Arct. Zool.* ii. No. 963.—Little Woodpecker, *Lath. Synop.* ii. 573, 19. *Id. Sup.* 106.—*Peale's Museum*, No. 1986.

DENDROCOPUS PUBESCENS.—SWAINSON.

Picus pubescens, *Bonap. Synop.* p. 46.—*Wagl. Syst. Av. Picus*, No. 23.—Dendrocopus, pubescens, *North. Zool.* ii. p. 307.

THIS is the smallest of our woodpeckers,* and so exactly resembles the former in its tints and markings, and in almost everything except its diminutive size, that I wonder how it passed through the Count de Buffon's hands without being branded as a "spurious race, degenerated by the influence of food, climate, or some unknown cause." But though it has escaped this infamy, charges of a much more heinous nature have been brought against it, not only by the writer above mentioned, but by the whole venerable body of zoologists in Europe, who have treated of its history, viz., that it is almost constantly boring and digging into apple trees, and that it is the most destructive of the whole genus to the orchards. The

* This species, as Wilson observes, is the smallest of the American woodpeckers, and it will fill the place in that country which is occupied in Europe and Great Britain by the *Picus minor*, or least woodpecker; unlike the latter, however, it is both abundant, and is familiar in its manners.

Mr Swainson, in a note to the "Northern Zoology," thinks that several American species are confounded under this. "We have no doubt," he says, "that two, if not three, species of these little woodpeckers, from different parts of North America, have been confounded under the common name of *pubescens.*" He proposes to distinguish them by the names of *Dendrocopus medianus*, inhabiting the middle parts of North America, chiefly different from *D. pubescens* in the greater portion of red on the hind head, relative length of the quills, and shape of the tail-feathers; and *Dendrocopus meridionalis*, inhabiting Georgia, less than *D. pubescens*, and with the under plumage hair brown.—ED.

first part of this charge I shall not pretend to deny; how far the other is founded in truth will appear in the sequel. Like the two former species, it remains with us the whole year. About the middle of May, the male and female look out for a suitable place for the reception of their eggs and young. An apple, pear, or cherry tree, often in the near neighbourhood of the farmhouse, is generally fixed upon for this purpose. The tree is minutely reconnoitered for several days previous to the operation, and the work is first begun by the male, who cuts out a hole in the solid wood as circular as if described with a pair of compasses. He is occasionally relieved by the female, both parties working with the most indefatigable diligence. The direction of the hole, if made in the body of the tree, is generally downwards, by an angle of thirty or forty degrees, for the distance of six or eight inches, and then straight down for ten or twelve more; within roomy, capacious, and as smooth as if polished by the cabinetmaker; but the entrance is judiciously left just so large as to admit the bodies of the owners. During this labour, they regularly carry out the chips, often strewing them at a distance to prevent suspicion. This operation sometimes occupies the chief part of a week. Before she begins to lay, the female often visits the place, passes out and in, examines every part both of the exterior and interior with great attention, as every prudent tenant of a new house ought to do, and at length takes complete possession. The eggs are generally six, pure white, and laid on the smooth bottom of the cavity. The male occasionally supplies the female with food while she is sitting; and about the last week in June the young are perceived making their way up the tree, climbing with considerable dexterity. All this goes on with great regularity where no interruption is met with; but the house wren, who also builds in the hollow of a tree, but who is neither furnished with the necessary tools nor strength for excavating such an apartment for himself, allows the woodpeckers to go on, till he thinks it will answer his purpose, then attacks them with violence, and generally succeeds in driving them off. I saw some weeks ago a striking example of this, where the woodpeckers we are now describing, after commencing in a cherry tree within a few yards of the house, and having made considerable progress, were turned out by the wren; the former began again on a pear tree in the garden, fifteen or twenty yards off, whence, after digging out a most complete apartment, and one egg being laid, they were once more assaulted by the same impertinent intruder, and finally forced to abandon the place.

The principal characteristics of this little bird are diligence, familiarity, perseverance, and a strength and energy in the head and muscles of the neck, which are truly astonishing. Mounted on the infected branch of an old apple tree, where insects have lodged their corroding and destructive brood in crevices between the bark and wood, he labours sometimes for half an hour incessantly at the same spot, before he has succeeded in dislodging and destroying them. At these times you may walk up pretty close to the tree, and even stand immediately below it, within five or six feet of the bird, without in the least embarrassing him; the strokes of his bill are distinctly heard several hundred yards off; and I have known him to be at work for two hours together on the same tree. Buffon calls this "incessant toil and slavery,"—their attitude "a painful posture,"—and their life "a dull and insipid existence;" expressions improper, because untrue; and absurd, because contradictory. The posture is that for which the whole organisation of his frame is particularly adapted; and though, to a wren or a humming bird, the labour would be both toil and slavery, yet to him it is, I am convinced, as pleasant and as amusing, as the sports of the chase to the hunter, or the sucking of flowers to the humming bird. The eagerness with which he traverses the upper and lower sides of the branches, the cheerfulness of his cry, and the liveliness of his motions while digging into the tree and dislodging the vermin, justify this belief. He has a single note, or *chink*, which, like the former species, he frequently repeats; and when he flies off, or alights on another tree, he utters a rather shriller cry, composed of nearly the same kind of note, quickly reiterated. In fall and winter, he associates with the titmouse, creeper, &c., both in their wood and orchard excursions; and usually leads the van. Of all our woodpeckers, none rid the apple trees of so many vermin as this, digging off the moss which the negligence of the proprietor had suffered to accumulate, and probing every crevice. In fact, the orchard is his favourite resort in all seasons; and his industry is unequalled, and almost incessant, which is more than can be said of any other species we have. In fall, he is particularly fond of boring the apple trees for insects, digging a circular hole through the bark, just sufficient to admit his bill; after that a second, third, &c., in pretty regular horizontal circles round the body of the tree; these parallel circles of holes are often not more than an inch or an inch and a half apart, and sometimes so close together, that I have covered eight or ten of them at once with a dollar. From nearly the surface of the ground up to the first fork, and sometimes far beyond it, the whole bark of many apple trees is perforated in this manner, so as to appear as if made by successive discharges of buckshot; and our little woodpecker, the subject of the present account, is the principal perpetrator of this supposed mischief,—I say, supposed; for so far from these perforations of the bark being ruinous, they are not only harmless, but, I have good reason to believe, really beneficial to the health and fertility of the tree. I leave it to the philosophical botanist to account for this; but the fact I am confident of. In more than fifty orchards which I have myself carefully examined, those trees which were marked by the woodpecker (for some trees they never touch, perhaps because not penetrated by insects) were uniformly the most thriving, and seemingly the most productive: many of these were upwards of sixty years old, their trunks completely covered with holes, while the branches were broad, luxuriant, and loaded with fruit. Of decayed trees, more than three-fourths were untouched by the woodpecker. Several intelligent farmers, with whom I have conversed, candidly acknowledge the truth of these observations, and with justice look upon these birds as beneficial; but the most common opinion is, that they bore the trees to suck the sap, and so destroy its vegetation; though pine and other resinous trees, on the juices of which it is not pretended they feed, are often found equally perforated. Were the sap of the tree their object, the saccharine juice of the birch, the sugar-maple, and several others, would be much more inviting, because more sweet and nourishing, than that of either the pear or apple tree; but I have not observed one mark on the former for ten thousand that may be seen on the latter; besides, the early part of spring is the season when the sap flows most abundantly; whereas it is only during the months of September, October, and November, that woodpeckers are seen so indefatigably engaged in orchards, probing every crack and crevice, boring through the bark, and, what is worth remarking, chiefly on the south and southwest sides of the tree, for the eggs and larvæ deposited there by the countless swarms of summer insects. These, if suffered to remain, would prey upon the very vitals, if I may so express it, of the tree, and in the succeeding summer give birth to myriads more of their race, equally destructive.

Here, then, is a whole species, I may say, genus, of birds, which Providence seems to have formed for the protection of our fruit and forest trees from the ravages of vermin, which every day destroy millions of those noxious insects that would otherwise blast the hopes of the husbandman, and which even promote the fertility of the tree; and, in return, are proscribed by those who ought to have been their protectors, and incitements and rewards held out for their destruction! Let us examine better into the operations of nature, and many of our mistaken opinions and groundless prejudices will be abandoned for more just, enlarged, and humane modes of thinking.

The length of the downy woodpecker is six inches and three quarters, and its extent twelve inches; crown, black; hind head, deep scarlet; stripe over the eye, white; nostrils, thickly

Drawn from Nature by A. Wilson. *Engraved by W. H. Lizars.*

1. Mocking Bird. 2. Egg. 3 & 4. Male & Female Humming Bird, nest and Eggs. 5. Towhé Bunting. 6. Egg.

10.

covered with recumbent hairs, or small feathers, of a cream colour; these, as in the preceding species, are thick and bushy, as if designed to preserve the forehead from injury during the violent action of digging; the back is black, and divided by a lateral strip of white, loose, downy, unwebbed feathers; wings, black, spotted with white; tail-coverts, rump, and four middle feathers of the tail, black; the other three on each side, white, crossed with touches of black; whole under parts, as well as the sides of the neck, white; the latter marked with a streak of black, proceeding from the lower mandible, exactly as in the hairy woodpecker; legs and feet, bluish green; claws, light blue, tipt with black; tongue formed like that of the preceding species, horny towards the tip, where, for one-eighth of an inch, it is barbed; bill, of a bluish horn colour, grooved, and wedge-formed, like most of the genus; eye, dark hazel. The female wants the red on the hind head, having that part white; and the breast and belly are of a dirty white.

This, and the two former species, are generally denominated *Sap-suckers*. They have also several other provincial appellations, equally absurd, which it may, perhaps, be more proper to suppress than to sanction by repeating.

MOCKING BIRD. (*Turdus polyglottus.*)

PLATE X.—FIG. 1.

Mimic Thrush, *Lath. Syn.* iii. p. 40, No. 42.—*Arct. Zool.* ii. No. 194.—Turdus polyglottus, *Linn. Syst.* i. p. 293, No. 10.—Le grand moqueur, *Briss. Orn.* ii. p. 266, 29.—*Buff. Ois.* iii. p. 325. *Pl. enl.* 558, fig. 1.—Singing Bird, Mocking Bird, or Nightingale, *Raii Syn.* p. 64, No. 5, p. 185, 31.—*Sloan. Jam.* ii. 306, No. 34.—The Mock Bird, *Catesby, Car.* i. pl. 27.—*Peale's Museum*, No. 5288.

ORPHEUS POLYGLOTTUS.—SWAINSON.

Turdus polyglottos, *Bonap. Synop.* p. 74. The Mocking Bird, *Aud.* pl. xxi. *Orn. Biog.* 108.

THIS celebrated and very extraordinary bird, in extent and variety of vocal powers, stands unrivalled by the whole feathered songsters of this or perhaps any other country; and shall receive from us, in this place, all that attention and respect which superior merit is justly entitled to.

Among the many novelties which the discovery of this part of the western continent first brought into notice, we may reckon that of the mocking bird; which is not only peculiar to the New World, but inhabits a very considerable extent of both North and South America; having been traced from the States of New England to Brazil, and also among many of the adjacent islands. They are, however, much more numerous in those States south, than in those north, of the river Delaware; being generally migratory in the latter, and resident (at least many of them) in the former. A warm climate and low country, not far from the sea, seems most congenial to their nature; accordingly, we find the species less numerous to the west than east of the great range of the Alleghany, in the same parallels of latitude. In the severe winter of 1808–9, I found these birds, occasionally, from Fredericksburg, in Virginia, to the southern parts of Georgia; becoming still more numerous the farther I advanced to the south. The berries of the red cedar, myrtle, holly, cassine shrub, many species of smilax, together with gum berries, gall berries, and a profusion of others with which the luxuriant swampy thickets of those regions abound, furnish them with a perpetual feast. Winged insects, also, of which they are very fond, and remarkably expert at catching, abound there even in winter, and are an additional inducement to residency. Though rather a shy bird in the northern States, here he appeared almost half-domesticated, feeding on the cedars, and among the thickets of smilax that lined the roads, while I passed within a few feet; playing around the planter's door, and hopping along the shingles. During the month of February, I sometimes heard a solitary one singing; but on the 2d of March, in the neighbourhood of Savannah, numbers of them were heard on every hand, vieing in song with each other, and with the brown thrush, making the whole woods vocal with their melody. Spring was at that time considerably advanced; and the thermometer ranging between 70 and 78 degrees. On arriving at New York, on the 22d of the same month, I found many parts of the country still covered with snow, and the streets piled with ice to the height of two feet; while neither the brown thrush nor mocking bird were observed, even in the lower parts of Pennsylvania, until the 20th of April.

The precise time at which the mocking bird begins to build his nest varies according to the latitude in which he resides. In the lower parts of Georgia, he commences building early in April; but in Pennsylvania, rarely before the 10th of May; and in New York and the States of New England, still later. There are particular situations to which he gives the preference. A solitary thorn bush, an almost impenetrable thicket, an orange tree, cedar, or holly bush, are favourite spots, and frequently selected. It is no great objection with him that these happen, sometimes, to be near the farm or mansion-house: always ready to defend, but never over-anxious to conceal, his nest, he very often builds within a small distance of the house; and not unfrequently in a pear or apple tree; rarely at a greater height than six or seven feet from the ground. The nest varies a little with different individuals, according to the conveniency of collecting suitable materials. A very complete one is now lying before me, and is composed of the following substances: First, a quantity of dry twigs and sticks, then withered tops of weeds of the preceding year, intermixed with fine straws, hay, pieces of wool and tow; and, lastly, a thick layer of fine fibrous roots, of a light brown colour, lines the whole. The eggs, one of which is represented at fig. 2, are four, sometimes five, of a cinereous blue, marked with large blotches of brown. The female sits fourteen days; and generally produces two broods in the season, unless robbed of her eggs, in which case she will even build and lay the third time. She is, however, extremely jealous of her nest, and very apt to forsake it if much disturbed. It is even asserted by some of our bird dealers, that the old ones will actually destroy the eggs, and poison the young, if either the one or the other have been handled. But I cannot give credit to this unnatural report. I know, from my own experience, at least, that it is not always their practice; neither have I ever witnessed a case of the kind above mentioned. During the period of incubation, neither cat, dog, animal, nor man, can approach the nest without being attacked. The cats, in particular, are persecuted whenever they make their appearance, till obliged to retreat. But his whole vengeance is most particularly directed against that mortal enemy of his eggs and young, the black snake. Whenever the insidious approaches of this reptile are discovered, the male darts upon it with the rapidity of an arrow, dexterously eluding its bite, and striking it violently and incessantly about the head, where it is very vulnerable. The snake soon becomes sensible of its danger, and seeks to escape; but the intrepid defender of his young redoubles his exertions, and, unless his antagonist be of great magnitude, often succeeds in destroying him. All its pretended powers of fascination avail it nothing against the vengeance of this noble bird. As the snake's strength begins to flag, the mocking bird seizes and lifts it up partly from the ground, beating it with his wings; and when the business is completed, he returns to the repository of his young, mounts the summit of the bush, and pours out a torrent of song in token of victory.

As it is of some consequence to be able to distinguish a young male bird from a female, the following marks may be attended to, by which some pretend to be able to distinguish them in less than a week after they are hatched. These are, the breadth and purity of the white on the wings, for that on the tail is not so much to be depended on. This white, in a full-grown male bird, spreads over the whole nine primaries, down to, and considerably below, their coverts, which are also white, sometimes slightly tipt with brown. The white of the primaries also extends equally far on both vanes of the feathers. In the female the white is less pure, spreads over only seven or eight of the primaries, does not descend so far, and extends considerably farther down on the broad than on the narrow side of the feathers. The black is also more of a brownish cast.

The young birds, if intended for the cage, ought not to be left till they are nearly ready to fly, but should be taken rather young than otherwise; and may be fed, every half hour, with milk thickened with Indian-meal; mixing occasionally with it a little fresh meat, cut or minced very fine. After they begin to eat of their own accord, they ought still to be fed by hand, though at longer intervals, and a few cherries, strawberries, &c., now and then thrown into them. The same sort of food, adding grasshoppers and fruit, particularly the various kinds of berries in which they delight, and plenty of clear fine gravel, is found very proper for them after they are grown up. Should the bird at any time appear sick or dejected, a few spiders thrown into him will generally remove these symptoms of disease.

If the young bird is designed to be taught by an old one, the best singer should be selected for this office, and no other allowed to be beside him. Or if by the bird-organ or mouth-whistling, it should be begun early, and continued, pretty constantly, by the same person, until the scholar, who is seldom inattentive, has completely acquired his lesson. The best singing birds, however, in my own opinion, are those that have been reared in the country, and educated under the tuition of the feathered choristers of the surrounding fields, groves, woods, and meadows.

The plumage of the mocking bird, though none of the homeliest, has nothing gaudy or brilliant in it; and had he nothing else to recommend him, would scarcely entitle him to notice; but his figure is well proportioned, and even handsome. The ease, elegance, and rapidity of his movements, the animation of his eye, and the intelligence he displays in listening and laying up lessons from almost every species of the feathered creation within his hearing, are really surprising, and mark the peculiarity of his genius. To these qualities we may add that of a voice full, strong, and musical, and capable of almost every modulation, from the clear mellow tones of the wood thrush to the savage scream of the bald eagle. In measure and accent, he faithfully follows his originals; in force and sweetness of expression, he greatly improves upon them. In his native groves, mounted on the top of a tall bush or half-grown tree, in the dawn of dewy morning, while the woods are already vocal with a multitude of warblers, his admirable song rises pre-eminent over every competitor. The ear can listen to *his* music alone, to which that of all the others seems a mere accompaniment. Neither is this strain altogether imitative. His own native notes, which are easily distinguishable by such as are well acquainted with those of our various song birds, are bold and full, and varied seemingly beyond all limits. They consist of short expressions of two, three, or, at the most, five or six syllables; generally interspersed with imitations, and all of them uttered with great emphasis and rapidity; and continued, with undiminished ardour, for half an hour or an hour at a time. His expanded wings and tail glistening with white, and the buoyant gaiety of his action arresting the eye, as his song most irresistibly does the ear, he sweeps round with enthusiastic ecstasy—he mounts and descends as his song swells or dies away; and, as my friend Mr Bartram has beautifully expressed it, "He bounds aloft with the celerity of an arrow, as if to recover or recall his very soul, expired in the last elevated strain." * While thus exerting himself, a bystander destitute of sight would suppose that the whole feathered tribes had assembled together on a trial of skill, each striving to produce his utmost effect; so perfect are his imitations. He many times deceives the sportsman, and sends him in search of birds that perhaps are not within

* Travels, p. 32, Introd.

miles of him, but whose notes he exactly imitates: even birds themselves are frequently imposed on by this admirable mimic, and are decoyed by the fancied calls of their mates, or dive with precipitation into the depth of thickets, at the scream of what they suppose to be the sparrow hawk.

The mocking bird loses little of the power and energy of his song by confinement. In his domesticated state, when he commences his career of song, it is impossible to stand by uninterested. He whistles for the dog,—Cæsar starts up, wags his tail, and runs to meet his master. He squeaks out like a hurt chicken,—and the hen hurries about with hanging wings, and bristled feathers, clucking to protect its injured brood. The barking of the dog, the mewing of the cat, the creaking of a passing wheelbarrow, follow with great truth and rapidity. He repeats the tune taught him by his master, though of considerable length, fully and faithfully. He runs over the quiverings of the canary, and the clear whistlings of the Virginia nightingale, or red bird, with such superior execution and effect, that the mortified songsters feel their own inferiority, and become altogether silent; while he seems to triumph in their defeat by redoubling his exertions.

This excessive fondness for variety, however, in the opinion of some, injures his song. His elevated imitations of the brown thrush are frequently interrupted by the crowing of cocks; and the warblings of the blue bird, which he exquisitely manages, are mingled with the screaming of swallows or the cackling of hens; amidst the simple melody of the robin, we are suddenly surprised by the shrill reiterations of the whip-poor-will; while the notes of the killdeer, blue jay, martin, baltimore, and twenty others, succeed, with such imposing reality, that we look round for the originals, and discover, with astonishment, that the sole performer in this singular concert is the admirable bird now before us. During this exhibition of his powers, he spreads his wings, expands his tail, and throws himself around the cage in all the ecstasy of enthusiasm, seeming not only to sing, but to dance, keeping time to the measure of his own music. Both in his native and domesticated state, during the solemn stillness of night, as soon as the moon rises in silent majesty, he begins his delightful solo, and serenades us the livelong night with a full display of his vocal powers, making the whole neighbourhood ring with his inimitable medley.*

Were it not to seem invidious in the eyes of foreigners, I might in this place, make a comparative statement between the powers of the mocking bird and the only bird, I believe, in the world, worthy of being compared with him,—the European nightingale. This, however, I am unable to do from my own observation, having never myself heard the song of the latter; and, even if I had, perhaps something might be laid to the score of partiality, which, as a faithful biographer, I am anxious to avoid. I shall, therefore, present the reader with the opinion of a distinguished English naturalist and curious observer on this subject, the Honourable Daines Barrington, who, at the time he made the communication, was Vice-president of the Royal Society, to which it was addressed.†

"It may not be improper here," says this gentleman, "to

* The hunters in the southern States, when setting out upon an excursion by night, as soon as they hear the mocking bird begin to sing, know that the moon is rising.

A certain anonymous author, speaking of the mocking birds in the island of Jamaica, and their practice of singing by moonlight, thus gravely philosophises, and attempts to account for the habit. "It is not certain," says he, "whether they are kept so wakeful by the clearness of the light, or by any extraordinary attention and vigilance, at such times, for the protection of their nursery from the piratical assaults of the owl and the night hawk. It is possible that fear may operate upon them, much in the same manner as it has been observed to affect some cowardly persons, who whistle stoutly in a lonesome place, while their mind is agitated with the terror of thieves or hobgoblins." —*History of Jamaica*, vol. iii. p. 894, 4to.

† Philosophical Transactions, vol. lxii. part ii. p. 284.

consider whether the nightingale may not have a very formidable competitor in the American mocking bird, though almost all travellers agree that the concert in the European woods is superior to that of the other parts of the globe." "I have happened, however, to hear the American mocking bird, in great perfection, at Messrs Vogels and Scotts, in Love Lane, Eastcheap. This bird is believed to be still living, and hath been in England these six years. During the space of a minute, he imitated the woodlark, chaffinch, blackbird, thrush, and sparrow; I was told also that he would bark like a dog; so that the bird seems to have no choice in his imitations, though his pipe comes nearest to our nightingale of any bird I have yet met with. With regard to the original notes, however, of this bird, we are still at a loss, as this can only be known by those who are accurately acquainted with the song of the other American birds. Kalm indeed informs us, that the natural song is excellent;* but this traveller seems not to have been long enough in America to have distinguished what were the genuine notes: with us, mimics do not often succeed but in imitations. I have little doubt, however, but that this bird would be fully equal to the song of the nightingale in its whole compass; but then, from the attention which the mocker pays to any other sort of disagreeable noise, these capital notes would be always debased by a bad mixture."

On this extract I shall make a few remarks. If, as is here conceded, the mocking bird be fully equal to the song of the nightingale, and, as I can with confidence add, not only to that, but to the song of almost every other bird, besides being capable of exactly imitating various other sounds and voices of animals,—his vocal powers are unquestionably superior to those of the nightingale, which possesses its own native notes alone. Further, if we consider, as is asserted by Mr Barrington, that "one reason of the nightingale's being more

* Travels, vol. i. p. 219.

attended to than others is, that it sings in the night;" and if we believe, with Shakespeare, that—

> The nightingale, if she should sing by day
> When every goose is cackling, would be thought
> No better a musician than a wren,

what must we think of that bird who, in the glare of day, when a multitude of songsters are straining their throats in melody, overpowers all competition, and, by the superiority of his voice, expression, and action, not only attracts every ear, but frequently strikes dumb his mortified rivals; when the silence of night, as well as the bustle of day, bear witness to his melody; and when, even in captivity, in a foreign country, he is declared, by the best judges in that country, to be fully equal to the song of their sweetest bird *in its whole compass?* The supposed degradation of his song by the introduction of extraneous sounds and unexpected imitations, is, in fact, one of the chief excellences of this bird; as these changes give a perpetual novelty to his strain, keep attention constantly awake, and impress every hearer with a deeper interest in what is to follow. In short, if we believe in the truth of that mathematical axiom, that the whole is greater than a part, all that is excellent or delightful, amusing or striking, in the music of birds, must belong to that admirable songster whose vocal powers are equal to the whole compass of their whole strains.

The native notes of the mocking bird have a considerable resemblance to those of the brown thrush, but may easily be distinguished, by their greater rapidity, sweetness, energy of expression, and variety. Both, however, have, in many parts of the United States, particularly in those to the south, obtained the name of mocking bird; the first, or brown thrush, from its inferiority of song, being called the French, and the other the English mocking bird,—a mode of expression probably originating in the prejudices of our forefathers, with whom everything French was inferior to everything English.*

The mocking bird is frequently taken in trap-cages, and, by proper management, may be made sufficiently tame to sing. The upper parts of the cage (which ought to be of wood) should be kept covered, until the bird becomes a little more reconciled to confinement. If placed in a wire cage, uncovered, he will soon destroy himself in attempting to get out. These birds, however, by proper treatment, may be brought to sing perhaps superior to those raised by hand, and cost less trouble. The opinion which the naturalists of Europe entertain of the great difficulty of raising the mocking bird, and that not one in ten survives, is very incorrect. A person called on me a few days ago with twenty-nine of these birds, old and young, which he had carried about the fields with him for several days, for the convenience of feeding them while engaged in trapping others. He had carried them thirty miles, and intended carrying them ninety-six miles farther, viz., to New York; and told me, that he did not expect to lose one out of ten of them. Cleanliness, and regularity in feeding, are the two principal things to be attended to; and these rarely fail to succeed.

The eagerness with which the nest of the mocking bird is sought after in the neighbourhood of Philadelphia, has rendered this bird extremely scarce for an extent of several miles round the city. In the country round Wilmington and Newcastle, they are very numerous, from whence they are frequently brought here for sale. The usual price of a singing bird is from seven to fifteen, and even twenty dollars. I have known fifty dollars paid for a remarkably fine singer, and one instance where one hundred dollars were refused for a still more extraordinary one.

* The observations of Mr Barrington, in the paper above referred to, make this supposition still more probable. "Some nightingales," says he, "are so vastly inferior, that the bird-catchers will not keep them, branding them with the name of Frenchmen."—P. 283.

Attempts have been made to induce these charming birds to pair, and rear their young in a state of confinement, and the result has been such as to prove it, by proper management, perfectly practicable. In the spring of 1808, a Mr Klein, living in North Seventh Street, Philadelphia, partitioned off about twelve feet square in the third story of his house. This was lighted by a pretty large wire-grated window. In the centre of this small room he planted a cedar bush, five or six feet high, in a box of earth, and scattered about a sufficient quantity of materials suitable for building. Into this place, a male and female mocking bird were put, and soon began to build. The female laid five eggs, all of which she hatched, and fed the young with great affection until they were nearly able to fly. Business calling the proprietor from home for two weeks, he left the birds to the care of his domestics; and on his return, found, to his great regret, that they had been neglected in food. The young ones were all dead, and the parents themselves nearly famished. The same pair have again commenced building this season, in the same place, and have at this time, July 4, 1809, three young, likely to do well. The place might be fitted up with various kinds of shrubbery, so as to resemble their native thickets; and ought to be as remote from noise and interruption of company as possible, and strangers rarely allowed to disturb, or even approach them.

The mocking bird is nine and a half inches long, and thirteen in breadth. Some individuals are, however, larger, and some smaller, those of the first hatch being uniformly the biggest and stoutest.* The upper parts of the head, neck, and back, are a dark brownish ash, and when new moulted, a fine light grey; the wings and tail are nearly black, the first

* Many people are of opinion that there are two sorts, the large and the small mocking bird; but, after examining great numbers of these birds in various regions of the United States, I am satisfied that this variation of size is merely accidental, or owing to the circumstance above mentioned.

and second rows of coverts tipt with white; the primary coverts, in some males, are wholly white, in others, tinged with brown. The three first primaries are white from their roots as far as their coverts; the white on the next six extends from an inch to one and three-fourths farther down, descending equally on both sides of the feather; the tail is cuneiform, the two exterior feathers wholly white, the rest, except the middle ones, tipt with white; the chin is white; sides of the neck, breast, belly, and vent, a brownish white, much purer in wild birds than in those that have been domesticated; iris of the eye, yellowish cream coloured, inclining to golden; bill, black, the base of the lower mandible, whitish; legs and feet, black, and strong. The female very much resembles the male; what difference there is, has been already pointed out in a preceding part of this account. The breast of the young bird is spotted like that of the thrush.*

Mr William Bartram observes of the mocking bird, that "formerly, say thirty or forty years ago, they were numerous, and often stayed all winter with us, or the year through, feeding on the berries of ivy, smilax, grapes, persimmons, and other berries. The ivy (*Hedera helix*) they were particularly fond of, though a native of Europe. We have an ancient plant adhering to the wall of the house, covering many yards of surface; this vine is very fruitful, and here many would feed and lodge during the winter, and, in very severe cold weather, sit on the top of the chimney to warm themselves." He also adds, "I have observed that the mocking bird ejects

* A bird is described in the "Northern Zoology" as the varied thrush of Pennant, the *Turdus nævius* of Latham, which will rank as an addition to the North American species of this genus, and has been named by Mr Swainson *O. meruloides*, thrush-like mocking bird. Mr Swainson has changed the name of Latham, to give it one expressive of its form, as he considers the structure intermediate between *Orpheus* and *Turdus*, though leaning most to the former. According to Dr Richardson, it was discovered by Captain Cook at Nootka Sound, and described by Latham from these specimens.—Ed.

from his stomach through his mouth the hard kernels of berries, such as smilax, grapes, &c., retaining the pulpy part." *

HUMMING BIRD.† (*Trochilus colubris.*)

PLATE X.—FIGS. 3 AND 4.

Trochilus colubris, *Linn. Syst.* i. p. 191, No. 12.—L'Oiseau mouche à gorge rouge de la Caroline, *Briss. Orn.* iii. p. 716, No. 13, t. 36, fig. 6.—Le Rubis, *Buff. Ois.* vi. p. 13.—Humming Bird, *Catesb. Car.* i. 65.—Red-throated Humming Bird, *Edw.* i. 38, male and female.—*Lath. Syn.* ii. 769, No. 35.—*Peale's Museum*, No. 2520.

TROCHILUS COLUBRIS.—LINNÆUS.

Trochilus colubris, *Bonap. Synop.* p. 98.—The Ruby-throated Humming Bird, *Aud.* pl. xlvii. *Orn. Biog.* i. 248.—Trochilus colubris, Northern Humming Bird, *North. Zool.* ii. p. 323.

NATURE, in every department of her work, seems to delight in variety; and the present subject of our history is almost as

* Letter from Mr Bartram to the author.

† The "fairy humming birds," "the jewels of ornithology,"

Least of the winged vagrants of the sky,

though amply dispersed over the southern continent of the New World, from their delicate and slender structure, being unable to bear the severities of a hardier climate, are, with two exceptions, withdrawn from its northern parts; and it is with wonder that we see creatures of such tiny dimensions occasionally daring to brave even the snows and frosts of a northern latitude. The present species, though sometimes exceeding its appointed time, is obliged to seek warmer abodes during winter; and it is another subject for astonishment and reflection how they are enabled to perform a lengthened migration, where the slightest gale would waft them far from their proper course. Mr Audubon is of opinion, that they migrate during the night, passing through the air in long undulations, raising themselves for some distance at an angle of about 40°, and then falling in a curve; but he adds, that the smallness of their size precludes the possibility of following them farther than fifty or sixty yards, even with a good glass.

The humming birds, or what are generally known by the genus *Trochilus* of Linnæus, have been, through the researches of late travellers and naturalists, vastly increased in their numbers; they form a large and closely-connected group, but show a considerable singular for its minuteness, beauty, want of song, and manner of feeding as the preceding is for unrivalled excellence of notes and plainness of plumage. Though this interesting and beautiful genus of birds comprehends upwards of seventy species, all of which, with a very few exceptions, are natives of America and its adjacent islands, it is yet singular that the species now before us should be the only one of its tribe that ever visits the territory of the United States.

According to the observations of my friend Mr Abbot, of Savannah, in Georgia, who has been engaged these thirty years in collecting and drawing subjects of natural history in that part of the country, the humming bird makes its first appearance there, from the south, about the 23d of March; two weeks earlier than it does in the county of Burke, sixty miles higher up the country towards the interior; and at least five weeks sooner than it reaches this part of Pennsylvania. As it passes on to the northward, as far as the interior of Canada, where it is seen in great numbers,* the wonder is excited how so feebly constructed and delicate a little creature can make its way over such extensive regions of lakes and

variety of form and character, and have been divided into different genera. They may be said to be strictly confined to the New World, with her islands; and although other countries possess many splendid and closely allied forms, "with gemmed frontlets and necks of verdant gold," which have been by some included, none we consider can properly range with any of those found in this division of the world. In India and the Asiatic continent, they may be represented by *Cœreba*, &c.; in Africa, by *Nectarinia* and *Cyniris;* and in Australia and in the Southern Pacific, by *Meliphaga*, *Myrzomela*, &c. Europe possesses no direct prototype.

The second northern species alluded to was discovered by Captain Cook in Nootka Sound, and first described by Dr Latham as the ruff-necked humming bird. Mr Swainson introduces it in the "Northern Zoology," under his genus *Selasphorus*. It ranges southwards to Real del Monte on the tableland of Mexico.—ED.

* Mr M'Kenzie speaks of seeing a "beautiful humming bird" near the head of the Unjigah, or Peace River, in lat. 54°; but has not particularised the species.

forests, among so many enemies, all its superiors in strength and magnitude. But its very minuteness, the rapidity of its flight, which almost eludes the eye, and that admirable instinct, reason, or whatever else it may be called, and daring courage, which Heaven has implanted in its bosom, are its guides and protectors. In these we may also perceive the reason why an all-wise Providence has made this little hero an exception to a rule which prevails almost universally through nature, viz., that the smallest species of a tribe are the most prolific. The eagle lays one, sometimes two, eggs; the crow, five; the titmouse, seven or eight; the small European wren, fifteen; the humming bird, *two:* and yet this latter is abundantly more numerous in America than the wren is in Europe.

About the 25th of April, the humming bird usually arrives in Pennsylvania, and about the 10th of May begins to build its nest. This is generally fixed on the upper side of a horizontal branch, not among the twigs, but on the body of the branch itself. Yet I have known instances where it was attached by the side to an old moss-grown trunk, and others where it was fastened on a strong rank stalk, or weed, in the garden; but these cases are rare. In the woods, it very often chooses a white oak sapling to build on; and in the orchard or garden, selects a pear tree for that purpose. The branch is seldom more than ten feet from the ground. The nest is about an inch in diameter, and as much in depth. A very complete one is now lying before me, and the materials of which it is composed are as follows:—The outward coat is formed of small pieces of a species of bluish gray lichen that vegetates on old trees and fences, thickly glued on with the saliva of the bird, giving firmness and consistency to the whole, as well as keeping out moisture. Within this are thick matted layers of the fine wings of certain flying seeds, closely laid together; and, lastly, the downy substance from the great mullein, and from the stalks of the common fern, lines the whole. The base of the nest is continued round the stem of the branch, to which it closely adheres; and, when viewed from below, appears a mere mossy knot or accidental protuberance. The eggs are two, pure white, and of equal thickness at both ends. The nest and eggs in the plate were copied with great precision, and by actual measurement, from one just taken in from the woods. On a person's approaching their nest, the little proprietors dart around with a humming sound, passing frequently within a few inches of one's head; and, should the young be newly hatched, the female will resume her place on the nest even while you stand within a yard or two of the spot. The precise period of incubation I am unable to give; but the young are in the habit, a short time before they leave the nest, of thrusting their bills into the mouths of their parents, and sucking what they have brought them. I never could perceive that they carried them any animal food; though, from circumstances that will presently be mentioned, I think it highly probable they do. As I have found their nests with eggs so late as the 12th of July, I do not doubt but that they frequently, and perhaps usually, raise two broods in the same season.

The humming bird is extremely fond of tubular flowers, and I have often stopped, with pleasure, to observe his manœuvres among the blossoms of the trumpet flower. When arrived before a thicket of these that are full blown, he poises, or suspends himself on wing, for the space of two or three seconds, so steadily, that his wings become invisible, or only like a mist, and you can plainly distinguish the pupil of his eye looking round with great quickness and circumspection; the glossy golden green of his back, and the fire of his throat, dazzling in the sun, form altogether a most interesting appearance. The position into which his body is usually thrown while in the act of thrusting his slender tubular tongue into the flower to extract its sweets, is exhibited in the figure on the plate. When he alights, which is frequently, he always prefers the small dead twigs of a tree or bush, where he dresses and arranges his plumage with great dexterity. His only note is a single chirp, not louder than that of a small cricket or grasshopper, generally uttered while passing from flower to flower, or when engaged in fight with his fellows; for, when two males meet at the same bush or flower, a battle instantly takes place; and the combatants ascend in the air, chirping, darting and circling around each other, till the eye is no longer able to follow them. The conqueror, however, generally returns to the place to reap the fruits of his victory. I have seen him attack, and for a few moments teaze, the king bird; and have also seen him, in his turn, assaulted by a humble bee, which he soon put to flight. He is one of those few birds that are universally beloved; and amidst the sweet dewy serenity of a summer's morning, his appearance among the arbours of honeysuckles and beds of flowers is truly interesting.

> When the morning dawns, and the blest sun again
> Lifts his red glories from the eastern main,
> Then through our woodbines, wet with glittering dews,
> The flower-fed humming bird his round pursues;
> Sips, with inserted tube, the honeyed blooms,
> And chirps his gratitude as round he roams;
> While richest roses, though in crimson drest,
> Shrink from the splendour of his gorgeous breast.
> What heavenly tints in mingling radiance fly!
> Each rapid movement gives a different dye;
> Like scales of burnished gold they dazzling show,
> Now sink to shade—now like a furnace glow!

The singularity of this little bird has induced many persons to attempt to raise them from the nest, and accustom them to the cage. Mr Coffer, of Fairfax county, Virginia, a gentleman who has paid great attention to the manners and peculiarities of our native birds, told me that he raised and kept two for some months in a cage, supplying them with honey dissolved in water, on which they readily fed. As the sweetness of the liquid frequently brought small flies and gnats about the cage and cup, the birds amused themselves by snapping at them on wing, and swallowing them with eagerness, so that these insects formed no inconsiderable part of their food. Mr

Charles Wilson Peale, proprietor of the museum, tells me that he had two young humming birds, which he raised from the nest. They used to fly about the room; and would frequently perch on Mrs Peale's shoulder to be fed. When the sun shone strongly in the chamber, he has observed them darting after the motes that floated in the light, as flycatchers would after flies. In the summer of 1803, a nest of young humming birds was brought me, that were nearly fit to fly. One of them actually flew out by the window the same evening, and, falling against a wall, was killed. The other refused food, and the next morning I could but just perceive that it had life. A lady in the house undertook to be its nurse, placed it in her bosom, and, as it began to revive, dissolved a little sugar in her mouth, into which she thrust its bill, and it sucked with great avidity. In this manner it was brought up until fit for the cage. I kept it upwards of three months, supplied it with loaf-sugar dissolved in water, which it preferred to honey and water, gave it fresh flowers every morning sprinkled with the liquid, and surrounded the space in which I kept it with gauze, that it might not injure itself. It appeared gay, active, and full of spirit, hovering from flower to flower as if in its native wilds, and always expressed by its motions and chirping, great pleasure at seeing fresh flowers introduced to its cage. Numbers of people visited it from motives of curiosity; and I took every precaution to preserve it, if possible, through the winter. Unfortunately, however, by some means it got at large, and flying about the room, so injured itself that it soon after died.

This little bird is extremely susceptible of cold, and, if long deprived of the animating influence of the sunbeams, droops and soon dies. A very beautiful male was brought me this season (1809), which I put into a wire cage, and placed in a retired shaded part of the room. After fluttering about for some time, the weather being uncommonly cool, it clung by the wires, and hung in a seemingly torpid state for a whole forenoon. No motion whatever of the lungs could be perceived, on the closest inspection, though, at other times, this is remarkably observable; the eyes were shut, and, when touched by the finger, it gave no signs of life or motion. I carried it out to the open air, and placed it directly in the rays of the sun, in a sheltered situation. In a few seconds, respiration became very apparent; the bird breathed faster and faster, opened its eyes, and began to look about, with as much seeming vivacity as ever. After it had completely recovered, I restored it to liberty: and it flew off to the withered top of a pear tree, where it sat for some time dressing its disordered plumage, and then shot off like a meteor.

The flight of the humming bird, from flower to flower, greatly resembles that of a bee; but is so much more rapid, that the latter appears a mere loiterer to him. He poises himself on wing, while he thrusts his long, slender, tubular tongue into the flowers in search of food. He sometimes enters a room by the window, examines the bouquets of flowers, and passes out by the opposite door or window. He has been known to take refuge in a hothouse during the cool nights of autumn, to go regularly out in the morning, and to return as regularly in the evening, for several days together.

The humming bird has hitherto been supposed to subsist altogether on the honey, or liquid sweets, which it extracts from flowers. One or two curious observers have indeed remarked, that they have found evident fragments of insects in the stomach of this species; but these have been generally believed to have been taken in by accident. The few opportunities which Europeans have to determine this point by observations made on the living bird, or by dissection of the newly killed one, have rendered this mistaken opinion almost general in Europe. For myself, I can speak decisively on this subject. I have seen the humming bird, for half an hour at a time, darting at those little groups of insects that dance in the air in a fine summer evening, retiring to an adjoining twig to rest, and renewing the attack with a dexterity that sets all our other flycatchers at defiance. I have opened, from time to time, great numbers of these birds; have examined the contents of the stomach with suitable glasses, and, in three cases out of four, have found these to consist of broken fragments of insects. In many subjects, entire insects of the coleopterous class, but very small, were found unbroken. The observations of Mr Coffer, as detailed above, and the remarks of my worthy friend Mr Peale, are corroborative of these facts. It is well known that the humming bird is particularly fond of tubular flowers, where numerous small insects of this kind resort to feed on the farina, &c.; and there is every reason for believing that he is as often in search of these insects as of honey; and that the former compose at least as great a portion of his usual sustenance as the latter. If this food be so necessary for the parents, there is no doubt but the young also occasionally partake of it.

To enumerate all the flowers of which this little bird is fond, would be to repeat the names of half our American flora. From the blossoms of the towering poplar or tulip tree, through a thousand intermediate flowers, to those of the humble larkspur, he ranges at will, and almost incessantly. Every period of the season produces a fresh multitude of new favourites. Towards the month of September, there is a yellow flower which grows in great luxuriance along the sides of creeks and rivers, and in low moist situations; it grows to the height of two or three feet, and the flower, which is about the size of a thimble, hangs in the shape of a cap of liberty above a luxuriant growth of green leaves. It is the *Balsamina noli me tangere* of botanists, and is the greatest favourite with the humming bird of all our other flowers. In some places, where these plants abound, you may see, at one time, ten or twelve humming birds darting about, and fighting with and pursuing each other. About the 20th of September they generally retire to the south. I have, indeed, sometimes seen a solitary individual on the 28th and 30th of that month, and sometimes even in October; but these cases are rare. About the beginning of November, they pass the southern boundary of the United States into Florida.

The humming bird is three inches and a half in length, and four and a quarter in extent; the whole back, upper part of the neck, sides under the wings, tail-coverts, and two middle feathers of the tail, are of a rich golden green; the tail is forked, and, as well as the wings, of a deep brownish purple; the bill and eyes are black; the legs and feet, both of which are extremely small, are also black; the bill is straight, very slender, a little inflated at the tip, and very incompetent to the exploit of penetrating the tough sinewy side of a crow, and precipitating it from the clouds to the earth, as Charlevoix would persuade his readers to believe.* The nostrils are two small oblong slits, situated at the base of the upper mandible, scarcely perceivable when the bird is dead, though very distinguishable and prominent when living; the sides of the belly, and belly itself, dusky white, mixed with green; but what constitutes the chief ornament of this little bird, is the splendour of the feathers of his throat, which, when placed in a proper position, glow with all the brilliancy of the ruby. These feathers are of singular strength and texture, lying close together like scales, and vary, when moved before the eye, from a deep black to a fiery crimson and burning orange. The female is destitute of this ornament; but differs little in other appearance from the male; her tail is tipt with white, and the whole lower parts are of the same tint. The young birds of the first season, both male and female, have the tail tipt with white, and the whole lower parts nearly white; in the month of September, the ornamental feathers on the throat of the young males begin to appear.

On dissection, the heart was found to be remarkably large, nearly as big as the cranium; and the stomach, though distended with food, uncommonly small, not exceeding the globe of the eye, and scarcely more than one-sixth part as large as

* Histoire de la Nouvelle France, iii. p. 185.

the heart; the fibres of the last were also exceedingly strong. The brain was in large quantity, and very thin; the tongue, from the tip to an extent equal with the length of the bill, was perforated, forming two closely attached parallel and cylindrical tubes; the other extremities of the tongue corresponded exactly to those of the woodpecker, passing up the hind head, and reaching to the base of the upper mandible. These observations were verified in five different subjects, all of whose stomachs contained fragments of insects, and some of them whole ones.

TOWHE BUNTING. (*Emberiza erythropthalma.*)

PLATE X.—FIG. 5.

Fringilla erythropthalma, *Linn. Syst.* p. 318, 6.—Le Pinson de la Caroline, *Briss. Orn.* iii. p. 169, 44.—*Buff. Ois.* iv. p. 141.—*Lath.* ii. p. 199, No. 43.—*Catesb.* Car. i. plate 34.—*Peale's Museum*, No. 5970.

PIPILO ERYTHROPTHALMA.—VIEILLOT.

Pipilo erythropthalma, *Vieill. Gal. des Ois.* plate 80.—Fringilla erythropthalma, *Bonap. Synop.* p. 112.—The Towhe Bunting, *Aud.* plate 29, male and female, *Orn. Biog.* i. p. 150.

THIS is a very common, but humble and inoffensive species, frequenting close sheltered thickets, where it spends most of its time in scratching up the leaves for worms, and for the larvæ and eggs of insects. It is far from being shy, frequently suffering a person to walk round the bush or thicket where it is at work, without betraying any marks of alarm, and when disturbed, uttering the notes *tow-he* repeatedly. At times the male mounts to the top of a small tree, and chants his few simple notes for an hour at a time. These are loud, not unmusical, something resembling those of the yellow-hammer of Britain, but more mellow and more varied. He is fond of thickets with a southern exposure, near streams of water, and where there is plenty of dry leaves; and is found generally over the whole United States. He is not gregarious, and you seldom see more than two together. About the middle or 20th of April, they arrive in Pennsylvania, and begin building about the first week in May. The nest is fixed on the ground among the dry leaves, near, and sometimes under, a thicket of briars, and is large and substantial. The outside is formed of leaves and dry pieces of grape-vine bark, and the inside of fine stalks of dried grass, the cavity completely sunk beneath the surface of the ground, and sometimes half covered above with dry grass or hay. The eggs are usually five, of a pale flesh colour, thickly marked with specks of rufous, most numerous near the great end (see fig. 6). The young are produced about the beginning of June, and a second brood commonly succeeds in the same season. This bird rarely winters north of the State of Maryland, retiring from Pennsylvania to the south about the 12th of October. Yet in the middle districts of Virginia, and thence south to Florida, I found it abundant during the months of January, February, and March. Its usual food is obtained by scratching up the leaves; it also feeds, like the rest of its tribe, on various hard seeds and gravel; but rarely commits any depredations on the harvest of the husbandman, generally preferring the woods, and traversing the bottom of fences sheltered with briars. He is generally very plump and fat; and, when confined in a cage, soon becomes familiar. In Virginia, he is called the bulfinch; in many places, the towhe bird; in Pennsylvania, the chewink; and by others, the swamp robin. He contributes a little to the harmony of our woods in spring and summer; and is remarkable for the cunning with which he conceals his nest. He shows great affection for his young, and the deepest marks of distress on the appearance of their mortal enemy, the black snake.

The specific name which Linnæus has bestowed on this bird is deduced from the colour of the iris of its eye, which, in those that visit Pennsylvania, is dark red. But I am suspicious that this colour is not permanent, but subject to a periodical change. I examined a great number of these birds in the month of March, in Georgia, every one of which had the iris of the eye white. Mr Abbot of Savannah assured me, that at this season, every one of these birds he shot had the iris white, while at other times it was red; and Mr Elliot of Beaufort, a judicious naturalist, informed me, that in the month of February he killed a towhe bunting with one eye red and the other white! It should be observed that the iris of the young bird's eye is of a chocolate colour during its residence in Pennsylvania: perhaps this may brighten into a white during winter, and these may have been all birds of the preceding year, which had not yet received the full colour of the eye.

The towhe bunting is eight inches and a half long, and eleven broad; above, black, which also descends, rounding on the breast, the sides of which are bright bay, spreading along under the wing; the belly is white; the vent, pale rufous; a spot of white marks the wings just below the coverts, and another a little below that extends obliquely across the primaries; the tail is long, nearly even at the end; the three exterior feathers white for an inch or so from the tips, the outer one wholly white, the middle ones black; the bill is black; the legs and feet, a dirty flesh colour, and strong, for scratching up the ground. The female differs in being of a light reddish brown in those parts where the male is black, and in having the bill more of a light horn colour.*

* Mr Swainson makes *Pipilo* a subgenus among the sparrows. Six species have been described, and the above-mentioned gentleman has lately received two in addition. They are confined to both continents of America, and the species of our author was considered as the only one belonging to the northern parts. The "Northern Zoology" will give to the public a second under the title *Pipilo arctica*, which was only met with on the plains of the Saskatchewan, where it was supposed to breed, from a specimen being killed late in July. It frequents shady and moist clumps of wood, and is generally seen on the ground. It feeds on grubs; is a solitary and retired, but not distrustful bird. It approaches nearest to the Mexican *Pipilo maculata*, Sw.

Mr Audubon says, "The haunts of the towhe bunting are dry barren

CARDINAL GROSBEAK. (*Loxia cardinalis.*)

PLATE XI.—Figs. 1 and 2.

Linn. Syst. i. p. 300, No. 5.—Le Gros-bec de Virginie, *Briss. Orn.* iii. p. 255, No. 17.—*Buff.* iii. p. 458, pl. 28. *Pl. enl.* 37.—*Lath. Syn.* ii. p. 118, No. 13. —Cardinal, *Brown's Jam.* p. 647.—*Peale's Museum*, No. 5668.

GUARICA CARDINALIS.—Swainson.

Fringilla cardinalis, *Bonap. Synop.* p. 113.

This is one of our most common cage birds; and is very generally known, not only in North America, but even in Europe, numbers of them having been carried over both to France and England, in which last country they are usually called Virginia nightingales. To this name, Dr Latham observes, "they are fully entitled," from the clearness and variety of their notes, which, both in a wild and domestic state, are very various and musical: many of them resemble the high notes of a fife, and are nearly as loud. They are in song from March to September, beginning at the first appearance of dawn, and repeating a favourite stanza, or passage, twenty or thirty times successively; sometimes, with little intermission, for a whole morning together, which, like a good story too often repeated, becomes at length tiresome and insipid. But the sprightly figure and gaudy plumage of the

tracts; but not, as others have said, low and swampy grounds, at least during the season of incubation." The name of *swamp robin* would indicate something the reverse of this, and provincial names are generally pretty correct in their application; different habits may perhaps be sought at different seasons. In "the Barrens of Kentucky they are found in the greatest abundance. They rest upon the ground at night. Their migrations are performed by day, from bush to bush; and they seem to be much at a loss when a large extent of forest is to be traversed by them. They perform these journeys almost singly. The females set out before the males in autumn, the males before the females in spring; the latter not appearing in the middle districts until the end of April, a fortnight after the males had arrived."—Ed.

Drawn from Nature by A. Wilson

Engraved by W.H. Lizars

1. Cardinal Grosbeak. 2. Female & egg. 3. Red Tanager. 4. Female & egg. II.

red bird, his vivacity, strength of voice, and actual variety of note, and the little expense with which he is kept, will always make him a favourite.

This species, like the mocking bird, is more numerous to the east of the great range of the Alleghany Mountains, and inhabits from New England to Carthagena. Michaux the younger, son to the celebrated botanist, informed me that he found this bird numerous in the Bermudas. In Pennsylvania and the northern States, it is rather a scarce species; but through the whole lower parts of the southern States, in the neighbourhood of settlements, I found them much more numerous; their clear and lively notes, in the months of January and February, being, at that time, almost the only music of the season. Along the roadsides and fences I found them hovering in half-dozens together, associated with snow birds, and various kinds of sparrows. In the northern States, they are migratory; but in the lower parts of Pennsylvania, they reside during the whole year, frequenting the borders of creeks and rivulets, in sheltered hollows, covered with holly, laurel, and other evergreens. They love also to reside in the vicinity of fields of Indian-corn, a grain that constitutes their chief and favourite food. The seeds of apples, cherries, and of many other sorts of fruit, are also eaten by them; and they are accused of destroying bees.

In the months of March and April, the males have many violent engagements for their favourite females. Early in May, in Pennsylvania, they begin to prepare their nest, which is very often fixed in a holly, cedar, or laurel bush. Outwardly, it is constructed of small twigs, tops of dry weeds, and slips of vine bark, and lined with stalks of fine grass. The female lays four eggs, thickly marked all over with touches of brownish olive, on a dull white ground, as represented in the figure; and they usually raise two broods in the season. These birds are rarely raised from the nest for singing, being so easily taken in trap cages, and soon domesticated. By long confinement, and perhaps unnatural food, they are found to fade in colour, becoming of a pale whitish red. If well taken care of, however, they will live to a considerable age. There is at present in Mr Peale's museum the stuffed skin of one of these birds, which is there said to have lived in a cage upwards of twenty-one years.

The opinion which so generally prevails in England, that the music of the groves and woods of America is far inferior to that of Europe, I, who have a thousand times listened to both, cannot admit to be correct. We cannot with fairness draw a comparison between the depth of the forest in America and the cultivated fields of England; because it is a well-known fact, that singing birds seldom frequent the former in any country. But let the latter places be compared with the like situations in the United States, and the superiority of song, I am fully persuaded, would justly belong to the western continent. The few of our song birds that have visited Europe extort admiration from the best judges. "The notes of the cardinal grosbeak," says Latham, "are almost equal to those of the nightingale." Yet these notes, clear and excellent as they are, are far inferior to those of the wood thrush, and even to those of the brown thrush, or thrasher. Our inimitable mocking bird is also acknowledged by themselves to be fully equal to the song of the nightingale, "in its whole compass." Yet these are not one-tenth of the number of our singing birds. Could these people be transported to the borders of our woods and settlements in the month of May, about half an hour before sunrise, such a ravishing concert would greet their ear as they have no conception of.

The males of the cardinal grosbeak, when confined together in a cage, fight violently. On placing a looking-glass before the cage, the gesticulations of the tenant are truly laughable; yet with this he soon becomes so well acquainted, that, in a short time, he takes no notice whatever of it; a pretty good proof that he has discovered the true cause of the appearance to proceed from himself. They are hardy birds, easily kept, sing six or eight months in the year, and are most lively in wet weather. They are generally known by the names, red bird, Virginia red bird, Virginia nightingale, and crested red bird, to distinguish them from another beautiful species, which is represented on the same plate.

I do not know that any successful attempts have been made to induce these birds to pair and breed in confinement; but I have no doubt of its practicability, by proper management. Some months ago, I placed a young unfledged cow bird (the *Fringilla pecoris* of Turton), whose mother, like the cuckoo of Europe, abandons her eggs and progeny to the mercy and management of other smaller birds, in the same cage with a red bird, which fed and reared it with great tenderness. They both continue to inhabit the same cage, and I have hopes that the red bird will finish his pupil's education by teaching him his song.

I must here remark, for the information of foreigners, that the story told by Le Page du Pratz, in his "History of Louisiana," and which has been so often repeated by other writers, that the cardinal grosbeak "collects together great hoards of maize and buckwheat, often as much as a bushel, which it artfully covers with leaves and small twigs, leaving only a small hole for entrance into the magazine," is entirely fabulous.

This species is eight inches long, and eleven in extent; the whole upper parts are a dull, dusky red, except the sides of the neck and head, which, as well as the whole lower parts, are bright vermilion; chin, front, and lores, black; the head is ornamented with a high, pointed crest, which it frequently erects in an almost perpendicular position, and can also flatten at pleasure, so as to be scarcely perceptible; the tail extends three inches beyond the wings, and is nearly even at the end; the bill is of a brilliant coralline colour, very thick and powerful, for breaking hard grain and seeds; the legs and feet, a light clay colour (not blood red, as Buffon describes them); iris of the eye, dark hazel. The female is less than the male, has the upper parts of a brownish olive, or drab colour, the tail, wings, and tip of the crest excepted, which are nearly as red as those of the male; the lores, front, and chin are light ash; breast, and lower parts, a reddish drab; bill, legs, and eyes, as those of the male; the crest is shorter, and less frequently raised.

One peculiarity in the female of this species is, that she often sings nearly as well as the male. I do not know whether it be owing to some little jealousy on this score or not, that the male, when both occupy the same cage, very often destroys the female.

SCARLET TANAGER. (*Tanagra rubra.*)

PLATE XI.—FIGS. 3 AND 4.

Tanagra rubra, *Linn. Syst.* i. p. 314, 3.—Cardinal de Canada, *Briss. Orn.* iii. p. 48, pl. 2, fig. 5.—*Lath.* ii. p. 217, No. 3.—Scarlet Sparrow, *Edw.* pl. 343.—Canada Tanager, and Olive Tanager, *Arct. Zool.* p. 369, No. 237, 238.—*Peale's Museum*, No. 6128.

PYRANGA RUBRA.*—SWAINSON.

Pyranga erythropis, *Vieill. Enc. Méthod.* p. 793.—Tanagra rubra, *Bonap. Synop.* p. 105.—Pyranga rubra, *North Zool.* ii. p. 273.

THIS is one of the gaudy foreigners (and perhaps the most showy) that regularly visit us from the torrid regions of the south. He is drest in the richest scarlet, set off with the most jetty black, and comes, over extensive countries, to sojourn

* *Pyranga* has been established for the reception of this bird as the type, and a few others, all natives of the New World, and more particularly inhabiting the warmer parts of it. The present species is, indeed, the only one which is common to the north and south continents; and, in the former, it ranks only as a summer visitant. They are all of very bright colours, and distinct markings. They are distinguished from the true tanagers by their stout and rounded bill, slightly notched, bent at the tip, and having a jutting out blunt tooth about the middle of the upper mandible. They are placed by Desmarest among his *Tangaras colluriens*, or shrike-like tanagers; and by Lesson among the *Tangaras cardinals*. The latter writer enumerates only three species belonging to his division.—ED.

for a time among us. While we consider him entitled to all the rights of hospitality, we may be permitted to examine a little into his character, and endeavour to discover whether he has anything else to recommend him, besides that of having a fine coat, and being a great traveller.

On or about the 1st of May, this bird makes his appearance in Pennsylvania. He spreads over the United States, and is found even in Canada. He rarely approaches the habitations of man, unless, perhaps, to the orchard, where he sometimes builds; or to the cherry trees, in search of fruit. The depth of the woods is his favourite abode. There, among the thick foliage of the tallest trees, his simple and almost monotonous notes, *chip, churr*, repeated at short intervals, in a pensive tone, may be occasionally heard, which appear to proceed from a considerable distance, though the bird be immediately above you,—a faculty bestowed on him by the beneficent Author of Nature, no doubt for his protection, to compensate, in a degree, for the danger to which his glowing colour would often expose him. Besides this usual note, he has, at times, a more musical chant, something resembling in mellowness that of the Baltimore oriole. His food consists of large winged insects, such as wasps, hornets, and humble bees, and also of fruit, particularly those of that species of *Vaccinium* usually called huckleberries, which, in their season, form almost his whole fare. His nest is built, about the middle of May, on the horizontal branch of a tree, sometimes an apple tree, and is but slightly put together; stalks of broken flax, and dry grass, so thinly woven together, that the light is easily perceivable through it, form the repository of his young. The eggs are three, of a dull blue, spotted with brown or purple. They rarely raise more than one brood in a season, and leave us for the south about the last week in August.

Among all the birds that inhabit our woods, there is none that strikes the eye of a stranger, or even a native, with so much brilliancy as this. Seen among the green leaves, with the light falling strongly on his plumage, he really appears beautiful. If he has little of melody in his notes to charm us, he has nothing in them to disgust. His manners are modest, easy, and inoffensive. He commits no depredations on the property of the husbandman, but rather benefits him by the daily destruction, in spring, of many noxious insects; and when winter approaches, he is no plundering dependant, but seeks, in a distant country, for that sustenance which the severity of the season denies to his industry in this. He is a striking ornament to our rural scenery, and none of the meanest of our rural songsters. Such being the true traits of his character, we shall always with pleasure welcome this beautiful, inoffensive stranger to our orchards, groves, and forests.

The male of this species, when arrived at his full size and colours, is six inches and a half in length, and ten and a half broad. The whole plumage is of a most brilliant scarlet, except the wings and tail, which are of a deep black; the latter, handsomely forked, sometimes minutely tipped with white, and the interior edges of the wing-feathers nearly white; the bill is strong, considerably inflated, like those of his tribe; the edge of the upper mandible, somewhat irregular, as if toothed, and the whole of a dirty gamboge, or yellowish horn colour; this, however, like that of most other birds, varies according to the season. About the 1st of August he begins to moult; the young feathers coming out, of a greenish yellow colour, until he appears nearly all dappled with spots of scarlet and greenish yellow. In this state of plumage he leaves us. How long it is before he recovers his scarlet dress, or whether he continues of this greenish colour all winter, I am unable to say. The iris of the eye is of a cream colour; the legs and feet, light blue. The female (now, I believe, for the first time figured) is green above and yellow below; the wings and tail, brownish black, edged with green. The young birds, during their residence here the first season, continue nearly of the same colour with the female. In this circumstance we again recognise the wise provision of the Deity, in thus clothing the female, and the inexperienced young, in a garb so favourable for concealment among the foliage; as the weakness of the one, and the frequent visits of the other to her nest, would greatly endanger the safety of all. That the young males do not receive their red plumage until the early part of the succeeding spring, I think highly probable, from the circumstance of frequently finding their red feathers, at that season, intermixed with green ones, and the wings also broadly edged with green. These facts render it also probable that the old males regularly change their colour, and have a summer and winter dress; but this further observations must determine.

There is in the Brazils a bird of the same genus with this, and very much resembling it, so much so as to have been frequently confounded with it by European writers. It is the *Tanagra Brazilia* of Turton; and, though so like, is yet a very distinct species from the present, as I have myself had the opportunity of ascertaining, by examining two very perfect specimens from Brazil, now in the possession of Mr Peale, and comparing them with this. The principal differences are these: The plumage of the Brazilian is almost black at bottom, very deep scarlet at the surface, and of an orange tint between; ours is ash coloured at bottom, white in the middle, and bright scarlet at top. The tail of ours is forked, that of the other cuneiform, or rounded. The bill of our species is more inflated, and of a greenish yellow colour; the other's is black above, and whitish below, towards the base. The whole plumage of the southern species is of a coarser, stiffer quality, particularly on the head. The wings and tail, in both, are black.

In the account which Buffon gives of the scarlet tanager, and cardinal grosbeak, there appears to be very great confusion, and many mistakes; to explain which, it is necessary to observe, that Mr Edwards, in his figure of the scarlet tanager, or scarlet sparrow, as he calls it, has given it a hanging crest, owing, no doubt, to the loose disordered state of the plumage of the stuffed or dried skin from which he made his drawing.

Buffon has afterwards confounded the two together, by applying many stories, originally related of the cardinal grosbeak, to the scarlet tanager; and the following he gravely gives as his reason for so doing: "We may presume," says he, "that when travellers talk of the warble of the cardinal, they mean the scarlet cardinal, for the other cardinal is of the genus of the grosbeaks, consequently a silent bird." * This silent bird, however, has been declared by an eminent English naturalist, to be almost equal to their own nightingale! The Count also quotes the following passage from Charlevoix to prove the same point, which, if his translator has done him justice, evidently proves the reverse:—"It is scarcely more than a hundred leagues," says this traveller, "south of Canada, that the cardinal begins to be seen. Their song is sweet, their plumage beautiful, and their head wears a crest." But the scarlet tanager is found even in Canada, as well as a hundred leagues to the south, while the cardinal grosbeak is not found in any great numbers north of Maryland. The latter, therefore, it is highly probable, was the bird meant by Charlevoix, and not the scarlet tanager. Buffon also quotes an extract of a letter from Cuba, which, if the circumstance it relates be true, is a singular proof of the estimation in which the Spaniards hold the cardinal grosbeak:—"On Wednesday arrived at the port of Havannah, a bark from Florida, loaded with cardinal birds, skins, and fruit. The Spaniards bought the cardinal birds at so high a price as ten dollars apiece; and, notwithstanding the public distress, spent on them the sum of 18,000 dollars!" †

With a few facts more I shall conclude the history of the scarlet tanager. When you approach the nest, the male keeps cautiously at a distance, as if fearful of being seen; while the female hovers around in the greatest agitation and distress. When the young leave the nest, the male parent takes a most active part in feeding and attending them, and is then altogether indifferent of concealment.

* Buffon, vol. iv. p. 209. † Gmelli Careri.

Passing through an orchard one morning, I caught one of these young birds, that had but lately left the nest. I carried it with me about half a mile, to show it to my friend Mr William Bartram; and having procured a cage, hung it up on one of the large pine trees in the botanic garden, within a few feet of the nest of an orchard oriole, which also contained young; hopeful that the charity or tenderness of the orioles would induce them to supply the cravings of the stranger. But charity with them, as with too many of the human race, began and ended at home. The poor orphan was altogether neglected, notwithstanding its plaintive cries; and, as it refused to be fed by me, I was about to return it back to the place where I found it, when, towards the afternoon, a scarlet tanager, no doubt its own parent, was seen fluttering round the cage, endeavouring to get in. Finding this impracticable, he flew off, and soon returned with food in his bill; and continued to feed it till after sunset, taking up his lodgings on the higher branches of the same tree. In the morning, almost as soon as day broke, he was again seen most actively engaged in the same affectionate manner; and, notwithstanding the insolence of the orioles, continued his benevolent offices the whole day, roosting at night as before. On the third or fourth day, he appeared extremely solicitous for the liberation of his charge, using every expression of distressful anxiety, and every call and invitation that nature had put in his power, for him to come out. This was too much for the feelings of my venerable friend; he procured a ladder, and, mounting to the spot where the bird was suspended, opened the cage, took out the prisoner, and restored him to liberty and to his parent, who, with notes of great exultation, accompanied his flight to the woods. The happiness of my good friend was scarcely less complete, and showed itself in his benevolent countenance; and I could not refrain saying to myself,—If such sweet sensations can be derived from a single circumstance of this kind, how exquisite—how unspeakably rapturous—must the delight of those individuals have been, who have rescued their fellow-beings from death, chains, and imprisonment, and restored them to the arms of their friends and relations! Surely, in such godlike actions, virtue is its own most abundant reward.

RICE BUNTING. (*Emberiza oryzivora.*)

PLATE XII.—FIGS. 1 AND 2.

Emberiza oryzivora, *Linn. Syst.* p. 311, 16.—Le Ortolan de la Caroline, *Briss. Orn.* iii. p. 282, 8, pl. 15, fig. 3. *Pl. enl.* 388. fig. 1.—L'Agripenne ou l'ortolan de Riz, *Buff. Ois.* iv. p. 337.—Rice Bird, *Catesb. Car.* i. pl. 14.—*Edw.* pl. 2. —*Latham,* ii. p. 188. No. 25.—*Peale's Museum,* No. 6026.

DOLYCHONYX ORYZIVORUS.—SWAINSON.

Icterus agripennis, *Bonap. Synop.* p. 53.—Dolychonyx oryzivorus, *Sw. Synop. Birds of Mexico,* 435.—*North. Zool.* ii. p. 278.—*Aud.* pl. 54, *Orn. Biog.* i. p. 283.

THIS is the *boblink* of the eastern and northern States, and the *rice* and *reed bird* of Pennsylvania and the southern States. Though small in size, he is not so in consequence; his coming is hailed by the sportsman with pleasure; while the careful planter looks upon him as a devouring scourge, and worse than a plague of locusts. Three good qualities, however, entitle him to our notice, particularly as these three are rarely found in the same individual,—his plumage is beautiful, his song highly musical, and his flesh excellent. I might also add, that the immense range of his migrations, and the havoc he commits, are not the least interesting parts of his history.*

* To Wilson's interesting account of the habits of this curious bird, Mr Audubon adds the following particulars :—In Louisiana, they pass under the name of *meadow birds*, and they arrive there in small flocks of males and females about the middle of March or beginning of April. Their song in spring is extremely interesting, and, emitted with a volubility bordering on the burlesque, is heard from a whole party at the same time, and it becomes amusing to hear thirty or forty of them beginning one after another, as if ordered to follow in quick succession, after the first notes are given by a leader, and producing such a medley as it is impossible to describe, although it is extremely pleasant to hear.

Drawn from Nature by A. Wilson.

Engraved by W. H. Lizars

1. Rice Bunting. 2. Female. 3. Red eyed Flycatcher. 4. Marsh Wren. 5. Great Carolina Wren. 6. Yellow throat Warbler.

12.

The winter residence of this species I suppose to be from Mexico to the mouth of the Amazon, from whence, in hosts innumerable, they regularly issue every spring ; perhaps to both hemispheres, extending their migrations northerly, as far as the banks of the Illinois, and the shores of the St Lawrence. Could the fact be ascertained, which has been asserted by some writers, that the emigration of these birds was altogether unknown in this part of the continent previous to the introduction of rice plantations, it would certainly be interesting. Yet, why should these migrations reach at least a thousand miles beyond those places where rice is now planted ; and this, not in occasional excursions, but regularly to breed and rear their young, where rice never was, and probably never will be, cultivated ? Their so recent arrival on this part of the continent, I believe to be altogether imaginary, because, though there were not a single grain of rice cultivated within the United States, the country produces an exuberance of food of which they are no less fond. Insects of various kinds, grubs, Mayflies, and caterpillars, the young ears of Indian-corn, and the seed of the wild oats, or, as it is called in Pennsylvania, reeds (the *Zizania aquatica* of Linnæus), which grows in prodigious abundance along the marshy shores of our large rivers, furnish, not only them, but millions of rail, with a delicious subsistence for several weeks. I do not doubt, however, that the introduction of rice, but more particularly the progress of agriculture, in this part of America, has greatly increased their numbers, by multiplying their sources of subsistence fiftyfold within the same extent of country.

In the month of April, or very early in May, the rice bunt-

While you are listening, the whole flock simultaneously ceases, which appears equally extraordinary. This curious exhibition takes place every time the flock has alighted on a tree.

Another curious fact mentioned by this gentleman is, that during their spring migrations eastward, they fly mostly at night ; whereas, in autumn, when they are returning southward, their flight is diurnal.—ED.

ing, male and female, in the dresses in which they are figured on the plate, arrive within the southern boundaries of the United States; and are seen around the town of Savannah in Georgia about the 4th of May, sometimes in separate parties of males and females, but more generally promiscuously. They remain there but a short time; and, about the 12th of May, make their appearance in the lower parts of Pennsylvania, as they did at Savannah. While here, the males are extremely gay and full of song; frequenting meadows, newly ploughed fields, sides of creeks, rivers, and watery places, feeding on Mayflies and caterpillars, of which they destroy great quantities. In their passage, however, through Virginia, at this season, they do great damage to the early wheat and barley while in its milky state. About the 20th of May, they disappear, on their way to the north. Nearly at the same time, they arrive in the State of New York, spread over the whole New England States, as far as the river St Lawrence, from Lake Ontario to the sea; in all of which places, north of Pennsylvania, they remain during the summer, building, and rearing their young. The nest is fixed in the ground, generally in a field of grass; the outside is composed of dry leaves and coarse grass, the inside is lined with fine stalks of the same, laid in considerable quantity. The female lays five eggs, of a bluish white, marked with numerous irregular spots of blackish brown. The song of the male, while the female is sitting, is singular, and very agreeable. Mounting and hovering on wing, at a small height above the field, he chants out such a jingling medley of short, variable notes, uttered with such seeming confusion and rapidity, and continued for a considerable time, that it appears as if half a dozen birds of different kinds were all singing together. Some idea may be formed of this song by striking the high keys of a pianoforte at random, singly, and quickly, making as many sudden contrasts of high and low notes as possible. Many of the tones are, in themselves, charming; but they succeed each other so rapidly, that the ear can hardly separate them. Nevertheless the general effect is good; and, when ten or twelve are all singing on the same tree, the concert is singularly pleasing. I kept one of these birds for a long time, to observe its change of colour. During the whole of April, May, and June, it sang almost continually. In the month of June, the colour of the male begins to change, gradually assimilating to that of the female, and before the beginning of August it is difficult to distinguish the one from the other, both being then in the dress of fig. 2. At this time, also, the young birds are so much like the female, or rather like both parents, and the males so different in appearance from what they were in spring, that thousands of people in Pennsylvania, to this day, persist in believing them to be a different species altogether; while others allow them, indeed, to be the same, but confidently assert that they are all females—none but females, according to them, returning in the fall; what becomes of the males they are totally at a loss to conceive. Even Mr Mark Catesby, who resided for years in the country they inhabit, and who, as he himself informs us, examined by dissection great numbers of them in the fall, and repeated his experiment the succeeding year, lest he should have been mistaken, declares that he uniformly found them to be females. These assertions must appear odd to the inhabitants of the eastern States, to whom the change of plumage in these birds is familiar, as it passes immediately under their eye; and also to those who, like myself, have kept them in cages, and witnessed their gradual change of colour.* That accurate observer, Mr William Bartram, appears, from the following

* The beautiful plumage of the male represented on the plate is that during the breeding-season, and is lost as soon as the duties incumbent thereon are completed. In this we have a striking analogy with some nearly allied African *Fringillidæ.*

The genus *Dolyconyx* has been made by Mr Swainson to contain this curious and interesting form: by that gentleman it is placed in the aberrant families of the *Sturnidæ.*—ED.

extract, to have taken notice of, or at least suspected, this change of colour in these birds, more than forty years ago. "Being in Charleston," says he, "in the month of June, I observed a cage full of rice birds, that is, of the yellow, or female colour, who were very merry and vociferous, having the same variable music with the pied, or male bird, which I thought extraordinary, and, observing it to the gentleman, he assured me that they were all of the male kind, taken the preceding spring; but had changed their colour, and would be next spring of the colour of the pied, thus changing colour with the seasons of the year. If this is really the case, it appears they are both of the same species intermixed, spring and fall." Without, however, implicating the veracity of Catesby, who, I have no doubt, believed as he wrote, a few words will easily explain why he was deceived. The internal organisation of undomesticated birds, of all kinds, undergoes a remarkable change every spring and summer; and those who wish to ascertain this point by dissection will do well to remember, that in this bird those parts that characterise the male are, in autumn, no larger than the smallest pin's head, and in young birds of the first year can scarcely be discovered; though in spring their magnitude in each is at least one hundred times greater. To an unacquaintance with this extraordinary circumstance, I am persuaded, has been owing the mistake of Mr Catesby, that the females only return in the fall; for the same opinion I long entertained myself, till a more particular examination showed me the source of my mistake. Since that, I have opened and examined many hundreds of these birds, in the months of September and October, and, on the whole, have found about as many males as females among them. The latter may be distinguished from the former by being of a rather more shining yellow on the breast and belly: it is the same with the young birds of the first season.

During the breeding season, they are dispersed over the country; but, as soon as the young are able to fly, they collect together in great multitudes, and pour down on the oatfields of New England like a torrent, depriving the proprietors of a good tithe of their harvest; but, in return, often supply his table with a very delicious dish. From all parts of the north and western regions, they direct their course towards the south; and, about the middle of August, revisit Pennsylvania, on their route to winter quarters. For several days, they seem to confine themselves to the fields and uplands; but, as soon as the seeds of the reed are ripe, they resort to the shores of the Delaware and Schuylkill in multitudes; and these places, during the remainder of their stay, appear to be their grand rendezvous. The reeds, or wild oats, furnish them with such abundance of nutritious food, that in a short time they become extremely fat; and are supposed, by some of our epicures, to be equal to the famous ortolans of Europe. Their note at this season is a single *chink*, and is heard overhead, with little intermission, from morning to night. These are halcyon days for our gunners of all descriptions, and many a lame and rusty gun-barrel is put in requisition for the sport. The report of musketry along the reedy shores of the Schuylkill and Delaware is almost incessant, resembling a running fire. The markets of Philadelphia, at this season, exhibit proofs of the prodigious havoc made among these birds; for almost every stall is ornamented with strings of reed birds. This sport, however, is considered inferior to that of rail-shooting, which is carried on at the same season and places, with equal slaughter. Of this, as well as of the rail itself, we shall give a particular account in its proper place.

Whatever apology the people of the eastern and southern States may have for the devastation they spread among the rice and reed birds, the Pennsylvanians—at least those living in this part of it—have little to plead in justification, but the pleasure of destruction, or the savoury dish they furnish their tables with; for the oat harvest is generally secured before the great body of these birds arrive, the Indian-corn too ripe and hard, and the reeds seem to engross all their attention.

But in the States south of Maryland, the harvest of early wheat and barley in spring, and the numerous plantations of rice in fall, suffer severely. Early in October, or as soon as the nights begin to set in cold, they disappear from Pennsylvania, directing their course to the south. At this time they swarm among the rice fields; and appear in the island of Cuba in immense numbers, in search of the same delicious grain. About the middle of October, they visit the island of Jamaica in equal numbers, where they are called butter birds. They feed on the seed of the guinea grass, and are also in high esteem there for the table.*

Thus it appears, that the regions north of the fortieth degree of latitude are the breeding places of these birds; that their migrations northerly are performed from March to May, and their return southerly from August to November; their precise winter quarters, or farthest retreat southerly, is not exactly known.

The rice bunting is seven inches and a half long, and eleven and a half in extent; his spring dress is as follows:— Upper part of the head, wings, tail, and sides of the neck, and whole lower parts, black; the feathers frequently skirted with brownish yellow, as he passes into the colours of the female; back of the head, a cream colour; back, black, seamed with brownish yellow; scapulars, pure white; rump and tail-coverts the same; lower part of the back, bluish white; tail, formed like those of the woodpecker genus, and often used in the same manner, being thrown in to support it while ascending the stalks of the reed; this habit of throwing in the tail it retains even in the cage; legs, a brownish flesh colour; hind heel, very long; bill, a bluish horn colour; eye, hazel; see fig. 1. In the month of June this plumage gradually changes to a brownish yellow, like that of the female, fig. 2, which has the back streaked with brownish black; whole lower parts, dull yellow; bill, reddish flesh colour; legs and eyes as in the male. The young birds retain the dress of the

* Rennel's Hist. Jam.

female until the early part of the succeeding spring; the plumage of the female undergoes no material change of colour.

RED-EYED FLYCATCHER. (*Muscicapa olivacea.*)

PLATE XII.—Fig. 3.

Linn. Syst. i. p. 327, 14.—Gobe-mouche de la Caroline et de la Jamaique, *Buff.* iv. p. 539. *Edw.* t. 253.—*Catesb.* t. 54. *Lath. Syn.* iii. p. 351, No. 52.—Muscicapa sylvicola, *Bartram*, p. 290.—*Peale's Museum*, No. 6675.

VIREO OLIVACEUS.—Bonaparte.

Vireo olivaceus, *Bonap. Synop.* p. 71.—Vireo olivaceus, Red-eyed Greenlet, *North. Zool.* ii. p. 233.

This is a numerous species, though confined chiefly to the woods and forests, and, like all the rest of its tribe that visit Pennsylvania, is a bird of passage. It arrives here late in April; has a loud, lively, and energetic song, which it continues, as it hunts among the thick foliage, sometimes for an hour with little intermission. In the months of May, June, and to the middle of July, it is the most distinguishable of all the other warblers of the forest; and even in August, long after the rest have almost all become mute, the notes of the red-eyed flycatcher are frequently heard with unabated spirit. These notes are in short, emphatical bars, of two, three, or four syllables. In Jamaica, where this bird winters, and is probably also resident, it is called, as Sloane informs us, whip-tom-kelly, from an imagined resemblance of its notes to these words. And, indeed, on attentively listening for some time to this bird in his full ardour of song, it requires but little of imagination to fancy that you hear it pronounce these words, "Tom-kelly, whip-tom-kelly!" very distinctly. It inhabits from Georgia to the river St Lawrence, leaving Pennsylvania about the middle of September.

This bird builds, in the month of May, a small, neat, pensile nest, generally suspended between two twigs of a young dogwood or other small sapling. It is hung by the two upper edges, seldom at a greater height than four or five feet from the ground. It is formed of pieces of hornets' nests, some flax, fragments of withered leaves, slips of vine bark, bits of paper, all glued together with the saliva of the bird, and the silk of caterpillars, so as to be very compact; the inside is lined with fine slips of grape-vine bark, fibrous grass, and sometimes hair. These nests are so durable, that I have often known them to resist the action of the weather for a year; and, in one instance, I have found the nest of the yellow bird built in the cavity of one of those of the preceding year. The mice very often take possession of them after they are abandoned by the owners. The eggs are four, sometimes five, pure white, except near the great end, where they are marked with a few small dots of dark brown or reddish. They generally raise two broods in the season.

The red-eyed flycatcher is one of the adopted nurses of the cow bird, and a very favourite one, showing all the symptoms of affection for the foundling, and as much solicitude for its safety, as if it were its own. The figure of that singular bird, accompanied by a particular account of its history, is given in Plate XVIII. of the present volume.

Before I take leave of this bird, it may not be amiss to observe that there is another, and a rather less species of flycatcher, somewhat resembling the red-eyed, which is frequently found in its company. Its eyes are hazel; its back more cinereous than the other, and it has a single light streak over the eye. The notes of this bird are low, somewhat plaintive, but warbled out with great sweetness; and form a striking contrast with those of the red-eyed flycatcher. I think it probable that Dr Barton had reference to this bird when he made the following remarks (see his "Fragments of the Natural History of Pennsylvania," page 19) :—"*Muscicapa olivacea.*—I do not think, with Mr Pennant, that this is the same bird as the whip-tom-kelly of the West Indies. Our bird has no such note; but a great variety of soft, tender, and agreeable notes. It inhabits forests; and does not, like the West India bird, build a pendulous nest." Had the learned professor, however, examined into this matter with his usual accuracy, he would have found, that the *Muscicapa olivacea*, and the soft and tender songster he mentions, are two very distinct species; and that both the one and the other actually build very curious pendulous nests.

This species is five inches and a half long, and seven inches in extent; crown, ash, slightly tinged with olive, bordered on each side with a line of black, below which is a line of white passing from the nostril over and a little beyond the eye; the bill is longer than usual with birds of its tribe, the upper mandible overhanging the lower considerably, and notched, dusky above, and light blue below; all the rest of the plumage above is of a yellow olive, relieved on the tail and at the tips of the wings with brown; chin, throat, breast, and belly, pure white; inside of the wings and vent-feathers, greenish yellow; the tail is very slightly forked; legs and feet, light blue; iris of the eye, red. The female is marked nearly in the same manner, and is distinguishable only by the greater obscurity of the colours.

MARSH WREN. (*Certhia palustris.*)

PLATE XII.—Fig. 4.

Lath. Syn. Suppl. p. 244.—Motacilla palustris (regulus minor), *Bartram*, p. 291. —*Peale's Museum*, No. 7282.

TROGLODYTES PALUSTRIS.—Bonaparte.

Troglodytes palustris, *Bonap. Synop.* p. 93.—The Marsh Wren, *Aud.* pl. 100, *Orn. Biog.* i. p. 500.—*North. Zool.* ii. p. 319.

This obscure but spirited little species has been almost overlooked by the naturalists of Europe, as well as by those of its own country. The singular attitude in which it is represented will be recognised, by those acquainted with its manners, as one of its most common and favourite ones, while skipping through among the reeds and rushes. The marsh wren arrives in Pennsylvania about the middle of May, or as soon as the reeds and a species of nymphea, usually called splatter-docks, which grow in great luxuriance along the tide water of our rivers, are sufficiently high to shelter it. To such places it almost wholly limits its excursions, seldom venturing far from the river. Its food consists of flying insects and their larvæ, and a species of green grasshoppers that inhabit the reeds. As to its notes, it would be mere burlesque to call them by the name of song. Standing on the reedy borders of the Schuylkill or Delaware, in the month of June, you hear a low, crackling sound, something similar to that produced by air bubbles forcing their way through mud or boggy ground when trod upon; this is the song of the marsh wren. But as, among the human race, it is not given to one man to excel in every thing, and yet each, perhaps, has something peculiarly his own; so, among birds, we find a like distribution of talents and peculiarities. The little bird now before us, if deficient and contemptible in singing, excels in the art of design, and constructs a nest, which, in durability, warmth, and convenience, is scarcely inferior to one, and far superior to many, of its more musical brethren. This is formed outwardly of wet rushes mixed with mud, well intertwisted, and fashioned into the form of a cocoa-nut. A small hole is left two-thirds up, for entrance, the upper edge of which projects like a penthouse over the lower, to prevent the admission of rain. The inside is lined with fine soft grass, and sometimes feathers; and the outside, when hardened by the sun, resists every kind of weather. This nest is generally suspended among the reeds, above the reach of the highest tides, and is tied so fast in every part to the surrounding reeds, as to bid defiance to the winds and the waves. The eggs are usually six, of a dark fawn colour, and very small. The young leave the nest about the 20th of June, and they generally have a second brood in the same season.

The size, general colour, and habit of this bird of erecting its tail, give it, to a superficial observer, something of the appearance of the common house wren, represented in Plate VIII.; and still more that of the winter wren, figured in the same plate; but with the former of these it never associates; and the latter has left us some time before the marsh wren makes his appearance. About the middle of August, they begin to go off; and on the 1st of September, very few of them are to be seen. How far north the migrations of this species extend, I am unable to say; none of them, to my knowledge, winter in Georgia, or any of the southern States.

The marsh wren is five inches long, and six in extent; the whole upper parts are dark brown, except the upper part of the head, back of the neck, and middle of the back, which are black, the two last streaked with white; the tail is short, rounded, and barred with black; wings, slightly barred; a broad strip of white passes over the eye half way down the neck; the sides of the neck are also mottled with touches of a light clay colour on a whitish ground; whole under parts, pure silvery white, except the vent, which is tinged with brown; the legs are light brown; the hind claw, large, semicircular, and very sharp; bill, slender, slightly bent; nostrils, prominent; tongue, narrow, very tapering, sharp pointed, and horny at the extremity; eye, hazel. The female almost exactly resembles the male in plumage.

From the above description and a view of the figure, the naturalist will perceive that this species is truly a *Certhia*, or creeper; and indeed its habits confirm this, as it is continually climbing along the stalks of reeds, and other aquatic plants, in search of insects.

GREAT CAROLINA WREN. (*Certhia Caroliniana.*)

PLATE XII.—Fig. 5.

Le Roitelet de la Louisiana, *Pl. enl.* 730, fig. 1.—*Lath. Syn.* vii. p. 507, var. B.—Le Troglodytes de la Louisiana, *Buff. Ois.* v. p. 361.—Motacilla Caroliniana (regulus magnus), *Bartram*, p. 291.—*Peale's Museum*, No. 7248.

TROGLODYTES LUDOVICIANUS.—Bonaparte.

Troglodytes Ludovicianus, *Bonap. Synop.* p. 93.—The Great Carolina Wren, *Aud.* pl. 78, male and female, *Orn. Biog.* i. p. 399.

This is another of those equivocal species that so often occur to puzzle the naturalist. The general appearance of this bird is such, that the most illiterate would at first sight call it a wren; but the common wren of Europe, and the winter wren of the United States, are both warblers, judging them according to the simple principle of Linnæus. The present species, however, and the preceding (the marsh wren), though possessing great family likeness to those above mentioned, are decisively creepers, if the bill, the tongue, nostrils, and claws, are to be the criteria by which we are to class them.

The colour of the plumage of birds is but an uncertain and inconstant guide; and though in some cases it serves to furnish a trivial or specific appellation, yet can never lead us to the generic one. I have, therefore, notwithstanding the general appearance of these birds, and the practice of former ornithologists, removed them to the genus *Certhia*, from that of *Motacilla*, where they have hitherto been placed.*

* Of this bird, and some others, Vieillot formed his genus *Tryothorus*, containing the larger wrens, with long and somewhat curved bills, and possessing, if possible, more of the habits of the creepers. This has, with almost universal consent, been laid aside even as a subgenus, and they are all included in *Troglodytes.* Read the descriptions of our author, or of Audubon, and the habits of the *wren* will be at once perceived. "Its tail," says the latter ornithologist, "is almost constantly erect; and before it starts to make the least flight, it uses a quick motion, which brings its body almost in contact with the object on which it stands. The quickness of the motions of this little bird is fully equal

This bird is frequently seen, early in May, along the shores of the Delaware, and other streams that fall into it on both sides, thirty or forty miles below Philadelphia; but is rather rare in Pennsylvania. This circumstance is a little extraordinary, since, from its size and stout make, it would seem more capable of braving the rigours of a northern climate than any of the others. It can, however, scarcely be called migratory. In the depth of winter I found it numerous in Virginia, along the shores and banks of the James River, and its tributary streams, and thence as far south as Savannah. I also observed it on the banks of the Ogechee. It seemed to be particularly attached to the borders of cypress swamps, deep hollows, among piles of old decaying timber, and by rivers and small creeks. It has all the restless, jerking manners of the wrens, skipping about with great nimbleness, hopping into caves, and disappearing into holes and crevices, like a rat, for several minutes, and then reappearing in another quarter. It occasionally utters a loud, strong, and singular twitter, resembling the word *chirr-rup*, dwelling long and strongly on the first syllable; and so loud, that I at first mistook it for the red bird (*L. cardinalis*). It has also another chant, rather more musical, like "*sweet William sweet William*," much softer than the former. Though I cannot positively say, from my own observations, that it builds in Pennsylvania, and have never yet been so fortunate as to find its nest, yet, from the circumstance of having several times observed it within a quarter of a mile of the Schuylkill, in the month of August, I have no doubt that some few breed here; and think it highly probable that Pennsylvania and New York may be the northern

to that of a mouse: it appears, and is out of sight in a moment; peeps into a crevice, passes rapidly through it, and shows itself at a different place the next instant. These wrens often sing from the roof of an abandoned flat-boat. When the song is finished, they creep from one board to another, thrust themselves through an auger hole, entering the boat's side at one place, and peeping out at another." In them we have exactly portrayed the manners of our British wren when engaged about a heap of rubbish, old stones, or barrels in a farmyard.—ED.

boundaries of their visits, having sought for it in vain among the States of New England. Its food appears to consist of those insects, and their larvæ, that frequent low, damp caves, piles of dead timber, old roots, projecting banks of creeks, &c. It certainly possesses the faculty of seeing in the dark better than day birds usually do; for I have observed it exploring the recesses of caves, where a good acute eye must have been necessary to enable it to distinguish its prey.

In the southern States, as well as in Louisiana, this species is generally resident; though in summer they are more numerous, and are found rather farther north than in winter. In this last season their chirruping is frequently heard in gardens soon after daybreak, and along the borders of the great rivers of the southern States, not far from the sea-coast.

The great wren of Carolina is five inches and a quarter long, and seven broad; the whole upper parts are reddish brown, the wings and tail being barred with black; a streak of yellowish white runs from the nostril over the eye, down the side of the neck, nearly to the back; below that, a streak of reddish brown extends from the posterior part of the eye to the shoulder; the chin is yellowish white; the breast, sides, and belly, a light rust colour, or reddish buff; vent feathers, white, neatly barred with black; in the female, plain; wing-coverts, minutely tipt with white; legs and feet, flesh coloured, and very strong; bill, three-quarters of an inch long, strong, a little bent, grooved, and pointed; the upper mandible, bluish black; lower, light blue; nostrils, oval, partly covered with a prominent convex membrane; tongue, pointed and slender; eyes, hazel; tail, cuneiform, the two exterior feathers on each side three-quarters of an inch shorter, whitish on their exterior edges, and touched with deeper black; the same may be said of the three outer primaries. The female wants the white on the wing-coverts, but differs little in colour from the male.

In this species I have observed a circumstance common to

the house and winter wren, but which is not found in the marsh wren; the feathers of the lower part of the back, when parted by the hand or breath, appear spotted with white, being at bottom deep ash, reddish brown at the surface, and each feather with a spot of white between these two colours. This, however, cannot be perceived without parting the feathers.

YELLOW-THROAT WARBLER. (*Sylvia flavicollis.*)

PLATE XII.—Fig. 6.

Yellow-throat Warbler,* *Arct. Zool.* p. 400, No. 286.—*Catesb.* i. 62.—*Lath.* ii. 437.—La Mesange grise à gorge jaune, *Buff.* v. 454.—La gorge jaune de St Domingue, *Pl. enl.* 686, fig. 1.

SYLVICOLA FLAVICOLLIS.—Swainson.

Sylvia pensilis, *Bonap. Synop.* p. 79.—S. pensilis, *Lath.*

The habits of this beautiful species, like those of the preceding, are not consistent with the shape and construction of

* As with many others, there has been some confusion in the synonyms of this species, and it has been described under different names by the same authors. That of *flavicollis*, adopted by our author, is characteristic of the markings; whereas *pensilis* of Latham and Vieillot is applicable to the whole group; and perhaps restoring Wilson's name will create less confusion than taking one less known. The genus *Sylvicola*, with the subgenus *Vermivora*, have been used by Mr Swainson to designate almost all those birds in North America which will represent the European *Sylvianæ*, or warblers. They are generally of a stronger make; the bill, though slender, is more conical, and the wings have the first and second quills of nearly equal length. The general dress is chaste and unobtrusive; but, at the same time, we have exceptions, showing great brilliancy and beauty of colouring. Their habits are precisely the same with our warblers. They frequent woods and thickets. They are in constant motion, creeping and clinging about the branches, and inspecting the crevices in the bark, or under sides of the leaves, in search of insects. When their duties of incubation are over, they become less retired, and, with their broods, assemble in the gardens and cultivated grounds, where they find sustenance in the various fruits and berries. The notes of all are sprightly and pleasant; and a few possess a melody hardly inferior to the best songsters of Europe.

its bill; the former would rank it with the titmouse, or with the creepers, the latter is decisively that of the warbler. The first opportunity I had of examining a living specimen of this bird was in the southern parts of Georgia, in the month of February. Its notes, which were pretty loud and spirited, very much resembled those of the indigo bird. It continued a considerable time on the same pine tree, creeping around the branches, and among the twigs, in the manner of the titmouse, uttering its song every three or four minutes. On flying to another tree, it frequently alighted on the body, and ran nimbly up or down, spirally and perpendicularly, in search of insects. I had afterwards many opportunities of seeing others of the same species, and found them all to correspond in these particulars. This was about the 24th of February, and the first of their appearance there that spring, for they leave the United States about three months during winter, and, consequently, go to no great distance. I had been previously informed that they also pass the summer in Virginia, and in the southern parts of Maryland; but they very rarely proceed as far north as Pennsylvania.

This species is five inches and a half in length, and eight

Mr Audubon has figured the following birds, which appear to rank under this genus, as hitherto undescribed:—*Sylvia Rathbonia*, Aud. male and female, plate lxv. He met with this species only once; it is entirely of a bright yellow colour, about four and a half inches in length. The bill appears more bent than in the typical species. *Sylvia Roscoe*, Aud. plate xxiv. male; looking more like a *Trichas*, shot on the Mississippi, the only one seen. The colours of the upper parts are dark olive, a slender white streak over each eye, and a broad black band from the eye downwards; the under parts, yellow. *Sylvia Childrenii*, Aud. plate xxxv.; killed in the State of Louisiana; only two specimens were met with. General colour of the plumage, yellowish green; length, about four inches and three-quarters.

We cannot but regret the want of specimens of these interesting and rare species. Their authority will rest upon Mr Audubon's plates. It is impossible, from them alone, to say, with precision, that they belong to this genus; and they are placed in it provisionally, with the view of making the list as complete as possible, and to point them out to others who may have the opportunity of examining them.—Ed.

1. Tyrant Flycatcher 2. Great Crested F. 3. Small Green Crested F. 4. Pe-we F. 5 Wood Pe-we F.

13.

and a half broad; the whole back, hind head, and rump, are a fine light slate colour; the tail is somewhat forked, black, and edged with light slate; the wings are also black, the three shortest secondaries, broadly edged with light blue; all the wing-quills are slightly edged with the same; the first row of wing-coverts are tipt and edged with white, the second, wholly white, or nearly so; the frontlet, ear-feathers, lores, and above the temple, are black; the line between the eye and nostril, whole throat, and middle of the breast, brilliant golden yellow; the lower eyelid, line over the eye, and spot behind the ear-feathers, as well as the whole lower parts, are pure white; the yellow on the throat is bordered with touches of black, which also extend along the sides, under the wings; the bill is black, and faithfully represented in the figure; the legs and feet, yellowish brown; the claws, extremely fine pointed; the tongue rather cartilaginous, and lacerated at the end. The female has the wings of a dingy brown, and the whole colours, particularly the yellow on the throat, much duller; the young birds of the first season are without the yellow.

TYRANT FLYCATCHER, OR KING BIRD.

(*Muscicapa tyrannus.**)

PLATE XIII.—Fig. 1.

Lanius tyrannus, *Linn. Syst.* 136.—*Lath. Syn.* i. 186.—*Catesb.* i. 55.—Le Tyran de la Caroline, *Buff.* iv. 577. *Pl. enl.* 676.—*Arct. Zool.* p. 384, No. 263.—*Peale's Museum,* No. 578.

TYRANNUS INTREPIDUS.—Vieillot.

Muscicapa tyrannus, *Bonap. Synop.* p. 66.—Tyrannus intrepidus, *Vieill. Gal. des Ois.* pl. 133.—*North. Zool.* ii. 137.—The Tyrant Flycatcher, *Aud.* pl. 79, male and female, *Orn. Biog.* i. 403.

This is the field martin of Maryland and some of the southern States, and the king bird of Pennsylvania and several of the

* Among the family of the *Lanaidæ*, North America possesses only two of the sub-families; the typical one, *Lanianæ*, represented by

northern districts. The epithet *tyrant*, which is generally applied to him by naturalists, I am not altogether so well satisfied with; some, however, may think the two terms pretty nearly synonymous.

Lanius, and an aberrant form, *Tyranninæ*, represented by Tyrannus. Of the former, we have already seen an example at page 73. These are comparatively few; the great bulk of that form being confined to Africa, and the warmer parts of Asia and India; and, with the latter, we enter into the great mass of American flycatchers, ranging over both the continents, particularly the southern.

"Tropical America," Mr Swainson remarks, "swarms with the *Tyranninæ*, so much so, that several individuals, of three or four species, may be seen in the surrounding trees at the same moment, watching for passing insects; each, however, looks out for its own particular prey, and does not interfere with such as appear destined by nature for its stronger and less feeble associates. It is only towards the termination of the rainy season, when myriads of the *Termites* and *Formicæ* emerge from the earth in their winged state, that the whole family of tyrants, of all sizes and species, commence a regular and simultaneous attack upon the thousands which then spring from the ground."

From their long-accepted name we have some idea of their manners. They possess extensive powers of locomotion, to enable them to secure a prey at once active and vigilant; and their long and sharp wings are beautifully formed for quick and rapid flight.* The tail, next in importance as a locomotive organ, is also generally of a form joining the greatest advantages,—that of a forked shape; in some with the exterior feathers extending to a considerable length, while, in others, certainly only slightly divaricating, or nearly square; but never, as among the *Thamnophilinæ*, or bush shrikes, of a graduated or rounded form, where the individuals seek their prey by stealth and prowling, and require no great extent of flight; on the other hand, those organs of less utility for securing the means of sustenance are of much inferior strength and power. The accessory members for seizing their insect prey are, in like manner, adapted to their other powers; the bill, though of considerable strength, is flattened; the rictus being ample, and furnished with bristles. The genus *Tyrannus*, however, does not entirely feed on insects when on wing, like the smaller *Tyrannulæ*, but, as shown by Mr Swainson, will also feed on small fish and aquatic insects; and if this fact be united with the weak formation of the tarsi, and, in several

* In many species the quills become suddenly emarginated at the tips. This also occurs in the subgenera *Milvulus* and *Negeta*, both much allied, and possessing great powers of flight.

The trivial name *king* as well as *tyrant* has been bestowed on this bird for its extraordinary behaviour, and the authority it assumes over all others, during the time of breeding. At

species, having the toes united at the base, there will be an evident connection between this group and the *Fissirostres*. That gentleman, in the second volume of the "Northern Zoology," relates a fact from his journal when resident in Brazil, most beautifully illustrative of this affinity, and shows the value of attending to all circumstances relative to the habits of individuals, which though, like the present, of no importance alone, will, when taken in connection with other views, be of the *very utmost consequence.* "*April* 7, 1817.—Sitting in the house this morning, I suddenly heard a splash in the lake close to the window; on looking out, I saw a common grey-breasted tyrant (*Tyrannus crudelis*) perched upon a dead branch hanging over the water, plunging and drying itself. Intent upon watching this bird, I saw it, within a quarter of an hour, dive into the lake two successive times, after some small fish or aquatic insects, precisely like a kingfisher; this action was done with amazing celerity, and it then took its former station to plume and dry its feathers." Here we have exactly the habits of the kingfisher; and I believe a contrariety of manner, equally worthy of remark, is observed among some of the *Dacelones*, frequenting woods, and darting by surprise on the larger insects. Both tribes have another similarity in their economy, and delight to sit motionless, either watching their prey, or pluming and resting on the extremity or top of some dead branch, pale, or peaked rock. With regard to the tyrant's being not only carnivorous, but preying also on the weaker reptiles, we have the authority of Azara, who mentions the common *Tyrannus sulphuratus*, or bentivo of Brazil, as "S'approchent des animaux morts pour l'emporter des debris et des petits morceaux de chair que laissent les Caraçaras." And Mr Swainson ("Northern Zoology," ii. 133) has himself taken from the stomach of this species lizards, in an entire state, sufficiently large to excite surprise how they possibly could have been swallowed by the bird; it is also here that we have the habits, and, in some respects, the form of the *Lanianæ*, serving at the other extremity as a connecting link. The North American species, coming under the definition which we would wish to adopt for this group, are comparatively few. A new and more northern species is added by the authors of the "Northern Zoology,"*—the *Tyrannus borealis*, Sw.

Only one specimen of this species, which Mr Swainson considers undescribed, was procured. It was shot on the banks of the Saskatchewan

* They are also baccivorous, as shown by our author in the description of this species and *T. crinitus*.

that season his extreme affection for his mate, and for his nest and young, makes him suspicious of every bird that happens to pass near his residence; so that he attacks, without discrimination, every intruder. In the months of May, June, and part of July, his life is one continued scene of broils and battles; in which, however, he generally comes off conqueror. Hawks and crows, the bald eagle and the great black eagle, all equally dread a rencounter with this dauntless little champion, who, as soon as he perceives one of these last approaching, launches into the air to meet him, mounts to a considerable height above him, and darts down on his back, sometimes fixing there, to the great annoyance of his sovereign, who, if no convenient retreat or resting-place be near, endeavours by various evolutions to rid himself of his merciless adversary. But the king bird is not so easily dismounted. He teases the eagle incessantly, sweeps upon him from right to left, remounts, that he may descend on his back with the greater violence; all the while keeping up a shrill and rapid twittering; and continuing the attack sometimes for more than a mile, till he is relieved by some other of his tribe equally eager for the contest.

There is one bird, however, which, by its superior rapidity of flight, is sometimes more than a match for him; and I have several times witnessed his precipitate retreat before this active antagonist. This is the purple martin, one whose food and disposition are pretty similar to his own, but who has greatly the advantage of him on wing, in eluding all his attacks, and teasing him as he pleases. I have also seen the

river. Like the king bird, it is found in the Fur Countries only in summer. It is considerably smaller than the *Tyrannus intrepidus*, and may at once be distinguished from it by the forked tail not tipped with white, and much shorter tarsi, as well as by very evident differences in the colours of the plumage. Its bill is rather more depressed at the base, and its lower mandible is dissimilar to the upper one; the relative length of the tail-feathers in the two species are also different; the first of *T. borealis*, *shorter* than the third, the fourth being farther apart from the latter than in *T. intrepidus*.—ED.

red-headed woodpecker, while clinging on a rail of the fence, amuse himself with the violence of the king bird, and play *bo-peep* with him round the rail, while the latter, highly irritated, made every attempt, as he swept from side to side, to strike him—but in vain. All this turbulence, however, vanishes as soon as his young are able to shift for themselves, and he is then as mild and peaceable as any other bird.

But he has a worse habit than all these, one much more obnoxious to the husbandman, and often fatal to himself. He loves, not the honey, but the *bees ;* and, it must be confessed, is frequently on the look-out for these little industrious insects. He plants himself on a post of the fence, or on a small tree in the garden, not far from the hives, and from thence sallies on them as they pass and repass, making great havoc among their numbers. His shrill twitter, so near to the house, gives intimation to the farmer of what is going on, and the gun soon closes his career for ever. Man arrogates to himself, in this case, the exclusive privilege of murder ; and, after putting thousands of these same little insects to death, seizes on the fruits of their labour.

The king birds arrive in Pennsylvania about the 20th of April, sometimes in small bodies of five and six together, and are at first very silent, until they begin to pair and build their nest. This generally takes place about the first week in May. The nest is very often built in the orchard, on the horizontal branch of an apple tree ; frequently also, as Catesby observes, on a sassafras tree, at no great height from the ground. The outside consists of small slender twigs, tops of withered flowers of the plant yarrow, and others, well wove together with tow and wool ; and is made large, and remarkably firm and compact. It is usually lined with fine dry fibrous grass and horse hair. The eggs are five, of a very pale cream colour, or dull white, marked with a few large spots of deep purple, and other smaller ones of light brown, chiefly, though not altogether, towards the great end (see fig. 1). They generally build twice in the season.

The king bird is altogether destitute of song, having only the shrill twitter above mentioned. His usual mode of flight is singular. The vibrations of his broad wings, as he moves slowly over the fields, resemble those of a hawk hovering and settling in the air to reconnoitre the ground below; and the object of the king bird is no doubt something similar, viz., to look out for passing insects, either in the air, or among the flowers and blossoms below him. In fields of pasture he often takes his stand on the tops of the mullein, and other rank weeds, near the cattle, and makes occasional sweeps after passing insects, particularly the large black gadfly, so terrifying to horses and cattle. His eye moves restlessly around him, traces the flight of an insect for a moment or two, then that of a second, and even a third, until he perceives one to his liking, when, with a shrill sweep, he pursues, seizes it, and returns to the same spot again, to look out for more. This habit is so conspicuous when he is watching the beehive, that several intelligent farmers of my acquaintance are of opinion that he picks out only the drones, and never injures the working bees. Be this as it may, he certainly gives a preference to one bee, and one species of insect, over another. He hovers over the river, sometimes for a considerable time, darting after insects that frequent such places, snatching them from the surface of the water, and diving about in the air like a swallow; for he possesses at will great powers of wing. Numbers of them are frequently seen thus engaged, for hours together, over the rivers Delaware and Schuylkill, in a calm day, particularly towards evening. He bathes himself by diving repeatedly into the water from the overhanging branches of some tree, where he sits to dry and dress his plumage.

Whatever antipathy may prevail against him for depredations on the drones, or, if you will, on the bees, I can assure the cultivator that this bird is greatly his friend, in destroying multitudes of insects, whose larvæ prey on the harvests of his fields, particularly his corn, fruit trees, cucumbers, and pumpkins. These noxious insects are the daily food of this bird; and he destroys, upon a very moderate average, some hundreds of them daily. The death of every king bird is therefore an actual loss to the farmer, by multiplying the numbers of destructive insects, and encouraging the depredations of crows, hawks, and eagles, who avoid as much as possible his immediate vicinity. For myself, I must say, that the king bird possesses no common share of my regard. I honour this little bird for his extreme affection for his young; for his contempt of danger, and unexampled intrepidity; for his meekness of behaviour when there are no calls on his courage, a quality which, even in the human race, is justly considered so noble:

> In peace there's nothing so becomes a man
> As modest stillness and humility;
> But when the blast of war, &c.

But, above all, I honour and esteem this bird for the millions of ruinous vermin which he rids us of; whose depredations, in *one* season, but for the services of this and other friendly birds, would far overbalance all the produce of the beehives in fifty.

As a friend to this persecuted bird, and an enemy to prejudices of every description, will the reader allow me to set this matter in a somewhat clearer and stronger light, by presenting him with a short poetical epitome of the king bird's history?

> FAR in the south, where vast Maragnon flows,
> And boundless forests unknown wilds enclose;
> Vine-tangled shores, and suffocating woods,
> Parched up with heat, or drowned with pouring floods;
> Where each extreme alternately prevails,
> And Nature sad their ravages bewails;
> Lo! high in air, above those trackless wastes,
> With Spring's return the king bird hither hastes;
> Coasts the famed Gulf,* and, from his height, explores
> Its thousand streams, its long indented shores,

* Of Mexico.

Its plains immense, wide opening on the day,
Its lakes and isles, where feathered millions play :
All tempt not him ; till, gazing from on high,
COLUMBIA'S regions wide below him lie ;
There end his wanderings and his wish to roam,
There lie his native woods, his fields, his *home ;*
Down, circling, he descends, from azure heights,
And on a full-blown sassafras alights.
  Fatigued and silent, for a while he views
His old frequented haunts and shades recluse,
Sees brothers, comrades, every hour arrive—
Hears, humming round, the tenants of the hive :
Love fires his breast ; he woos, and soon is blest ;
And in the blooming orchard builds his nest.
  Come now, ye cowards ! ye whom Heaven disdains,
Who boast the happiest home—the richest plains ;
On whom, perchance, a wife, an infant's eye,
Hang as their hope, and on your arm rely ;
Yet, when the hour of danger and dismay
Comes on your country, sneak in holes away,
Shrink from the perils ye were bound to face,
And leave those babes and country to disgrace;
Come here (if such we have), ye dastard herd !
And kneel in dust before this noble bird.
  When the specked eggs within his nest appear,
Then glows affection, ardent and sincere ;
No discord sours him when his mate he meets ;
But each warm heart with mutual kindness beats.
For her repast he bears along the lea
The bloated gadfly, and the balmy bee ;
For her repose scours o'er the adjacent farm,
Whence hawks might dart, or lurking foes alarm ;
For now abroad a band of ruffians prey,
The crow, the cuckoo, and the insidious jay ;
These, in the owner's absence, all destroy,
And murder every hope and every joy.
  Soft sits his brooding mate, her guardian he,
Perched on the top of some tall neighbouring tree ;
Thence, from the thicket to the concave skies,
His watchful eye around unceasing flies.
Wrens, thrushes, warblers, startled at his note,
Fly in affright the consecrated spot.
He drives the plundering *jay*, with honest scorn,
Back to his woods ; the *mocker*, to his thorn ;
Sweeps round the *cuckoo*, as the thief retreats;
Attacks the *crow;* the diving *hawk* defeats ;
Darts on the *eagle* downwards from afar,
And, 'midst the clouds, prolongs the whirling war.

All danger o'er, he hastens back elate,
To guard his post, and feed his faithful mate.
Behold him now, his little family flown,
Meek, unassuming, silent, and alone;
Lured by the well-known hum of favourite bees,
As slow he hovers o'er the garden trees
(For all have failings, passions, whims that lead,
Some favourite wish, some appetite to feed);
Straight he alights, and, from the pear tree, spies
The circling stream of humming insects rise;
Selects his prey; darts on the busy brood,
And shrilly twitters o'er his savoury food.
  Ah! ill-timed triumph! direful note to thee,
That guides thy murderer to the fatal tree;
See where he skulks! and takes his gloomy stand,
The deep-charged musket hanging in his hand;
And, gaunt for blood, he leans it on a rest,
Prepared, and pointed at thy snow-white breast.
Ah, friend! good friend! forbear that barbarous deed,
Against it valour, goodness, pity, plead;
If e'er a family's griefs, a widow's woe,
Have reached thy soul, in mercy let him go!
Yet, should the tear of pity nought avail,
Let *interest* speak, let *gratitude* prevail;
Kill not thy friend, who thy whole harvest shields,
And sweeps ten thousand vermin from thy fields;
Think how this dauntless bird, thy poultry's guard,
Drove every hawk and eagle from thy yard;
Watched round thy cattle as they fed, and slew
The hungry blackening swarms that round them flew;
Some small return—some little right resign,
And spare *his* life whose services are thine!
—— I plead in vain! Amid the bursting roar,
The poor lost king bird welters in his gore!

This species is eight inches long, and fourteen in extent; the general colour above is a dark slaty ash; the head and tail are nearly black; the latter *even* at the end, and tipt with white; the wings are more of a brownish cast; the quills and wing-coverts are also edged with dull white; the upper part of the breast is tinged with ash; the throat, and all the rest of the lower parts, are pure white; the plumage on the crown, though not forming a crest, is frequently erected, as represented in the plate, and discovers a rich bed of brilliant orange, or flame colour, called by the country people his crown: when the feathers lie close, this is altogether concealed. The bill is very broad at the base, overhanging at the point, and notched, of a glossy black colour, and furnished with bristles at the base; the legs and feet are black, seamed with gray; the eye, hazel. The female differs in being more brownish on the upper parts, has a smaller streak of paler orange on the crown, and a narrower border of duller white on the tail. The young birds do not receive the orange on the head during their residence here the first season.

This bird is very generally known, from the Lakes to Florida. Besides insects, they feed, like every other species of their tribe with which I am acquainted, on various sorts of berries, particularly blackberries, of which they are extremely fond. Early in September they leave Pennsylvania, on their way to the south.

A few days ago I shot one of these birds, the whole plumage of which was nearly white, or a little inclining to a cream colour; it was a bird of the present year, and could not be more than a month old. This appeared also to have been its original colour, as it issued from the egg. The skin was yellowish white; the eye, much lighter than usual; the legs and bill, blue. It was plump, and seemingly in good order. I presented it to Mr Peale. Whatever may be the cause of this loss of colour, if I may so call it, in birds, it is by no means uncommon among the various tribes that inhabit the United States. The sparrow-hawk, sparrow, robin, red-winged blackbird, and many others, are occasionally found in white plumage; and I believe that such birds do not become so by climate, age, or disease, but that they are universally hatched so. The same phenomena are observable not only among various sorts of animals, but even among the human race; and a white negro is no less common, in proportion to their numbers, than a white blackbird; though the precise cause of this in either is but little understood.

GREAT CRESTED FLYCATCHER. (*Muscicapa crinita.*)

PLATE XIII.—FIG. 2.

Linn. Syst. 325.—*Lath.* ii. 357.—*Arct. Zool.* p. 386, No. 267.—Le Moucherolle de Virginie à huppe verte, *Buff.* iv. 565, *Pl. enl.* 569.—*Peale's Museum*, No. 6645.

TYRANNUS CRINITUS.—SWAINSON.

Tyrannus crinitus, *Swain. Monog. Journ. of Science*, vol. xx. p. 271.—Muscicapa crinita, *Bonap. Synop.* p. 67.

BY glancing at the physiognomy of this bird, and the rest of the figures on the same plate, it will readily be observed, that they all belong to one particular family of the same genus. They possess strong traits of their particular *cast*, and are all remarkably dexterous at their profession of fly-catching. The one now before us is less generally known than the preceding, being chiefly confined to the woods. There his harsh *squeak*—for he has no song—is occasionally heard above most others. He also visits the orchard, is equally fond of bees, but wants the courage and magnanimity of the king bird. He arrives in Pennsylvania early in May, and builds his nest in a hollow tree deserted by the blue bird or woodpecker. The materials of which this is formed are scanty, and rather novel. One of these nests, now before me, is formed of a little loose hay, feathers of the guinea fowl, hogs' bristles, pieces of cast snake skins, and dogs' hair. Snake skins with this bird appear to be an indispensable article, for I have never yet found one of his nests without this material forming a part of it.* Whether he surrounds his nest with this by way of *terrorem*, to prevent other birds or animals from entering, or whether it be that he finds its silky softness suitable for his young, is uncertain;

* As I have mentioned at page 143, this forms the lining to the nests of other birds also; and as the number of snakes is considerable in those uncultivated and woody countries, their castings may form a more frequent substitute than is generally supposed.—ED.

the fact, however, is notorious. The female lays four eggs of a dull cream colour, thickly scratched with purple lines of various tints, as if done with a pen (see fig. 2).

This species is eight inches and a half long, and thirteen inches in extent; the upper parts are of a dull greenish olive; the feathers on the head are pointed, centred with dark brown, ragged at the sides, and form a kind of blowsy crest; the throat and upper parts of the breast, delicate ash; rest of the lower parts, a sulphur yellow; the wing-coverts are pale drab, crossed with two bars of dull white; the primaries are of a bright ferruginous, or sorrel colour; the tail is slightly forked, its interior vanes of the same bright ferruginous as the primaries; the bill is blackish, very much like that of the king bird, furnished also with bristles; the eye is hazel; legs and feet, bluish black. The female can scarcely be distinguished by its colours from the male.

This bird also feeds on berries towards the end of summer, particularly on huckleberries, which, during the time they last, seem to form the chief sustenance of the young birds. I have observed this species here as late as the 10th of September; rarely later. They do not, to my knowledge, winter in any of the southern States.

SMALL GREEN-CRESTED FLYCATCHER.*
(*Muscicapa querula.*)

PLATE XIII.—Fig. 3.

Muscicapa subviridis, *Bartram*, p. 289.—*Arct. Zool.* p. 386, No. 268.—*Peale's Museum*, No. 6825.

TYRANNULA ACADICA.—Swainson.

Muscicapa acadica, *Bonap. Synop.* p. 68.

This bird is but little known. It inhabits the deepest, thick shaded, solitary parts of the woods, sits generally on the lower

* This species, with the two following of our author, have been separated from the tyrants, and placed in a subgenus, *Tyrannula.* They

branches, utters, every half minute or so, a sudden sharp squeak, which is heard a considerable way through the woods; and, as it flies from one tree to another, has a low, querulous

are, however, in reality, *little tyrants*, and agree in their habits, as far as their smaller size and weaker powers enable them. Their food is nearly the same, more confined, however, to insects, sufficient power being wanting to overcome any stronger prey. *Tyrannula* will contain a great many species most closely allied to each other in form, size, and colour; so much so, that it is nearly impossible to distinguish them without a comparison of many together. When they are carefully analysed, they seem distinct, and the characters being constant, are also of sufficient specific importance. They are natives of both North and South America, and the adjacent islands. The North American known species are those described by our author, which will be found in Vol. III., one or two figured by Bonaparte, with two new species discovered in the course of the last Overland Arctic Expedition, and described by Mr Swainson in the second volume of the "Northern Zoology." South America, however, possesses the great host of species, where we may yet expect many novelties. The extent and the closely allied features of the group render them most difficult of distinction.*

Both this form and the tyrants are confined to the New World, and the latter may be said to represent the great mass of our flycatchers.

The new species described by Mr Swainson are, *Tyrannula pusilla*, Sw., very closely allied to *Muscicapa querula* of Wilson, but satisfactorily proved distinct; the wings are much shorter,so mewhat rounded, and the comparative proportion of the quills differ; the colours, however, nearly agree: the species brought home by the expedition was killed at Carlton House in 53° N. lat., and it extends southward to Mexico.—*T. Richardsonii*, closely resembling *T. fusca;* it differs in the form of the bill and size of the feet; the crest is thick and lengthened; the upper plumage is more olive, while the under has an olive whitish tint; the tail is more forked: it was found in the neighbourhood of Cumberland House, frequenting moist shady woods by the banks of rivers and lakes.

Mr Audubon also figures a species as new, and dedicates it to Dr Trail of Liverpool; but, as I have remarked before, it is impossible to decide from a plate, however accurate. *Tyrannula Trailii* will come nearest to the wood pewee, but differs as well in some parts of the plumage as in the habits. It is found in the woods which skirt the prairie lands of the Arkansas river.—Ed.

* It may be here remarked, that the Prince of Musignano, in his Synopsis, evidently recognises this form as a subgenus, though he has not characterised it.—Ed.

note, something like the twitterings of chickens nestling under the wings of the hen. On alighting, this sound ceases, and it utters its note as before. It arrives from the south about the middle of May; builds on the upper side of a limb, in a low, swampy part of the woods, and lays five white eggs. It leaves us about the beginning of September. It is a rare and very solitary bird, always haunting the most gloomy, moist, and unfrequented parts of the forest. It feeds on flying insects, devours bees, and, in the season of huckleberries, they form the chief part of its food. Its northern migrations extend as far as Newfoundland.

The length of this species is five inches and a half; breadth, nine inches; the upper parts are of a green olive colour, the lower, pale greenish yellow, darkest on the breast; the wings are deep brown, crossed with two bars of yellowish white, and a ring of the same surrounds the eye, which is hazel. The tail is rounded at the end; the bill is remarkably flat and broad, dark brown above, and flesh colour below; legs and feet, pale ash. The female differs little from the male in colour.

PEWIT FLYCATCHER. (*Muscicapa nunciola.*)

PLATE XIII.—FIG. 4.

Bartram, p. 289.—Blackcap Flycatcher, *Lath. Syn.* ii. 353.—Phœbe Flycatcher, *Id. Sup.* p. 173.—Le Gobe-mouche noirâtre de la Caroline, *Buff.* iv. 541.—*Arct. Zool.* p. 387, No. 269.—*Peale's Museum*, No. 6618.

TYRANNULA FUSCA.—JARDINE.

Muscicapa fusca, *Bonap. Synop.* p. 68.

THIS well-known bird is one of our earliest spring visitants, arriving in Pennsylvania about the first week in March, and continuing with us until October. I have seen them here as late as the 12th of November. In the month of February, I overtook these birds lingering in the low swampy woods of

North and South Carolina. They were feeding on smilax berries, and chanting, occasionally, their simple notes. The favourite resort of this bird is by streams of water, under or near bridges, in caves, &c. Near such places he sits on a projecting twig, calling out, *pe-wée*, *pe-wittitee pe-wée*, for a whole morning; darting after insects, and returning to the same twig; frequently flirting his tail, like the wagtail, though not so rapidly. He begins to build about the 20th or 25th of March, on some projecting part under a bridge, in a cave, in an open well, five or six feet down, among the interstices of the side walls, often under a shade in the low eaves of a cottage, and such-like places.* The outside is composed of mud mixed with moss, is generally large and solid, and lined with flax and horse hair. The eggs are five, pure white, with two or three dots of red near the great end (see fig. 4). I have known them rear three broods in one season.

In a particular part of Mr Bartram's woods, with which I am acquainted, by the side of a small stream, in a cave five or six feet high, formed by the undermining of the water below and the projection of two large rocks above—

> There down smooth glist'ning rocks the rivulet pours,
> Till in a pool its silent waters sleep;

* The general manners of this species, and indeed of the greater part of the smaller *Tyrannulæ*, bear a considerable resemblance to those of the common spotted flycatcher of this country, which the dilatation at the base of the bill, and the colour of the plumage, render still greater. The peculiar droop of the tail, and occasional rise and depression of the feathers on the crown, which are somewhat elongated—the motionless perch on some bare branch—the impatient call—the motion of the tail —and the sudden dart after some insect, and return to the same spot— are all close resemblances to the manners delineated by our author; and the resort by streams, bridges, or caves, with the manner and place of building—even the colour of the eggs—are not to be mistaken. In one instance our flycatcher and the *Tyrannulæ* disagree; the former possess no pleasing notes; its only cries are a single, rather harsh and monotonous *click* and a shrill *peep*. The song of the *Tyrannulæ* is "simple," but "lively."—Ed.

> A dark-browed cliff, o'ertopped with fern and flowers,
> Hangs, grimly low'ring, o'er the glassy deep ;
> Above through every chink the woodbines creep,
> And smooth-barked beeches spread their arms around,
> Whose roots cling twisted round the rocky steep :
> A more sequestered scene is nowhere found,
> For contemplation deep, and silent thought profound,—

in this cave I knew the pewit to build for several years. The place was solitary, and he was seldom disturbed. In the month of April, one fatal Saturday, a party of boys from the city, armed with guns, dealing indiscriminate destruction among the feathered tribes around them, directed their murderous course this way, and, within my hearing, destroyed both parents of this old and peaceful settlement. For two successive years, and I believe to this day, there has been no pewee seen about this place. This circumstance almost convinces me that birds, in many instances, return to the same spots to breed; and who knows but, like the savage nations of Indians, they may usurp a kind of exclusive right of tenure to particular districts, where they themselves have been reared?

The notes of the pewee, like those of the blue bird, are pleasing, not for any melody they contain, but from the ideas of spring and returning verdure, with all the sweets of this lovely season, which are associated with his simple but lively ditty. Towards the middle of June, he becomes nearly silent; and late in the fall gives us a few farewell and melancholy repetitions, that recall past imagery, and make the decayed and withered face of nature appear still more melancholy.

The pewit is six inches and a half in length, and nine and a half broad ; the upper parts are of a dark dusky olive ; the plumage of the head, like that of the two preceding, is loose, subcrested, and of a deep brownish black ; wings and tail, deep dusky ; the former edged, on every feather, with yellowish white, the latter forked, and widening remarkably towards the end ; bill, formed exactly like that of the king bird ; whole lower parts, a pale delicate yellow ; legs and bill, wholly black; iris, hazel. The female is almost exactly like the male, except in having the crest somewhat more brown. This species inhabits from Canada to Florida; great numbers of them usually wintering in the two Carolinas and Georgia. In New York, they are called the phœby bird, and are accused of destroying bees. With many people in the country, the arrival of the pewee serves as a sort of almanack, reminding them that now it is time such and such work should be done. "Whenever the pewit appears," says Mr Bartram, "we may plant peas and beans in the open grounds, French beans, sow radishes, onions, and almost every kind of esculent garden seeds, without fear or danger from frosts; for, although we have sometimes frosts after their first appearance for a night or two, yet not so severe as to injure the young plants."*

WOOD PEWÈE FLYCATCHER. (*Muscicapa rapax.*)

PLATE XIII.—Fig. 5.

Muscicapa virens, *Linn. Syst.* 327.—*Lath. Syn.* ii. 350, *Id. Sup.* p. 174, No. 82.—*Catesb.* i. 54, fig. 1.—Le Gobe-mouche brun de la Caroline, *Buff.* iv. 543. —Muscicapa acadica, *Gmel. Syst.* i. p. 947.—*Arct. Zool.* 387, No. 270.—*Peale's Museum*, No. 6660.

TYRANNULA VIRENS.—Jardine.

Muscicapa virens, *Linn. Syst.*—*Bonap. Synop.*

I have given the name wood pewèe to this species, to discriminate it from the preceding, which it resembles so much in form and plumage as scarcely to be distinguished from it but by an accurate examination of both. Yet in manners, mode of building, period of migration, and notes, the two species differ greatly. The pewèe is among the first birds that visit us in spring, frequenting creeks, building in caves, and under arches of bridges; the wood pewèe, the subject

* Travels, p. 288.

of our present account, is among the latest of our summer birds, seldom arriving before the 12th or 15th of May; frequenting the shadiest high timbered woods, where there is little underwood, and abundance of dead twigs and branches shooting across the gloom; generally in low situations; builds its nest on the upper side of a limb or branch, forming it outwardly of moss, but using no mud, and lining it with various soft materials. The female lays five white eggs, and the first brood leaves the nest about the middle of June.

This species is an exceeding expert fly-catcher. It loves to sit on the high dead branches, amid the gloom of the woods, calling out in a feeble plaintive tone, *peto wāy, peto wāy, pee way;* occasionally darting after insects; sometimes making a circular sweep of thirty or forty yards, snapping up numbers in its way with great adroitness; and returning to its position and chant as before. In the latter part of August, its notes are almost the only ones to be heard in the woods; about which time, also, it even approaches the city, where I have frequently observed it busily engaged under trees, in solitary courts, gardens, &c., feeding and training its young to their profession. About the middle of September it retires to the south, a full month before the other.

Length, six inches; breadth, ten; back, dusky olive, inclining to greenish; head, subcrested, and brownish black; tail, forked, and widening towards the tips, lower parts, pale yellowish white. The only discriminating marks between this and the preceding are the size and the colour of the lower mandible, which in this is yellow, in the pewèe black. The female is difficult to be distinguished from the male.

This species is far more numerous than the preceding, and probably winters much farther south. The pewèe was numerous in North and South Carolina in February; but the wood pewèe had not made its appearance in the lower parts of Georgia, even so late as the 16th of March.

Drawn from Nature by A. Wilson — *Engraved by W. H. Lizars*

1. Brown Thrush. 2. Golden-crowned Th. 3. Cat Bird. 4. Bay breasted Warbler. 5. Chesnut sided W. 6. Mourning W.

14.

FERRUGINOUS THRUSH.* (*Turdus rufus.*)

PLATE XIV.—Fig. 1.

Fox-coloured Thrush, *Catesb.* i. 28.—Turdus rufus, *Linn. Syst.* 293.—*Lath.* iii. 39. —La Grive de la Caroline, *Briss.* ii. 223.—Le Moquer Francois, *Buff.* iii. 323, *Pl. enl.* 645.—*Arct. Zool.* p. 335, No. 195.—*Peale's Museum*, No. 5285.

ORPHÆUS RUFUS.—Swainson.

Turdus rufus, *Bonap. Synop.* p. 75.—Orphæus rufus, Fox-coloured Mock Bird, *North. Zool.* ii. p. 190.

This is the brown thrush, or thrasher, of the middle and eastern States, and the French mocking bird of Maryland,

* This species, with *O. polyglottos*, is the typical form of Mr Swainson's genus *Orphæus*, differing from *Turdus* in its longer form, chiefly apparent from the greater length of its tail, its rounded and shorter wings, its long and bending, and, in proportion, more slender bill. The form is confined to the New World, and will be represented in Africa by *Crateropus* and *Donocobius*, Swain. ; and in Asia and Australia by *Pomatorhinus*, Horsf. They appear to live nearer the ground than the true thrushes, frequenting the lower brushwood ; and it is only during the spring and breeding season that they mount aloft, to serenade their mates. The cries or notes are generally loud ; some possess considerable melody, which, however, is only exercised as above mentioned ; but many of the aberrant species possess only harsh and grating notes, incessantly kept up ; in which respect they resemble the more typical African form, and many of the aquatic warblers.

In the account given by our author of the manners of *O. rufus*, we perceive a very close resemblance to our common blackbird. The blackbird is seldom seen on lofty trees, except during the season of incubation, or occasionally in search of a roosting place ; its true habitat is brushwood or shrubbery, and, unless at one season, its only note is that of alarm, shrill and rapid, or a kind of chuck. The manner of flight, when raised from cover, along a hedge, or among bushes, with the tail expanded, is also similar ; we have thus two types of very nearly allied genera, varying decidedly in form, but agreeing almost entirely in habit. The gregarious thrushes, again, possess much more activity, enjoy lofty forests or the open country, and protect themselves by vigilance, not by stealth and concealment.

This species was met by Dr Richardson at Carlton House. It ex-

Virginia, and the Carolinas.* It is the largest of all our thrushes, and is a well known and very distinguished songster. About the middle or 20th of April, or generally about the time the cherry trees begin to blossom, he arrives in Pennsylvania; and from the tops of our hedgerows, sassafras, apple or cherry trees, he salutes the opening morning with his charming song, which is loud, emphatical, and full of variety. At that serene hour, you may plainly distinguish his voice fully half a mile off. These notes are not imitative, as his name would import, and as some people believe, but seem solely his own; and have considerable resemblance to the notes of the song thrush (*Turdus musicus*) of Britain. Early in May he builds his nest, choosing a thorn bush, low cedar, thicket of briers, dogwood sapling, or cluster of vines, for its situation, generally within a few feet of the ground. Outwardly, it is constructed of small sticks; then layers of dry leaves, and, lastly, lined with fine fibrous roots, but without any plaster. The eggs are five, thickly sprinkled with ferruginous grains, on a very pale bluish ground. They generally have two broods in a season. Like all birds that build near the ground, he shows great anxiety for the safety of his nest and young, and often attacks the black snake in their defence, generally, too, with success, his strength being greater, and his bill stronger and more powerful, than any other of his tribe within the United States. His food consists of worms, which he scratches from the ground, caterpillars, and many kinds of berries. Beetles, and the whole race of coleopterous insects, wherever he can meet with them, are sure to suffer. He is accused, by some people, of scratching up the hills of Indian-corn in planting time; this may be partly true; but, for every grain of maize he

tends from Pennsylvania to the Saskatchewan; but Dr Richardson thinks it probable that it does not extend its range beyond the 54th parallel of latitude. It quits the Fur Countries, with the other migratory birds, early in September.—Ed.

* See p. 173 for the supposed origin of this name.

pilfers, I am persuaded he destroys five hundred insects; particularly a large dirty-coloured grub, with a black head, which is more pernicious to the corn and other grain and vegetables than nine-tenths of the whole feathered race. He is an active, vigorous bird, flies generally low, from one thicket to another, with his long broad tail spread like a fan; is often seen about brier and bramble bushes, along fences; and has a single note or chuck when you approach his nest. In Pennsylvania, they are numerous, but never fly in flocks. About the middle of September, or as soon as they have well recovered from moulting, in which they suffer severely, they disappear for the season. In passing through the southern parts of Virginia, and south as far as Georgia, in the depth of winter, I found them lingering in sheltered situations, particularly on the borders of swamps and rivers. On the 1st of March, they were in full song round the commons at Savannah, as if straining to outstrip the mocking bird, that prince of feathered musicians.

The thrasher is a welcome visitant in spring to every lover of rural scenery and rural song. In the months of April and May, when our woods, hedgerows, orchards, and cherry trees, are one profusion of blossoms, when every object around conveys the sweet sensations of joy, and Heaven's abundance is, as it were, showering around us, the grateful heart beats in unison with the varying elevated strains of this excellent bird; we listen to its notes with a kind of devotional ecstasy, as a morning hymn to the great and most adorable Creator of all. The human being who, amidst such scenes, and in such seasons of rural serenity and delight, can pass them with cold indifference, and even contempt, I sincerely pity; for abject must that heart be, and callous those feelings, and depraved that taste, which neither the charms of nature, nor the melody of innocence, nor the voice of gratitude or devotion, can reach.

This bird inhabits North America, from Canada to the point of Florida. They are easily reared, and become very amiliar when kept in cages; and though this is rarely done, yet I have known a few instances where they sang in confinement with as much energy as in their native woods. They ought frequently to have earth and gravel thrown in to them, and have plenty of water to bathe in.

The ferruginous thrush is eleven inches and a half long, and thirteen in extent; the whole upper parts are of a bright reddish brown; wings, crossed with two bars of white, relieved with black; tips and inner vanes of the wings, dusky; tail, very long, rounded at the end, broad, and of the same reddish brown as the back; whole lower parts, yellowish white; the breast, and sides under the wings, beautifully marked with long pointed spots of black, running in chains; chin, white; bill, very long and stout, not notched, the upper mandible overhanging the lower a little, and beset with strong bristles at the base, black above, and whitish below, near the base; legs, remarkably strong, and of a dusky clay colour; iris of the eye, brilliant yellow. The female may be distinguished from the male by the white on the wing being much narrower, and the spots on the breast less. In other respects, their plumage is nearly alike.

Concerning the sagacity and reasoning faculty of this bird, my venerable friend Mr Bartram writes me as follows:—"I remember to have reared one of these birds from the nest; which, when full grown, became very tame and docile. I frequently let him out of his cage to give him a taste of liberty; after fluttering and dusting himself in dry sand and earth, and bathing, washing, and dressing himself, he would proceed to hunt insects, such as beetles, crickets, and other shelly tribes; but, being very fond of wasps, after catching them, and knocking them about to break their wings, he would lay them down, then examine if they had a sting, and, with his bill, squeeze the abdomen to clear it of the reservoir of poison before he would swallow his prey. When in his cage, being very fond of dry crusts of bread, if, upon trial, the corners of the crumbs were too hard and sharp for his throat, he would throw them up, carry, and put them in his water dish to soften; then take them out and swallow them. Many other remarkable circumstances might be mentioned that would fully demonstrate faculties of *mind;* not only innate, but acquired ideas (derived from necessity in a state of domestication), which we call understanding and knowledge. We see that this bird could associate those ideas, arrange and apply them in a rational manner, according to circumstances. For instance, if he knew that it was the hard sharp corners of the crumb of bread that hurt his gullet, and prevented him from swallowing it, and that water would soften, and render it easy to be swallowed, this knowledge must be acquired by observation and experience; or some other bird taught him. Here the bird perceived, by the effect, the cause, and then took the quickest, the most effectual, and agreeable method to remove that cause. What could the wisest man have done better? Call it reason, or instinct, it is the same that a sensible man would have done in this case.

"After the same manner this bird reasoned with respect to the wasps. He found, by experience and observation, that the first he attempted to swallow hurt his throat, and gave him extreme pain; and, upon examination, observed that the extremity of the abdomen was armed with a poisonous sting; and, after this discovery, never attempted to swallow a wasp until he first pinched his abdomen to the extremity, forcing out the sting, with the receptacle of poison."

It is certainly a circumstance highly honourable to the character of birds, and corroborative of the foregoing sentiments, that those who have paid the most minute attention to their manners are uniformly their advocates and admirers. "He must," said a gentleman to me the other day, when speaking of another person,—"he must be a good man; for those who have long known him, and are most intimate with him, respect him greatly, and always speak well of him."

GOLDEN-CROWNED THRUSH.* (*Turdus aurocapillus.*)

PLATE XIV.—Fig. 2.

Edw. 252.—*Lath.* iii. 21.—La Figuier à tête d'or, *Briss.* iii. 504.—La Grivelette de St Domingue, *Buff.* iii. 317, *Pl. enl.* 398.—*Arct. Zool.* p. 339, No. 203.—Turdus minimus, vertice aureo, The Least Golden-crown Thrush, *Bartram*, p. 290.—*Peale's Museum*, No. 7122.

SEIURUS AUROCAPILLUS.—Swainson.

Sylvia aurocapilla, *Bonap. Synop.* p. 77.—Seiürus aurocapillus, *North. Zool.* ii. 227.

Though the epithet *golden-crowned* is not very suitable for this bird, that part of the head being rather of a brownish orange, yet, to avoid confusion, I have retained it.

This is also a migratory species, arriving in Pennsylvania late in April, and leaving us again late in September. It is altogether an inhabitant of the woods, runs along the ground like a lark, and even along the horizontal branches, frequently moving its tail in the manner of the wagtails. It has no song; but a shrill energetic twitter, formed by the rapid reiteration of two notes, *peche, peche, peche,* for a quarter of a minute at a time. It builds a snug, somewhat singular nest, on the ground, in the woods, generally on a declivity

* This curious species, with the *S. aquaticus* of Plate XXIII., and some others, differs materially in economy from the thrushes, notwithstanding their general form and colours; and, to judge from the account of the manners of our present species given by Wilson, it will approach very closely to *Anthus* and our *A. arboreus*, and in form and structure to some of the warblers. The manners of *S. aquaticus*, again, resemble more those of the wagtails; but has somewhat of the true thrush in perching high, and in possessing a sweet and pensive song. We have, therefore, in shape, colour, and some of the habits, an alliance to the thrushes, while the colours and their distribution agree both with *Merula* and *Anthus*, and in their principal economy a combination of the *Sylvianæ* and *Motacillanæ*,—altogether a most interesting form; while, in the structure of their nest, and the colour of the eggs, they agree with the wrens. Mr Swainson has made from it his genus *Seiürus*.—Ed.

facing the south. This is formed of leaves and dry grass, and lined with hair. Though sunk below the surface, it is arched over, and only a small hole left for entrance: the eggs are four, sometimes five, white, irregularly spotted with reddish brown, chiefly near the great end. When alarmed, it escapes from the nest with great silence and rapidity, running along the ground like a mouse, as if afraid to tread too heavily on the leaves; if you stop to examine its nest, it also stops, droops its wings, flutters, and tumbles along, as if hardly able to crawl, looking back now and then to see whether you are taking notice of it. If you slowly follow, it leads you fifty or sixty yards off, in a direct line from its nest, seeming at every advance to be gaining fresh strength; and when it thinks it has decoyed you to a sufficient distance, it suddenly wheels off and disappears. This kind of deception is practised by many other species of birds that build on the ground; and is sometimes so adroitly performed, as actually to have the desired effect of securing the safety of its nest and young.

This is one of those birds frequently selected by the cow-pen bunting to be the foster parent of its young. Into the nest of this bird the cow bird deposits its egg, and leaves the result to the mercy and management of the thrush, who generally performs the part of a faithful and affectionate nurse to the foundling.

The golden-crowned thrush is six inches long, and nine in extent; the whole upper parts, except the crown and hind head, are a rich yellow olive; the tips of the wings, and inner vanes of the quills, are dusky brown; from the nostrils, a black strip passes to the hind head on each side, between which lies a bed of brownish orange; the sides of the neck are whitish; the whole lower parts, white, except the breast, which is handsomely marked with pointed spots of black, or deep brown, as in the figure; round the eye is a narrow ring of yellowish white; legs, pale flesh colour; bill, dusky above, whitish below. The female has the orange on the crown considerably paler.

This bird might with propriety be ranged with the wagtails, its notes, manners, and habit of building on the ground being similar to these. It usually hatches twice in the season; feeds on small bugs and the larvæ of insects, which it chiefly gathers from the ground. It is very generally diffused over the United States, and winters in Jamaica, Hispaniola, and other islands of the West Indies.

CAT BIRD.* (*Turdus lividus.*)

PLATE XIV.—Fig. 3.

Muscicapa Carolinensis, *Linn. Syst.* 328.—Le Gobe-mouche brun de Virginie, *Briss.* ii. 365.—Cat Bird, *Catesb.* i. 66.—*Latham*, ii. 353.—Le Moucherolle de Virginie, *Buff.* iv. 562.—Lucar lividus, apice nigra, The Cat Bird, or Chicken Bird, *Bartram*, p. 290.—*Peale's Museum*, No. 6770.

ORPHÆUS FELIVOX.—Swainson.

Turdus felivox, *Bonap. Synop.* p. 75.

We have here before us a very common and very numerous species in this part of the United States; and one as well known to all classes of people as his favourite briers or blackberry bushes. In spring or summer, on approaching thickets of brambles, the first salutation you receive is from the cat bird; and a stranger, unacquainted with its note, would instantly conclude that some vagrant orphan kitten had got bewildered among the briers, and wanted assistance; so exactly does the call of the bird resemble the voice of that

* At first sight, this species, singular both in habits and structure, appears to range with *Brachypus;* but a more minute inspection shows that it will rather stand as an aberrant form with *Orphæus.* The structure of the bill, feet, and tail, are all of the latter; while the colours, and their distribution, agree with *Brachypus*, particularly the rufous vent. That part is a nearly constant mark among the *Brachipi*, being of a different and brighter colour, and very generally red or yellow. The true *Brachipi* do not seem to extend to North America; they are chiefly confined to Africa and the warmer countries of India.

animal. Unsuspicious, and extremely familiar, he seems less apprehensive of man than almost any other of our summer visitants; for whether in the woods or in the garden, where he frequently builds his nest, he seldom allows you to pass without approaching to pay his respects in his usual way. This humble familiarity and deference, from a stranger, too, who comes to rear his young, and spend the summer with us, ought to entitle him to a full share of our hospitality. Sorry I am, however, to say, that this, in too many instances, is cruelly the reverse. Of this I will speak more particularly in the sequel.

About the 28th of February, the cat bird first arrives in the lower parts of Georgia from the south, consequently winters not far distant, probably in Florida. On the second week in April, he usually reaches this part of Pennsylvania; and about the beginning of May, has already succeeded in building his nest. The place chosen for this purpose is generally a thicket of briers or brambles, a thorn bush, thick vine, or the fork of a small sapling; no great solicitude is shown for concealment, though few birds appear more interested for the safety of their nest and young. The materials are dry leaves and weeds, small twigs, and fine dry grass; the inside is lined with the fine black fibrous roots of some plant. The female lays four, sometimes five eggs, of a uniform greenish blue colour, without any spots. They generally raise two, and sometimes three broods in a season.

In passing through the woods in summer, I have sometimes amused myself with imitating the violent chirping or squeaking of young birds, in order to observe what different species were around me; for such sounds, at such a season, in the woods, are no less alarming to the feathered tenants of the bushes, than the cry of fire or murder in the streets is to the inhabitants of a large and populous city. On such occasions of alarm and consternation, the cat bird is the first to make his appearance, not singly, but sometimes half a dozen at a time, flying from different quarters to the spot. At this time, those who are disposed to play with his feelings may almost throw him into fits, his emotion and agitation are so great, at the distressful cries of what he supposes to be his suffering young. Other birds are variously affected; but none show symptoms of such extreme suffering. He hurries backwards and forwards, with hanging wings and open mouth, calling out louder and faster, and actually screaming with distress, till he appears hoarse with his exertions. He attempts no offensive means; but he bewails—he implores—in the most pathetic terms with which nature has supplied him, and with an agony of feeling which is truly affecting. Every feathered neighbour within hearing hastens to the place, to learn the cause of the alarm, peeping about with looks of consternation and sympathy. But their own powerful parental duties and domestic concerns soon oblige each to withdraw. At any other season, the most perfect imitations have no effect whatever on him.

The cat bird will not easily desert its nest. I took two eggs from one which was sitting, and in their place put two of the brown thrush or thrasher, and took my stand at a convenient distance, to see how she would behave. In a minute or two the male made his approaches, stooped down, and looked earnestly at the strange eggs, then flew off to his mate, who was not far distant, with whom he seemed to have some conversation, and instantly returning, with the greatest gentleness took out both the thrasher's eggs, first one and then the other, carried them singly about thirty yards, and dropt them among the bushes. I then returned the two eggs I had taken, and, soon after, the female resumed her place on the nest as before.

From the nest of another cat bird I took two half-fledged young, and placed them in that of another, which was sitting on five eggs. She soon turned them both out. The place where the nest was not being far from the ground, they were little injured, and the male observing their helpless situation, began to feed them with great assiduity and tenderness.

I removed the nest of a cat bird, which contained four eggs, nearly hatched, from a fox-grape vine, and fixed it firmly and carefully in a thicket of briers close by, without injuring its contents. In less than half an hour I returned, and found it again occupied by the female.

The cat bird is one of our earliest morning songsters, beginning generally before break of day, and hovering from bush to bush, with great sprightliness, when there is scarce light sufficient to distinguish him. His notes are more remarkable for singularity than for melody. They consist of short imitations of other birds and other sounds ; but, his pipe being rather deficient in clearness and strength of tone, his imitations fail where these are requisite. Yet he is not easily discouraged, but seems to study certain passages with great perseverance ; uttering them at first low, and, as he succeeds, higher and more free, nowise embarrassed by the presence of a spectator even within a few yards of him. On attentively listening for some time to him, one can perceive considerable variety in his performance, in which he seems to introduce all the odd sounds and quaint passages he has been able to collect. Upon the whole, though we cannot arrange him with the grand leaders of our vernal choristers, he well merits a place among the most agreeable *general* performers.

This bird, as has been before observed, is very numerous in summer in the middle States. Scarcely a thicket in the country is without its cat birds ; and were they to fly in flocks, like many other birds, they would darken the air with their numbers. But their migrations are seldom observed, owing to their gradual progress and recession, in spring and autumn, to and from their breeding places. They enter Georgia late in February, and reach New England about the beginning of May. In their migrations, they keep pace with the progress of agriculture; and the first settlers in many parts of the Gennesee country have told me, that it was several years after they removed there before the cat bird made his appearance among them. With all these amiable qualities to recommend him, few people in the country respect the cat bird; on the contrary, it is generally the object of dislike; and the boys of the United States entertain the same prejudice and contempt for this bird, its nest, and young, as those of Britain do for the yellow hammer, and its nest, eggs, and young. I am at a loss to account for this cruel prejudice. Even those by whom it is entertained can scarcely tell you why; only they "hate cat birds;" as some persons tell you they hate Frenchmen, they hate Dutchmen, &c.—expressions that bespeak their own narrowness of understanding and want of liberality. Yet, after ruminating over in my own mind all the probable causes, I think I have at last hit on some of them; the principal of which seems to me to be a certain similarity of taste, and clashing of interest, between the cat bird and the farmer. The cat bird is fond of large ripe garden strawberries; so is the farmer, for the good price they bring in market: the cat bird loves the best and richest early cherries; so does the farmer, for they are sometimes the most profitable of his early fruit: the cat bird has a particular partiality for the finest ripe mellow pears; and these are also particular favourites with the farmer. But the cat bird has frequently the advantage of the farmer, by snatching off the first fruits of these delicious productions; and the farmer takes revenge, by shooting him down with his gun, as he finds old hats, windmills, and scarecrows, are no impediments in his way to these forbidden fruits; and nothing but this resource—the ultimatum of farmers as well as kings—can restrain his visits. The boys are now set to watch the cherry trees with the gun: and thus commences a train of prejudices and antipathies, that commonly continue through life. Perhaps, too, the common note of the cat bird, so like the mewing of the animal whose name it bears, and who itself sustains no small share of prejudice, the homeliness of his plumage, and even his familiarity, so proverbially known to beget contempt, may also contribute to this mean, illiberal, and persecuting prejudice; but, with the generous and the good, the lovers of nature and of rural charms, the confidence which this familiar bird places in man by building in his garden, under his eye, the music of his song, and the interesting playfulness of his manners, will always be more than a recompense for all the little stolen morsels he snatches.

The cat bird measures nine inches in length; at a small distance he appears nearly black; but, on a closer examination, is of a deep slate colour above, lightest on the edges of the primaries, and of a considerably lighter slate colour below, except the under tail-coverts, which are very dark red; the tail, which is rounded, and upper part of the head, as well as the legs and bill, are black. The female differs little in colour from the male. Latham takes notice of a bird, exactly resembling this, being found at Kamtschatka, only it wanted the red under the tail. Probably it might have been a young bird, in which the red is scarcely observable.

This bird has been very improperly classed among the fly-catchers. As he never seizes his prey on wing, has none of their manners, feeds principally on fruit, and seems to differ so little from the thrushes, I think he more properly belongs to the latter tribe, than to any other genus we have. His bill, legs, and feet, place and mode of building, the colour of the eggs, his imitative notes, food, and general manners, all justify me in removing him to this genus.

The cat bird is one of those unfortunate victims, and indeed the principal, against which credulity and ignorance have so often directed the fascinating quality of the black snake. A multitude of marvellous stories have been told me by people who have themselves seen the poor cat birds drawn, or sucked, as they sometimes express it, from the tops of the trees (which, by the by, the cat bird rarely visits), one by one into the yawning mouth of the immovable snake. It has so happened with me, that in all the adventures of this kind that I have personally witnessed, the cat bird was actually the assailant, and always the successful one. These rencounters never take place but during the breeding time of birds, for whose eggs and young the snake has a particular partiality. It is no wonder that those species whose nests are usually built near the ground should be the greatest sufferers, and the most solicitous for their safety: hence the cause why the cat bird makes such a distinguished figure in most of these marvellous narrations. That a poisonous snake will strike a bird or mouse, and allow it to remain till nearly expiring before he begins to devour it, our observations on the living rattlesnake at present (1811) kept by Mr Peale satisfy us is a fact; but that the same snake, with eyes, breath, or any other known quality he possesses, should be capable of drawing a bird, reluctantly, from the tree tops to its mouth, is an absurdity too great for me to swallow.

I am led to these observations by a note which I received this morning from my worthy friend Mr Bartram:—"Yesterday," says this gentleman, "I observed a conflict or contest between a cat bird and a snake. It took place in a gravel walk in the garden, near a dry wall of stone. I was within a few yards of the combatants. The bird pounced or darted upon the snake, snapping his bill; the snake would then draw himself quickly into a coil, ready for a blow; but the bird would cautiously circumvent him at a little distance, now and then running up to, and snapping at him; but keeping at a sufficient distance to avoid a blow. After some minutes, it became a running fight, the snake retreating; and, at last, he took shelter in the wall. The cat bird had young ones in the bushes near the field of battle.

"This may show the possibility of poisonous snakes biting birds; the operation of the poison causing them to become, as it were, fascinated."

BAY-BREASTED WARBLER. (*Sylvia castanea.*)

PLATE XIV.—Fig. 4.

Parus peregrinus, The Little Chocolate-breasted Titmouse, *Bartram*, p. 292.—*Peale's Museum*, No. 7311.

SYLVICOLA CASTANEA.—Swainson.

Sylvia castanea, *Bonap. Synop.** p. 81.

This very rare species passes through Pennsylvania about the beginning of May, and soon disappears. It has many of the habits of the titmouse, and all its activity; hanging among the extremities of the twigs, and darting about from place to place, with restless diligence, in search of various kinds of the larvæ of insects. It is never seen here in summer, and very rarely on its return, owing, no doubt, to the greater abundance of foliage at that time, and to the silence and real scarcity of the species. Of its nest and eggs we are altogether uninformed.

The length of this bird is five inches, breadth eleven; throat, breast, and sides under the wings, pale chestnut, or bay; forehead, cheeks, line over and strip through the eye, black; crown, deep chestnut; lower parts, dull yellowish white; hind head and back, streaked with black, on a grayish buff ground; wings, brownish black, crossed with two bars of white; tail, forked, brownish black, edged with ash, the three exterior feathers marked with a spot of white on the inner edges; behind the eye is a broad oblong spot of yellowish white. The female has much less of the bay colour on the breast; the black on the forehead is also less, and of a brownish tint. The legs and feet, in both, are dark ash, the claws extremely sharp for climbing and hanging; the bill is black; irides, hazel.

The ornithologists of Europe take no notice of this species,

* According to Bonaparte, discovered and first described by Wilson.—Ed.

and have probably never met with it. Indeed, it is so seldom seen in this part of Pennsylvania, that few even of our own writers have mentioned it.

I lately received a very neat drawing of this bird, done by a young lady in Middleton, Connecticut, where it seems also to be a rare species.

CHESTNUT-SIDED WARBLER. (*Sylvia Pennsylvanica.*)

PLATE XIV.—Fig. 5.

Linn. Syst. 333.—Red-throated Flycatcher, *Edw.* 301.—Bloody-sided Warbler, *Turton, Syst.* i. p. 596.—Le Figuier à poitrine rouge, *Buff.* v. 308.—*Briss. Add.* 105.—*Lath.* ii. 489.—*Arct. Zool.* p. 405, No. 298.—*Peale's Museum*, No. 7006.

SYLVICOLA ICTEROCEPHALA.—Swainson.

Sylvia icterocephala, *Bonap. Synop.* p. 80.—The Chestnut-sided Warbler, *Aud.* pl. 59, *Orn. Biog.* p. 306.

Of this bird I can give but little account. It is one of those transient visitors that pass through Pennsylvania in April and May, on their way further north to breed. During its stay here, which seldom exceeds a week or ten days, it appears actively engaged among the opening buds and young leaves, in search of insects; has no song but a feeble chirp, or twitter; and is not numerous. As it leaves us early in May, it probably breeds in Canada, or perhaps some parts of New England, though I have no certain knowledge of the fact. In a whole day's excursion it is rare to meet with more than one or two of these birds, though a thousand individuals of some species may be seen in the same time. Perhaps they may be more numerous on some other part of the continent.

The length of this species is five inches; the extent, seven and three-quarters. The front, line over the eye, and ear-feathers, are pure white; upper part of the head, brilliant yellow; the lores and space immediately below are marked with a triangular patch of black; the back and hind head are streaked with gray, dusky black, and dull yellow; wings, black; primaries edged with pale blue, the first and second row of coverts, broadly tipt with pale yellow; secondaries, broadly edged with the same; tail, black, handsomely forked, exteriorly edged with ash; the inner webs of the three exterior feathers with each a spot of white; from the extremity of the black at the lower mandible, on each side, a streak of deep reddish chestnut descends along the sides of the neck, and under the wings, to the root of the tail; the rest of the lower parts are pure white; legs and feet, ash; bill, black; irides, hazel. The female has the hind head much lighter, and the chestnut on the sides is considerably narrower, and not of so deep a tint.

Turton, and some other writers, have bestowed on this little bird the singular epithet of "bloody-sided," for which I was at a loss to know the reason, the colour of that part being a plain chestnut; till, on examining Mr Edward's coloured figure of this bird in the public library of Philadelphia, I found its side tinged with a brilliant blood colour. Hence, I suppose, originated the name!

MOURNING WARBLER. (*Sylvia Philadelphia.*)

PLATE XIV.—FIG. 6.

TRICHAS? PHILADELPHIA.—JARDINE.

Sylvia Philadelphia, *Bonap. Synop.* p. 85.

I HAVE now the honour of introducing to the notice of naturalists and others a very modest and neat little species, which has hitherto eluded their research. I must also add, with regret, that it is the only one of its kind I have yet met with. The bird from which the figure in the plate was taken was shot in the early part of June, on the border of a marsh, within a few miles of Philadelphia. It was flitting from one low bush to another, very busy in search of insects; and had a sprightly and pleasant warbling song, the novelty of which first attracted my attention. I have traversed the same and many such places, every spring and summer since, in expectation of again meeting with some individual of the species, but without success. I have, however, the satisfaction to say, that the drawing was done with the greatest attention to peculiarity of form, markings, and tint of plumage; and the figure on the plate is a good resemblance of the original. I have yet hopes of meeting, in some of my excursions, with the female, and, should I be so fortunate, shall represent her in some future volume of the present work, with such farther remarks on their manners, &c., as I may then be enabled to make.

There are two species mentioned by Turton, to which the present has some resemblance, viz., *Motacilla mitrata,* or mitred warbler, and *M. cucullata,* or hooded warbler; both birds of the United States, or, more properly, a single bird; for they are the same species twice described, namely, the hooded warbler. The difference, however, between that and the present is so striking, as to determine this at once to be a very distinct species. The singular appearance of the head, neck, and breast, suggested the name.

The mourning warbler is five inches long, and seven in extent; the whole back, wings, and tail, are of a deep greenish olive, the tips of the wings and the centre of the tail-feathers excepted, which are brownish; the whole head is of a dull slate colour; the breast is ornamented with a singular crescent of alternate transverse lines of pure glossy white and very deep black; all the rest of the lower parts are of a brilliant yellow; the tail is rounded at the end; legs and feet, a pale flesh colour; bill, deep brownish black above, lighter below; eye, hazel.*

* Wilson saw this bird only once, and I have met with no one who has since seen it. From the general appearance of the representation, it seems to approach nearest to the generic appellation we have given, but which must rest yet undecided. Bonaparte observes, "The excessive rarity might lead us to suppose it an accidental variety of some other,—*perhaps S. trichas.*"—ED.

Drawn from Nature by A.Wilson.　　Engraved by W.H.Lizars

1 Red-cocaded Woodpecker. 2. Brownheaded Nuthatch. 3. Pigeon Hawk. 4. Blue-winged Yellow Warbler. 5. Golden-winged W. 6. Blue-eyed Yellow W. 7. Black breasted Blue W.

15.

RED-COCKADED WOODPECKER. (*Picus querulus.*)

PLATE XV.—Fig. 1.

Peale's Museum, No. 2027.

DENDROCOPUS QUERULUS.—Koch.

Picus querulus, *Bonap. Synop.* p. 46.

This new species I first discovered in the pine woods of North Carolina. The singularity of its voice, which greatly resembles the chirping of young nestlings, and the red streak on the side of its head, suggested the specific name I have given it. It also extends through South Carolina and Georgia, at least as far as the Altamaha river. Observing the first specimen I found to be so slightly marked with red, I suspected it to be a young bird, or imperfect in its plumage; but the great numbers I afterwards shot satisfied me that this is a peculiarity of the species. It appeared exceedingly restless, active, and clamorous; and everywhere I found its manners the same.

This bird seems to be an intermediate link between the red-bellied and the hairy woodpecker, represented in Plates VII. and IX. It has the back of the former, and the white belly and spotted neck of the latter; but wants the breadth of red in both, and is less than either. A preserved specimen has been deposited in the Museum of Philadelphia.

This woodpecker is seven inches and a half long, and thirteen broad; the upper part of the head is black; the back barred with twelve white transversely semicircular lines, and as many of black, alternately; the cheeks and sides of the neck are white; whole lower parts the same; from the lower mandible, a list of black passes towards the shoulder of the wing, where it is lost in small black spots on each side of the breast; the wings are black, spotted with white; the four middle tail-feathers, black; the rest white, spotted with black; rump, black, variegated with white; the vent, white, spotted with black; the hairs that cover the nostrils are of a pale cream colour; the bill, deep slate. But what forms the most distinguishing peculiarity of this bird is a fine line of vermilion on each side of the head, seldom occupying more than the edge of a single feather. The female is destitute of this ornament; but, in the rest of her plumage, differs in nothing from the male. The iris of the eye, in both, was hazel.

The stomachs of all those I opened were filled with small black insects and fragments of large beetles. The posterior extremities of the tongue reached nearly to the base of the upper mandible.

BROWN-HEADED NUTHATCH. (*Sitta pusilla.*)

PLATE XV.—Fig. 2.

Small Nuthatch, *Catesby*, *Car.* i. 22, upper figure.—La Petite Sitelle à tête brune, *Buff.* v. 474.—*Peale's Museum*, No. 2040.—*Briss.* iii. 958.—*Lath.* i. 651, c.

SITTA PUSILLA.—Latham.

Sitta pusilla, *Bonap. Synop.* p. 97.

This bird is chiefly an inhabitant of Virginia and the southern States, and seems particularly fond of pine trees. I have never yet discovered it either in Pennsylvania or any of the regions north of this. Its manners are very similar to those of the red-bellied nuthatch, represented in Plate II.; but its notes are more shrill and chirping. In the countries it inhabits it is a constant resident; and in winter associates with parties of eight or ten of its own species, who hunt busily from tree to tree, keeping up a perpetual screeping. It is a frequent companion of the woodpecker figured beside it; and you rarely find the one in the woods without observing or hearing the other not far off. It climbs equally in every direction, on the smaller branches as well as on the body of the tree, in search of its favourite food, small insects and their larvæ. It also feeds on the seeds of the pine tree. I have never met with its nest.

This species is four inches and a quarter long, and eight broad; the whole upper part of the head and neck, from the bill to the back, and as far down as the eyes, is light brown or pale ferruginous, shaded with darker touches, with the exception of a spot of white near the back; from the nostril through the eyes, the brown is deepest, making a very observable line there; the chin, and sides of the neck under the eyes are white; the wings, dusky; the coverts and three secondaries next the body, a slate or lead colour; which is also the colour of the rest of the upper parts; the tail is nearly even at the end, the two middle feathers slate colour, the others black, tipped with slate, and crossed diagonally with a streak of white; legs and feet, dull blue; upper mandible, black; lower, blue at the base; iris, hazel. The female differs in having the brown on the head rather darker, and the line through the eye less conspicuous.

This diminutive bird is little noticed in history, and what little has been said of it by Europeans is not much to its credit. It is characterised as "a very stupid bird," which may easily be knocked down from the sides of the tree with one's cane. I confess I found it a very dexterous climber; and so rapid and restless in its motions as to be shot with difficulty. Almost all very small birds seem less suspicious of man than large ones; but that activity and restless diligence should constitute stupidity, is rather a new doctrine. Upon the whole, I am of opinion, that a person who should undertake the destruction of these birds, at even a dollar ahead for all he knocked down with his cane, would run a fair chance of starving by his profession.*

* In our note at page 36 of this volume, we mentioned that the American nuthatches and that of Europe were the only species known. M. Vigors has since described, in the proceedings of the Committee of Science of the Zoological Society, one under the name of *Sitta castaneoventris* from India, which, if true to the type, may prove an addition. In the same place, that gentleman also describes a second species of *Certhia* (*C. spilonata*), but adds, "The tail of this bird is soft and flexible." We have noticed, in a former note, the *C. familiaris* as the only known species, and we doubt if that now mentioned can rank with it.—ED.

PIGEON-HAWK. (*Falco columbarius.*)

PLATE XV.—FIG. 3, MALE.

Linn. Syst. p. 128, No. 21.—*Lath. Syn.* i. p. 101, No. 86.—L'Epervier de la Caroline, *Briss. Orn.* i. p. 238.—*Catesb.* i. p. 3, t. 3.—*Bartram*, p. 290.—*Turton, Syst.* i. p. 162.—*Peale's Museum*, No. 352.

FALCO COLUMBARIUS.—LINNÆUS.

Pigeon-Hawk, *Penn. Arct. Zool.* ii. 222.—Falco columbarius, *Bonap. Synop.* p. 28.—*North. Zool.* ii. p. 35.

THIS small hawk possesses great spirit and rapidity of flight. He is generally migratory in the middle and northern States, arriving in Pennsylvania early in spring, and extending his migrations as far north as Hudson's Bay. After building, and rearing his young, he retires to the south early in November. Small birds and mice are his principal food. When the reed birds, grakles, and red-winged blackbirds congregate in large flights, he is often observed hovering in their rear, or on their flanks, picking up the weak, the wounded, or stragglers, and frequently making a sudden and fatal sweep into the very midst of their multitudes. The flocks of robins and pigeons are honoured with the same attentions from this marauder, whose daily excursions are entirely regulated by the movements of the great body on whose unfortunate members he fattens. The individual from which the drawing on the plate was taken was shot in the meadows below Philadelphia in the month of August. He was carrying off a blackbird (*Oriolus phœniceus*) from the flock, and, though mortally wounded and dying, held his prey fast till his last expiring breath, having struck his claws into its very heart. This was found to be a male. Sometimes when shot at, and not hurt, he will fly in circles over the sportsman's head, shrieking out with great violence, as if highly irritated. He frequently flies low, skimming a little above the field. I have never seen his nest.*

The pigeon-hawk is eleven inches long, and twenty-three broad; the whole upper parts are of a deep dark brown, except the tail, which is crossed with bars of white; the inner vanes of the quill-feathers are marked with round spots of reddish brown; the bill is short, strongly toothed, of a light blue colour, and tipt with black; the skin surrounding the eye, greenish; cere, the same; temples and line over the eye, lighter brown; the lower parts, brownish white, streaked laterally with dark brown; legs, yellow; claws, black. The female is an inch and a half longer, of a still deeper colour, though marked nearly in the same manner, with the exception of some white on the hind head. The femoral, or thigh feathers, in both are of a remarkable length, reaching nearly to the feet, and are also streaked longitudinally with dark brown. The irides of the eyes of this bird have been hitherto described as being of a brilliant yellow; but every specimen I have yet met with had the iris of a deep hazel. I must therefore follow nature, in opposition to very numerous and respectable authorities.

I cannot, in imitation of European naturalists, embellish the history of this species with anecdotes of its exploits in falconry. This science, if it may be so called, is among the few that have never yet travelled across the Atlantic; neither does it appear that the idea of training our hawks or eagles to the chase ever suggested itself to any of the Indian nations of North America. The Tartars, however, from whom, according to certain writers, many of these nations

* Mr Hutchins, in his notes on the Hudson's Bay birds, informs us, that this species makes its nest in hollow rocks and trees, of sticks and grass, lined with feathers, laying from two to four white eggs, thinly marked with red spots.

This species has the form of the falcons, with the bill strongly toothed, but somewhat of the plumage of the sparrow-hawks. The colour of the eggs is also that of the latter.—Ed.

originated, have long excelled in the practice of this sport; which is indeed better suited to an open country than to one covered with forest. Though once so honourable and so universal, it is now much disused in Europe, and in Britain is nearly extinct. Yet I cannot but consider it as a much more noble and princely amusement than horse-racing and cock-fighting, cultivated in certain States with so much care; or even than pugilism, which is still so highly patronised in some of those enlightened countries.

BLUE-WINGED YELLOW WARBLER. (*Sylvia solitaria.*)

PLATE XV.—Fig. 4.

Parus aureus alis cœruleis, *Bartram*, p. 292.—*Edw.* pl. 277, upper figure.—Pine Warbler, *Arct. Zool.* p. 412, No. 318.—*Peale's Museum*, No. 7307.

VERMIVORA SOLITARIA.—Swainson.

Sylvia solitaria, *Bonap. Synop.* p. 87.—The Blue-winged Yellow Warbler, *Aud.* pl. 20, *Orn. Biog.* i. 102.

This bird has been mistaken for the pine creeper of Catesby. It is a very different species. It comes to us early in May from the south; haunts thickets and shrubberies, searching the branches for insects; is fond of visiting gardens, orchards, and willow trees, of gleaning among blossoms and currant bushes; and is frequently found in very sequestered woods, where it generally builds its nest. This is fixed in a thick bunch or tussock of long grass, sometimes sheltered by a brier bush. It is built in the form of an inverted cone or funnel, the bottom thickly bedded with dry beech leaves, the sides formed of the dry bark of strong weeds, lined within with fine dry grass. These materials are not placed in the usual manner, circularly, but shelving downwards on all sides from the top; the mouth being wide, the bottom very narrow, filled with leaves, and the eggs or young occupying the middle. The female lays five eggs, pure white, with a few very faint dots of reddish near the great end; the young appear the first week in June. I am not certain whether they raise a second brood in the same season.

I have met with several of these nests, always in a retired, though open, part of the woods, and very similar to each other.

The first specimen of this bird taken notice of by European writers was transmitted, with many others, by Mr William Bartram to Mr Edwards, by whom it was drawn and etched in the 277th plate of his "Ornithology." In his remarks on this bird, he seems at a loss to determine whether it is not the pine creeper of Catesby; * a difficulty occasioned by the very imperfect colouring and figure of Catesby's bird. The pine creeper, however, is a much larger bird; is of a dark yellow olive above, and orange yellow below; has all the habits of a creeper, alighting on the trunks of the pine trees, running nimbly round them, and, according to Mr Abbot, builds a pensile nest. I observed thousands of them in the pine woods of Carolina and Georgia, where they are resident, but have never met with them in any part of Pennsylvania.

This species is five inches and a half long, and seven and a half broad; hind head and whole back a rich green olive; crown and front, orange yellow; whole lower parts, yellow, except the vent-feathers, which are white; bill, black above, lighter below; lores, black; the form of the bill approximates a little to that of the finch; wings and tail, deep brown, broadly edged with pale slate, which makes them appear wholly of that tint, except at the tips; first and second row of coverts, tipt with white slightly stained with yellow; the three exterior tail feathers have their inner vanes nearly all white; legs, pale bluish; feet, dirty yellow; the two middle tail-feathers are pale slate. The female differs very little in colour from the male.

* Catesby, Car. vol. i. pl. 61.

This species very much resembles the prothonotary warbler of Pennant and Buffon; the only difference I can perceive, on comparing specimens of each, is, that the yellow of the prothonotary is more of an orange tint, and the bird somewhat larger.

BLUE-EYED YELLOW WARBLER. (*Sylvia citrinella.*)

PLATE XV.—Fig. 6.

Yellow-poll Warbler, *Lath. Syn.* vol. ii. No. 148.—*Arct. Zool.* p. 402, No. 292.—Le Figuier tacheté, *Buff. Ois.* v. p. 285.—Motacilla æstiva, *Turton's Syst.* p. 615.—Parus luteus, Summer Yellow Bird, *Bartram*, p. 292.—*Peale's Museum*, No. 7266.

SYLVICOLA ÆSTIVA.—Swainson.

Sylvia æstiva, *Bonap. Synop.* p. 83.—Sylvicola æstiva, *North. Zool.* ii. p. 212.

This is a very common summer species, and appears almost always actively employed among the leaves and blossoms of the willows, snowball shrub, and poplars, searching after small green caterpillars, which are its principal food. It has a few shrill notes, uttered with emphasis, but not deserving the name of song. It arrives in Pennsylvania about the beginning of May, and departs again for the south about the middle of September. According to Latham, it is numerous in Guiana, and is also found in Canada. It is a very sprightly, unsuspicious, and familiar little bird; is often seen in and about gardens, among the blossoms of fruit trees and shrubberies; and, on account of its colour, is very noticeable. Its nest is built with great neatness, generally in the triangular fork of a small shrub, near or among brier bushes. Outwardly it is composed of flax or tow, in thick circular layers, strongly twisted round the twigs that rise through its sides, and lined within with hair and the soft downy substance from the stalks of the fern. The eggs are four or five, of a dull white, thickly sprinkled near the great end with specks of pale brown.

They raise two broods in the season. This little bird, like many others, will feign lameness, to draw you away from its nest, stretching out his neck, spreading and bending down his tail, until it trails along the branch, and fluttering feebly along, to draw you after him; sometimes looking back, to see if you are following him, and returning back to repeat the same manœuvres, in order to attract your attention. The male is most remarkable for this practice.

The blue-eyed warbler is five inches long and seven broad; hind head and back, greenish yellow; crown, front, and whole lower parts, rich golden yellow; breast and sides, streaked laterally with dark red; wings and tail, deep brown, except the edges of the former, and the inner vanes of the latter, which are yellow; the tail is also slightly forked; legs, a pale clay colour; bill and eyelids, light blue. The female is of a less brilliant yellow, and the streaks of red on the breast are fewer and more obscure. Buffon is mistaken in supposing No. 1. of Pl. enl. plate lviii. to be the female of this species.

GOLDEN-WINGED WARBLER. (*Sylvia chrysoptera.*)

PLATE XV.—Fig. 5.

Edw. 299.—Le Figuier aux ailes dorées, *Buff.* v. 311.—*Lath.* ii. 492.—*Arct. Zool.* 403, No. 295, *Ib.* No. 296.—Motacilla chrysoptera, *Turt. Syst.* i. 597.—Mot. flavifrons, Yellow-fronted Warbler, *Id.* 601.—Parus alis aureis, *Bartram*, p. 292.—*Peale's Museum*, No. 7010.

VERMIVORA CHRYSOPTERA.—Swainson.

Sylvia chrysoptera, *Bonap. Synop.* p. 87.

This is another spring passenger through the United States to the north. This bird, as well as fig. 4, from the particular form of its bill, ought rather to be separated from the warblers; or, along with several others of the same kind, might be arranged as a 'subgenera, or particular family of that tribe, which might with propriety be called worm-eaters, the *Motacilla vermivora* of Turton having the bill exactly of this form. The habits of these birds partake a good deal of those of the titmouse; and in their language and action they very much resemble them. All that can be said of this species is, that it appears in Pennsylvania for a few days about the last of April or beginning of May, darting actively among the young leaves and opening buds, and is rather a scarce species.

The golden-winged warbler is five inches long, and seven broad; the crown, golden yellow; the first and second row of wing coverts, of the same rich yellow; the rest of the upper parts, a deep ash or dark slate colour; tail, slightly forked, and, as well as the wings, edged with whitish; a black band passes through the eye, and is separated from the yellow of the crown by a fine line of white; chin and throat, black, between which and that passing through the eye runs a strip of white, as in the figure; belly and vent, white; bill, black, gradually tapering to a sharp point; legs, dark ash; irides, hazel.

Pennant has described this species twice, first, as the golden-winged warbler, and immediately after as the yellow-fronted warbler. See the synonyms at the beginning of this article.

BLACK-THROATED BLUE WARBLER. (*Sylvia Canadensis.*)

PLATE XV.—FIG. 7.

Motacilla Canadensis, *Linn. Syst.* 336.—Le Figuier bleu, *Buff.* v. 304, *Pl. enl.* 685, fig. 2.—*Lath. Syn.* ii. p. 487, No. 113.—*Edw.* 252.—*Arct. Zool.* p. 399, No. 285.—*Peale's Museum*, No. 7222.

SYLVICOLA CANADENSIS.—SWAINSON.

Sylvia Canadensis, *Bonap. Synop.* p. 84.

I KNOW little of this bird. It is one of those transient visitors that, in the month of April, pass through Pennsylvania on its way to the north to breed. It has much of the flycatcher in its manners, though the form of its bill is decisively that of the warbler. These birds are occasionally seen for about a week or ten days, viz., from the 25th of April to the end of the first week in May. I sought for them in the southern States in winter, but in vain. It is highly probable that they breed in Canada; but the summer residents among the feathered race on that part of the continent are little known or attended to. The habits of the bear, the deer, and beaver are much more interesting to those people, and for a good substantial reason too, because more lucrative; and unless there should arrive an order from England for a cargo of skins of warblers and flycatchers, sufficient to make them an object worth speculation, we are likely to know as little of them hereafter as at present.

This species is five inches long, and seven and a half broad, and is wholly of a fine light slate colour above; the throat, cheeks, front and upper part of the breast, are black; wings and tail, dusky black, the primaries marked with a spot of white immediately below their coverts; tail, edged with blue; belly and vent, white; legs and feet, dirty yellow; bill, black, and beset with bristles at the base. The female is more of a dusky ash on the breast, and, in some specimens, nearly white.

They, no doubt, pass this way on their return in autumn, for I have myself shot several in that season; but as the woods are then still thick with leaves, they are much more difficult to be seen, and make a shorter stay than they do in spring.

AMERICAN SPARROW-HAWK. (*Falco sparverius.*)

PLATE XVI.—FIG. 1, FEMALE.

Emerillon de St Domingue, *Buff.* i. 291, *Pl. enl.* 465.—*Arct. Zool.* 212.—Little Falcon, *Lath. Syn.* i. p. 110, No. 94, *Ib.* 95.—*Peale's Museum*, No. 389.

FALCO SPARVERIUS.—LINNÆUS.

Falco sparverius, *Bonap. Synop.* p. 27.—Falco sparverius, Little Rusty-crowned Falcon, *North. Zool.* ii. p. 31.

IN no department of ornithology has there been greater confusion or more mistakes made than among this class of birds of prey. The great difference of size between the male and female, the progressive variation of plumage to which, for several years, they are subject, and the difficulty of procuring a sufficient number of specimens for examination; all these causes conspire to lead the naturalist into almost unavoidable mistakes. For these reasons, and in order, if possible, to ascertain each species of this genus distinctly, I have determined, where any doubt or ambiguity prevails, to represent both male and female, as fair and perfect specimens of each may come into my possession. According to fashionable etiquette, the honour of precedence, in the present instance, is given to the *female* of this species; both because she is the most courageous, the largest and handsomest of the two, best ascertained, and less subject to change of colour than the male, who will require some further examination and more observation before we can venture to introduce him.

This bird is a constant resident in almost every part of the United States, particularly in the States north of Maryland. In the southern States there is a smaller species found, which is destitute of the black spots on the head; the legs are long and very slender, and the wings light blue. This has been supposed by some to be the male of the present species; but this is an error. The eye of the present species is dusky;

Drawn from Nature by A. Wilson *Engraved by W.H.Lizars*

1. American Sparrow Hawk. 2. Field Sparrow. 3. Tree Sp. 4. Song Sp. 5. Chipping Sp. 6. Snow Bird.

16.

that of the smaller species a brilliant orange; the former has the tail *rounded* at the end, the latter slightly *forked*. Such essential differences never take place between two individuals of the same species. It ought, however, to be remarked, that in all the figures and descriptions I have hitherto met with of the bird now before us, the iris is represented of a bright golden colour; but, in all the specimens I have shot, I uniformly found the eye very dark, almost black, resembling a globe of black glass. No doubt the golden colour of the iris would give the figure of the bird a more striking appearance; but, in works of natural history, to sacrifice truth to mere picturesque effect is detestable, though, I fear, but too often put in practice.

The nest of this species is usually built in a hollow tree; generally pretty high up, where the top, or a large limb, has been broken off. I have never seen its eggs; but have been told that the female generally lays four or five, which are of a light brownish yellow colour, spotted with a darker tint; the young are fed on grasshoppers, mice, and small birds, the usual food of the parents.

The habits and manners of this bird are well known. It flies rather irregularly, occasionally suspending itself in the air, hovering over a particular spot for a minute or two, and then shooting off in another direction. It perches on the top of a dead tree or pole, in the middle of a field or meadow, and, as it alights, shuts its long wings so suddenly, that they seem instantly to disappear; it sits here in an almost perpendicular position, sometimes for an hour at a time, frequently jerking its tail, and reconnoitring the ground below in every direction for mice, lizards, &c. It approaches the farmhouse, particularly in the morning, skulking about the barnyard for mice or young chickens. It frequently plunges into a thicket after small birds, as if by random; but always with a particular, and generally a fatal, aim. One day I observed a bird of this species perched on the highest top of a large poplar, on the skirts of the wood, and was in the act of raising the gun to my eye, when he swept down, with the rapidity of an arrow, into a thicket of briers about thirty yards off, where I shot him dead, and, on coming up, found the small field sparrow (fig. 2) quivering in his grasp. Both our aims had been taken in the same instant, and, unfortunately for him, both were fatal. It is particularly fond of watching along hedgerows and in orchards, where those small birds represented in the same plate usually resort. When grasshoppers are plenty, they form a considerable part of its food.

Though small snakes, mice, lizards, &c., be favourite morsels with this active bird, yet we are not to suppose it altogether destitute of delicacy in feeding. It will seldom or never eat of anything that it has not itself killed, and even that, if not (as epicures would term it) *in good eating order*, is sometimes rejected. A very respectable friend, through the medium of Mr Bartram, informs me that one morning he observed one of these hawks dart down on the ground and seize a mouse, which he carried to a fence post, where, after examining it for some time, he left it, and, a little while after, pounced upon another mouse, which he instantly carried off to his nest, in the hollow of a tree hard by. The gentleman, anxious to know why the hawk had rejected the first mouse, went up to it, and found it to be almost covered with lice, and greatly emaciated! Here was not only delicacy of taste, but sound and prudent reasoning :—If I carry this to my nest, thought he, it will fill it with vermin, and hardly be worth eating.

The blue jays have a particular antipathy to this bird, and frequently insult it by following and imitating its notes so exactly, as to deceive even those well acquainted with both. In return for all this abuse, the hawk contents himself with now and then feasting on the plumpest of his persecutors, who are, therefore, in perpetual dread of him; and yet, through some strange infatuation, or from fear that, if they lose sight of him, he may attack them unawares, the sparrow-hawk no sooner appears than the alarm is given, and the whole posse of jays follow.

The female of this species, which is here faithfully represented from a very beautiful living specimen, furnished by a particular friend, is eleven inches long, and twenty-three from tip to tip of the expanded wings. The cere and legs are yellow; bill, blue, tipt with black; space round the eye, greenish blue; iris, deep dusky; head, bluish ash; crown, rufous; seven spots of black on a white ground surround the head, in the manner represented in the figure; whole upper parts reddish bay, transversely streaked with black; primary and secondary quills, black, spotted on their inner vanes with brownish white; whole lower parts yellowish white, marked with longitudinal streaks of brown, except the chin, vent, and femoral feathers, which are white; claws, black.

The male of this species (which is an inch and a half shorter, has the shoulder of the wings blue, and also the black marks on the head, but is, in other respects, very differently marked from the female) will appear in an early part of the present work, with such other particulars as may be thought worthy of communicating.*

FIELD SPARROW.† (*Fringilla pusilla.*)

PLATE XVI.—FIG. 2.

Passer agrestis, *Bartram*, p. 291.—*Peale's Museum*, No. 6560.

EMBERIZA PUSILLA.—JARDINE, SW. MSS.

Fringilla pusilla, *Bonap. Synop.* p. 110.

THIS is the smallest of all our sparrows, and, in Pennsylvania, is generally migratory. It arrives early in April, frequents dry fields covered with long grass, builds a small

* See description of male, and note, Vol. II.

† The American *bunting finches* are most puzzling, the forms being constantly intermediate, and never assuming the true type. Mr Swainson has also felt this, and has been obliged to form a new genus, to contain one portion nearly inadmissible to any of the others. The present species will rank as allied nearest to the reed bunting of Europe, *E.*

nest on the ground, generally at the foot of a brier; lines it with horse hair; lays six eggs, so thickly sprinkled with ferruginous, as to appear altogether of that tint; and raises two, and often three, broods in a season. It is more frequently found in the middle of fields and orchards than any of the other species, which usually lurk along hedgerows. It has no song, but a kind of chirruping, not much different from the chirpings of a cricket. Towards fall, they assemble in loose flocks, in orchards and cornfields, in search of the seeds of various rank weeds, and are then very numerous. As the weather becomes severe, with deep snow, they disappear. In the lower parts of North and South Carolina, I found this species in multitudes in the months of January and February. When disturbed, they take to the bushes, clustering so close together, that a dozen may easily be shot at a time. I continued to see them equally numerous through the whole lower parts of Georgia, from whence, according to Mr Abbot, they all disappear early in the spring.

None of our birds have been more imperfectly described than that family of the finch tribe usually called sparrows. They have been considered as too insignificant for particular notice, yet they possess distinct characters, and some of them peculiarities, well worthy of notice. They are innocent in their habits, subsisting chiefly on the small seeds of wild plants, and seldom injuring the property of the farmer. In the dreary season of winter, some of them enliven the prospect by hopping familiarly about our doors, humble pensioners on the sweepings of the threshold.

The present species has never before, to my knowledge, been figured. It is five inches and a quarter long, and eight inches broad; bill and legs, a reddish cinnamon colour; upper

schœniculus. Another, mentioned neither by Wilson nor Bonaparte, has been added by the Overland Expedition,—*Emberiza pallida,* clay coloured bunting, Sw. and Richard., North. Zool. It approaches nearest to *E. socialis,* but differs in wanting the bright rufous crown, and having the ear-feathers brown, margined above and below with a dark edge.—Ed.

part of the head, deep chestnut, divided by a slight streak of drab, widening as it goes back; cheeks, line over the eye, breast, and sides under the wings, a brownish clay colour, lightest on the chin, and darkest on the ear-feathers; a small streak of brown at the lower angle of the bill; back, streaked with black, drab, and bright bay, the latter being generally centred with the former; rump, dark drab, or cinereous; wings, dusky black, the primaries edged with whitish, the secondaries bordered with bright bay; greater wing coverts, black edged and broadly tipt with brownish white; tail, dusky black, edged with clay colour: male and female nearly alike in plumage; the chestnut on the crown of the male rather brighter.

TREE SPARROW. (*Fringilla arbcrea.*)

PLATE XVI.—FIG. 3.

Le Soulciet, *Buff.* iii. 500.—Moineau de Canada, *Briss.* iii. 101, *Pl. enl.* 223.—*Lath.* ii. 252.—*Edw.* 269.—*Arct. Zool.* p. 373, No. 246.—*Peale's Museum,* No. 6575.

EMBERIZA CANADENSIS.—SWAINSON.

Fringilla Canadensis, *Bonap. Synop.* p. 109.—Emberiza Canadensis, *North. Zool.* ii. p. 252.

THIS sparrow is a native of the north, who takes up his winter quarters in Pennsylvania, and most of the northern States, as well as several of the southern ones. He arrives here about the beginning of November, and leaves us again early in April; associates in flocks with the snow birds; frequents sheltered hollows, thickets, and hedgerows, near springs of water; and has a low warbling note, scarcely audible at the distance of twenty or thirty yards. If disturbed, he takes to trees, like the white-throated sparrow, but contrary to the habit of most of the others, who are inclined rather to dive into thickets. Mr Edwards has erroneously represented this as the female of the mountain sparrow; but that judicious and excellent naturalist, Mr Pennant, has given a more correct account of it, and informs us that it inhabits the country bordering on Hudson's Bay during summer ; comes to Severn settlement in May ; advances farther north to breed ; and returns in autumn on its way southward. It also visits Newfoundland.*

By some of our own naturalists, this species has been confounded with the chipping sparrow (fig. 5), which it very much resembles, but is larger and handsomer, and is never found with us in summer. The former departs for the south about the same time that the latter arrives from the north ; and from this circumstance, and their general resemblance, has arisen the mistake.

The tree sparrow is six inches and a half long, and nine and a half in extent ; the whole upper part of the head is of a bright reddish chestnut, sometimes slightly skirted with gray ; from the nostrils, over the eye, passes a white strip, fading into pale ash, as it extends back ; sides of the neck, chin, and breast, very pale ash ; the centre of the breast marked with an obscure spot of dark brown : from the lower angle of the bill proceeds a slight streak of chestnut ; sides, under the wings, pale brown ; back, handsomely streaked with pale drab, bright bay, and black ; lower part of the back and rump, brownish drab ; lesser wing-coverts, black, edged with pale ash ; wings, black, broadly edged with bright bay ; the first and second row of coverts, tipt with pure white ; tail, black, forked, and exteriorly edged with dull white ; belly and vent, brownish white ; bill, black above, yellow below ; legs, a brownish clay colour ; feet, black. The female is about half an inch shorter ; the chestnut or bright bay on the wings, back, and crown, is less brilliant ; and the white on the coverts narrower, and not so pure. These are all the differences I can perceive.†

* Arctic Zoology, vol. ii. p. 373.

† Peculiar to America, and we should say, going more off from the group than *F. socialis*, Wils., as mentioned by Swainson in the "Northern Zoology."—ED.

SONG SPARROW. (*Fringilla melodia.*)

PLATE XVI.—FIG. 4.

Fasciated Finch? *Arct. Zool.* p. 375, No. 252.—*Peale's Museum*, No. 6573.

EMBERIZA MELODIA.*—JARDINE.

Bonap. Synop. p. 108.—The Song Sparrow, *Aud.* pl. 25, *Orn. Biog.* i. p. 126.

So nearly do many species of our sparrows approximate to each other in plumage, and so imperfectly have they been taken notice of, that it is absolutely impossible to say, with certainty, whether the present species has ever been described or not. And yet, of all our sparrows, this is the most numerous, the most generally diffused over the United States, and by far the earliest, sweetest, and most lasting songster. It may be said to be partially migratory, many passing to the south in the month of November; and many of them still remaining with us, in low, close, sheltered meadows and swamps, during the whole of winter. It is the first singing bird in spring, taking precedence even of the pewee and blue bird. Its song continues occasionally during the whole summer and fall, and is sometimes heard even in the depth of winter. The notes or chant are short, but very sweet, resembling the beginning of the canary's song, and frequently repeated, generally from the branches of a bush or small tree, where it sits chanting for an hour together. It is fond of frequenting the borders of rivers, meadows, swamps, and such like watery places; and, if wounded, and unable to fly, will readily take to the water, and swim with considerable rapidity. In the great cypress swamps of the southern States, in the depth of winter, I

* I have been puzzled where to place this bird—in *Emberiza*, or as a subgenus of it. There seems much difference in the form of the bill, though it has "a rudiment of the knob." I have been unable to obtain a specimen for comparison. Mr Swainson thinks it connects the American bunting with his *Zonotrichia.*—ED.

observed multitudes of these birds mixed with several other species; for these places appear to be the grand winter rendezvous of almost all our sparrows. I have found this bird in every district of the United States, from Canada to the southern boundaries of Georgia; but Mr Abbot informs me, that he knows of only one or two species that remain in that part of Georgia during the summer.

The song sparrow builds on the ground, under a tuft of grass; the nest is formed of fine dry grass, and lined with horse hair; the eggs are four or five, thickly marked with spots of reddish brown, on a white, sometimes bluish white ground; if not interrupted, raises three broods in the season. I have found his nest with young as early as the 26th of April, and as late as the 12th of August. What is singular, the same bird often fixes his nest in a cedar tree, five or six feet from the ground. Supposing this to have been a variety, or different species, I have examined the bird, nest, and eggs, with particular care several times, but found no difference. I have observed the same accidental habit in the red-winged blackbird, which sometimes builds among the grass, as well as on alder bushes.

This species is six inches and a half long, and eight and a half in extent; upper part of the head, dark chestnut, divided laterally by a line of pale dirty white; spot at each nostril, yellow ochre; line over the eye, inclining to ash; chin, white; streak from the lower mandible, slit of the mouth, and posterior angle of the eye, dark chestnut; breast, and sides, under the wings, thickly marked with long pointed spots of dark chestnut, centred with black, and running in chains; belly, white; vent, yellow ochre, streaked with brown; back, streaked with black, bay, and pale ochre; tail, brown, rounded at the end, the two middle feathers streaked down their centres with black; legs, flesh coloured; wing-coverts, black, broadly edged with bay, and tipt with yellowish white; wings, dark brown. The female is scarcely distinguishable by its plumage from the male. The bill in both, horn coloured.

CHIPPING SPARROW. (*Fringilla socialis.*)

PLATE XVI.—Fig. 5.

Passer domesticus, the Little House Sparrow, or Chipping Bird, *Bartram*, p. 291.—*Peale's Museum*, No. 6571.

EMBERIZA SOCIALIS.—Swainson.

Fringilla socialis, *Bonap. Synop.* p. 109.

This species, though destitute of the musical talents of the former, is, perhaps, more generally known, because more familiar, and even domestic. He inhabits, during summer, the city, in common with man, building in the branches of the trees with which our streets and gardens are ornamented; and gleaning up crumbs from our yards, and even our doors, to feed his more advanced young with. I have known one of these birds attend regularly every day, during a whole summer, while the family were at dinner, under a piazza, fronting the garden, and pick up the crumbs that were thrown to him. This sociable habit, which continues chiefly during the summer, is a singular characteristic. Towards the end of summer he takes to the fields and hedges, until the weather becomes severe, with snow, when he departs for the south.

The chipping bird builds his nest most commonly in a cedar bush, and lines it thickly with cow hair. The female lays four or five eggs, of a light blue colour, with a few dots of purplish black near the great end.

This species may easily be distinguished from the four preceding ones by his black bill and frontlet, and by his familiarity in summer; yet, in the month of August and September, when they moult their feathers, the black on the front, and partially on the bill, disappears. The young are also without the black during the first season.

The chipping sparrow is five inches and a quarter long, and eight inches in extent; frontlet, black; chin, and line over the eye, whitish; crown, chestnut; breast and sides of the neck, pale ash; bill, in winter, black; in summer, the lower mandible flesh coloured; rump, dark ash; belly and vent, white; back, variegated with black and bright bay; wings, black, broadly edged with bright chestnut; tail, dusky, forked, and slightly edged with pale ochre; legs and feet, a pale flesh colour. The female differs in having less black on the frontlet, and the bay duller. Both lose the black front in moulting.

SNOW BIRD. (*Fringilla Hudsonia.*)*

PLATE XVI.—FIG. 6.

Fringilla Hudsonia, *Turton, Syst.* i. 568.—Emberiza hyemalis, *Id.* 531.—*Lath.* i. 66.—*Catesb.* i. 36.—*Arct. Zool.* p. 359, No. 223.—Passer nivalis, *Bartram*, p. 291.—*Peale's Museum*, No. 6532.

FRINGILLA HYEMALIS.—LINNÆUS.

Fringilla hyemalis, *Bonap. Synop.* p. 109.—*North. Zool.* ii. p. 259.—The Snow Bird, *Aud.* pl. 13, *Orn. Biog.* i. p. 72.

THIS well-known species, small and insignificant as it may appear, is by far the most numerous, as well as the most extensively disseminated, of all the feathered tribes that visit us from the frozen regions of the north,—their migrations extending from the arctic circle, and probably beyond it, to the shores of the Gulf of Mexico, spreading over the whole breadth of the United States, from the Atlantic Ocean to Louisiana; how much farther westward, I am unable to say. About the 20th of October, they make their first appearance in those parts of Pennsylvania east of the Alleghany Mountains. At first they are most generally seen on the borders of woods among the falling and decayed leaves, in loose flocks of thirty or forty together, always taking to the trees when disturbed. As the weather sets in colder, they approach nearer the farmhouse and villages; and on the appearance of what

* Nivalis of first edition.

is usually called falling weather, assemble in larger flocks, and seem doubly diligent in searching for food. This increased activity is generally a sure prognostic of a storm. When deep snows cover the ground, they become almost half domesticated. They collect about the barn, stables, and other outhouses, spread over the yard, and even round the steps of the door; not only in the country and villages, but in the heart of our large cities, crowding around the threshold early in the morning, gleaning up the crumbs, appearing very lively and familiar. They have also recourse, at this severe season, when the face of the earth is shut up from them, to the seeds of many kinds of weeds that still rise above the snow, in corners of fields, and low, sheltered situations along the borders of creeks and fences, where they associate with several species of sparrows, particularly those represented on the same plate. They are, at this time, easily caught with almost any kind of trap; are generally fat, and, it is said, are excellent eating.

I cannot but consider this bird as the most numerous of its tribe of any within the United States. From the northern parts of the district of Maine, to the Ogeechee river in Georgia, a distance, by the circuitous route in which I travelled, of more than 1800 miles, I never passed a day, and scarcely a mile, without seeing numbers of these birds, and frequently large flocks of several thousands. Other travellers with whom I conversed, who had come from Lexington in Kentucky, through Virginia, also declared that they found these birds numerous along the whole road. It should be observed, that the roadsides are their favourite haunts, where many rank weeds that grow along the fences furnish them with food, and the road with gravel. In the vicinity of places where they were most numerous, I observed the small hawk, represented in the same plate, and several others of his tribe, watching their opportunity, or hovering cautiously around, making an occasional sweep among them, and retiring to the bare branches of an old cypress to feed on their victim. In the month of

April, when the weather begins to be warm, they are observed to retreat to the woods, and to prefer the shaded sides of hills and thickets; at which time the males warble out a few very low sweet notes, and are almost perpetually pursuing and fighting with each other. About the 20th of April they take their leave of our humble regions, and retire to the north, and to the high ranges of the Alleghany, to build their nests and rear their young. In some of those ranges, in the interior of Virginia, and northward about the waters of the west branch of the Susquehanna, they breed in great numbers. The nest is fixed in the ground, or among the grass, sometimes several being within a small distance of each other. According to the observations of the gentlemen residing at Hudson Bay Factory, they arrive there about the beginning of June, stay a week or two, and proceed farther north to breed. They return to that settlement in the autumn on their way to the south.

In some parts of New England, I found the opinion pretty general, that the snow bird, in summer, is transformed into the small chipping sparrow, which we find so common in that season, and which is represented in the same plate. I had convinced a gentleman of New York of his mistake in this matter by taking him to the house of a Mr Gautier there, who amuses himself by keeping a great number of native as well as foreign birds. This was in the month of July, and the snow bird appeared there in the same coloured plumage he usually has. Several individuals of the chipping sparrow were also in the same apartment. The evidence was, therefore, irresistible; but, as I had not the same proofs to offer to the eye in New England, I had not the same success.

There must be something in the temperature of the blood or constitution of this bird which unfits it for residing during summer in the lower parts of the United States, as the country here abounds with a great variety of food of which during its stay it appears to be remarkably fond. Or perhaps its habit of associating in such numbers to breed, and building its nest with so little precaution, may, to ensure its

Drawn from Nature by A. Wilson. Engraved by W.H. Lizars

1. American Siskin. 2. Rose breasted Grosbeak. 3. Green black throated Warbler. 4. Yellow rump. W.
5. Cærulean W. 6. Solitary Flycatcher.

17.

safety, require a solitary region, far from the intruding footsteps of man.

The snow bird is six inches long, and nine in extent; the head, neck, and upper parts of the breast, body, and wings, are of a deep slate colour; the plumage sometimes skirted with brown, which is the colour of the young birds; the lower parts of the breast, the whole belly, and vent, are pure white; the three secondary quill-feathers next the body are edged with brown, the primaries with white; the tail is dusky slate, a little forked, the two exterior feathers wholly white, which are flirted out as it flies, and appear then very prominent; the bill and legs are of a reddish flesh colour; the eye, bluish black. The female differs from the male in being considerably more brown. In the depth of winter, the slate colour of the male becomes more deep and much purer, the brown disappearing nearly altogether.

PINE FINCH. (*Fringilla pinus.*)

PLATE XVII.—Fig. 1.

Peale's Museum, No. 6577.

CARDUELIS PINUS.—Swainson.

Fringilla pinus (subgenus Carduelis), *Bonap. Synop.* p. 111.

This little northern stranger visits us in the month of November, and seeks the seeds of the black alder, on the borders of swamps, creeks, and rivulets. As the weather becomes more severe, and the seeds of the *Pinus Canadensis* are fully ripe, these birds collect in larger flocks, and take up their residence almost exclusively among these trees. In the gardens of Bush Hill, in the neighbourhood of Philadelphia, a flock of two or three hundred of these birds has regularly wintered many years, where a noble avenue of pine trees, and walks covered with fine white gravel, furnish them with abundance through the winter. Early in March they disappear, either to the north, or to the pine woods that cover many lesser ranges of the Alleghany. While here, they are often so tame as to allow you to walk within a few yards of the spot where a whole flock of them are sitting. They flutter among the branches, frequently hanging by the cones, and uttering a note almost exactly like that of the goldfinch (*F. tristis*). I have not a doubt but this bird appears in a richer dress in summer in those places where he breeds, as he has so very great a resemblance to the bird above mentioned, with whose changes we are well acquainted.

The length of this species is four inches; breadth, eight inches; upper part of the head, the neck, and back, a dark flaxen colour, streaked with black; wings black, marked with two rows of dull white or cream colour; whole wing-quills, under the coverts, rich yellow, appearing even when the wings are shut; rump and tail-coverts, yellowish streaked with dark brown; tail-feathers, rich yellow from the roots half way to the tips, except the two middle ones, which are blackish brown, slightly edged with yellow; sides under the wings, of a cream colour, with long streaks of black; breast, a light flaxen colour, with small streaks or pointed spots of black; legs, purplish brown; bill, a dull horn colour; eyes, hazel. The female was scarcely distinguishable by its plumage from the male. The New York siskin of Pennant * appears to be only the yellow bird (*Fringilla tristis*) in his winter dress.

This bird has a still greater resemblance to the siskin of Europe (*F. spinus*), and may perhaps be the species described by Turton † as the black Mexican siskin, which he says is varied above with black and yellowish, and is white beneath, and which is also said to sing finely. This change from flaxen to yellow is observable in the goldfinch; and no other two birds of our country resemble each other more than these do in their winter dresses. Should these surmises be found correct, a figure of this bird, in his summer dress, shall appear in some future part of our work. ‡

* Arctic Zoology, p. 372, No. 243. † Turton, vol. i. p. 560.

‡ This is a true siskin; and we have a very accurate description of the general manners of the group in those of the individual now described

ROSE-BREASTED GROSBEAK.* (*Loxia rosea.*)

PLATE XVII.—Fig. 2.

Loxia Ludoviciana, *Turton, Syst.*—Red-breasted Grosbeak, *Arct. Zool.* p. 350, No. 212.—Red-breasted Finch, *Id.* 372, No. 245.—Le Rose gorge, *Buff.* iii. 460.—Gros-bec de la Louisiane, *Pl. enl.* 153, fig. 2.—*Lath.* ii. 126.—*Peale's Museum*, No. 5806, male ; 5807, female ; 5806 A, male of one year old.

GUIRACA LUDOVICIANA.—Swainson.

Fringilla (subgenus Coccothraustes) Ludoviciana, *Bonap. Synop.* p. 113.—Coccothraustes (Guiraca) Ludoviciana, *North. Zool.* i. p. 271.

This elegant species is rarely found in the lower parts of Pennsylvania; in the State of New York, and those of New England, it is more frequently observed, particularly in fall, when the berries of the sour gum are ripe, on the kernels

by Wilson. Little seems to be known of their summer haunts ; and, indeed, the more northern species remain in the same obscurity. They generally all migrate, go north to breed, and winter in southern latitudes. The species of Great Britain and Europe performs a like migration, assembling in very large flocks during winter, feeding upon seeds, &c., and retiring north to breed. A few pairs, not performing the migration to its utmost northern extent, breed in the larger pine woods in the Highlands of Scotland. In 1829, they were met with in June in a large fir wood at Killin, evidently breeding; last year they were known to breed in an extensive wood at New Abbey, in Galloway. In their winter migrations they are not regular, particular districts being visited by them at uncertain periods. In Annandale, Dumfriesshire, they were always accounted rare, and the first pair I ever saw there was shot in 1827. Early in October, as the winter advanced, very large flocks arrived, and fed chiefly upon the ragweed, and under some large beech trees, turning over the fallen mast, and eating part of the kernels, as well as any seeds they could find among them. In 1828, they again appeared ; but in 1829, not one was seen ; and the present winter (1830), they are equally wanting. The plate of our author is that of the bird in its winter dress. As he justly observes, the plumage becomes much richer during the season of incubation. The black parts become brighter and deeper, and the olive of a yellower green.—Ed.

* This species seems to have been described under various specific names by various authors. Wilson, in the body of his work, calls it *L. rosea ;* but he corrects that name afterwards in the index, and restores

of which it eagerly feeds. Some of its trivial names would import that it is also an inhabitant of Louisiana; but I have not heard of its being seen in any of the southern States. A gentleman of Middleton, Connecticut, informed me that he kept one of these birds for some considerable time in a cage, and observed that it frequently sang at night, and all night; that its notes were extremely clear and mellow, and the sweetest of any bird with which he is acquainted.

The bird from which the figure on the plate was taken was shot, late in April, on the borders of a swamp, a few miles from Philadelphia. Another male of the same species was killed at the same time, considerably different in its markings; a proof that they do not acquire their full colours until at least the second spring or summer.

The rose-breasted grosbeak is eight inches and a half long, and thirteen inches in extent; the whole upper parts are black, except the second row of wing-coverts, which are broadly tipt with white; a spot of the same extends over the primaries, immediately below their coverts; chin, neck, and upper part of the breast, black; lower part of the breast, middle of the belly, and lining of the wings, a fine light carmine or rose colour; tail, forked, black, the three exterior feathers on each side white on their inner vanes for an inch or more from the tips; bill, like those of its tribe, very thick and strong, and pure white; legs and feet, light blue; eyes, hazel. The young male of the first spring has the plumage of the back variegated with light brown, white, and black;

that by which it must now stand. The generic appellation has also been various, and the necessity of some decided one cannot be better shown than in the different opinions expressed by naturalists, who have placed it in three or four of the known genera, without being very well satisfied with any of its situations. Gmelin and Latham have even placed the young and old in different genera, *Loxia* and *Fringilla;* by Brisson, it is a *Coccothraustes;* and by Sabine, a *Phyrrhula.* It appears a form exclusively American, supplanting the *Coccothraustes* of Asia and the Indian continent; and *Guiraca* has been appropriated to it by Mr Swainson, in which will also range the cardinal and blue grosbeaks of our author.—ED.

a line of white extends over the eye; the rose colour also reaches to the base of the bill, where it is speckled with black and white. The female is of a light yellowish flaxen colour, streaked with dark olive and whitish; the breast is streaked with olive, pale flaxen, and white; the lining of the wings is pale yellow; the bill more dusky than in the male, and the white on the wing less.

BLACK-THROATED GREEN WARBLER. (*Sylvia virens.*)

PLATE XVII.—FIG. 3.

Motacilla virens, *Gmel. Syst.* i. p. 985.—Le Figuier à cravate noire, *Buff.* v. p. 298.—Black-throated Green Flycatcher, *Edw.* t. 300.—Green Warbler, *Arct. Zool.* ii. No. 297.—*Lath. Syn.* iv. p. 484, 108.—*Turton, Syst.* p. 607.—Parus viridis gutture nigro, the Green Black-throated Flycatcher, *Bartram,* p. 292.

SYLVICOLA VIRENS.—SWAINSON.

Sylvia virens, *Bonap. Synop.* p. 80.

THIS is one of those transient visitors that pass through Pennsylvania in the latter part of April and beginning of May, on their way to the north to breed. It generally frequents the high branches and tops of trees in the woods, in search of the larvæ of insects that prey on the opening buds. It has a few singular chirruping notes, and is very lively and active. About the 10th of May it disappears. It is rarely observed on its return in the fall, which may probably be owing to the scarcity of its proper food at that season obliging it to pass with greater haste; or to the foliage, which prevents it and other passengers from being so easily observed. Some few of these birds, however, remain all summer in Pennsylvania, having myself shot three this season (1809) in the month of June; but I have never yet seen their nest.

This species is four inches and three-quarters long, and seven broad; the whole back, crown, and hind head is of a rich yellowish green; front, cheeks, sides of the breast, and line over the eye, yellow; chin and throat, black; sides, under the wings, spotted with black; belly and vent, white; wings, dusky black, marked with two white bars; bill, black; legs and feet brownish yellow; tail, dusky, edged with light ash; the three exterior feathers spotted on their inner webs with white. The female is distinguished by having no black on the throat.

YELLOW-RUMPED WARBLER. (*Sylvia coronata.*)

PLATE XVII.—Fig. 4.

Motacilla maculosa, *Gmel. Syst.* i. p. 984.—Motacilla coronata, *Linn. Syst.* i. p. 332, No. 31.—Le Figuier à tête cendrée, *Buff.* v. p. 291.—Le Figuier couronné d'or, *Id.* v. p. 312.—Yellow-rump Flycatcher, *Edw.* t. 255.—Golden-crowned Flycatcher, *Id.* t. 298.—Yellow-rump Warbler, *Arct. Zool.* ii. No. 288.—Golden-crowned Warbler, *Id.* ii. No. 294.—*Lath. Syn.* iv. p. 481, No. 104, *Id. Supp.* p. 182, *Id. Syn.* iv. p. 486, No. 11.—*Turton*, p. 599, *Id.* 606.—Parus cedrus uropygio flavo.—The Yellow Rump, *Bartram*, p. 292.—Parus aurio vertice.—The Golden-crowned Flycatcher, *Id.* 292.—*Peale's Museum*, No. 7134.

SYLVICOLA CORONATA.—Swainson.

Sylvia coronata, *Bonap. Synop.* p. 77 (summer plumage.*)—Sylvicola coronata, *North. Zool.* ii. p. 216.

In this beautiful little species we have another instance of the mistakes occasioned by the change of colour to which many of our birds are subject. In the present case, this change is both progressive and periodical. The young birds of the first season are of a brown olive above, which continues until the month of February and March; about which time it gradually changes into a fine slate colour, as in the figure on the plate. About the middle of April this change is completed. I have shot them in all their gradations of change. While in their brown olive dress, the yellow on the sides of the breast and crown is scarcely observable, unless the feathers be parted with the hand; but that on the rump is still vivid; the spots of black on the cheek are then also obscured. The difference of appearance, however, is so great, that we need scarcely wonder that

* Winter plumage, Vol. II. Plate XLV.

foreigners, who have no opportunity of examining the progress of these variations, should have concluded them to be two distinct species, and designated them as in the above synonyms.

This bird is also a passenger through Pennsylvania. Early in October he arrives from the north, in his olive dress, and frequents the cedar trees, devouring the berries with great avidity. He remains with us three or four weeks, and is very numerous wherever there are trees of the red cedar covered with berries. He leaves us for the south, and spends the winter season among the myrtle swamps of Virginia, the Carolinas, and Georgia. The berries of the *Myrica cerifera*, both the large and dwarf kind, are his particular favourites. On those of the latter I found them feeding in great numbers, near the sea-shore, in the district of Maine, in October; and through the whole of the lower parts of the Carolinas, wherever the myrtles grew, these birds were numerous, skipping about, with hanging wings, among the bushes. In those parts of the country, they are generally known by the name of myrtle birds. Round Savannah, and beyond it as far as the Alatamaha, I found him equally numerous, as late as the middle of March, when his change of colour had considerably progressed to the slate hue. Mr Abbot, who is well acquainted with this change, assured me that they attain this rich slate colour fully before their departure from thence, which is about the last of March, and to the 10th of April. About the middle or 20th of the same month, they appear in Pennsylvania, in full dress, as represented in the plate; and after continuing to be seen for a week or ten days, skipping among the high branches and tops of the trees after those larvæ that feed on the opening buds, they disappear until the next October. Whether they retire to the north, or to the high ranges of our mountains to breed, like many other of our passengers, is yet uncertain. They are a very numerous species, and always associate together in considerable numbers, both in spring, winter, and fall.

This species is five inches and a half long, and eight inches broad; whole back, tail-coverts, and hind head, a fine slate colour streaked with black; crown, sides of the breast, and rump, rich yellow; wings and tail, black, the former crossed with two bars of white, the three exterior feathers of the latter spotted with white; cheeks and front, black; chin, line over and under the eye, white; breast, light slate, streaked with black extending under the wings; belly and vent, white, the latter spotted with black; bill and legs, black. This is the spring and summer dress of the male; that of the female of the same season differs but little, chiefly in the colours being less vivid, and not so strongly marked with a tincture of brownish on the back.

In the month of October the slate colour has changed to a brownish olive; the streaks of black are also considerably brown, and the white is stained with the same colour; the tail-coverts, however, still retain their slaty hue, the yellow on the crown and sides of the breast becomes nearly obliterated. Their only note is a kind of chip, occasionally repeated; their motions are quick, and one can scarcely ever observe them at rest.

Though the form of the bill of this bird obliges me to arrange him with the warblers, yet in his food and all his motions he is decidedly a flycatcher.

On again recurring to the descriptions in Pennant of the "yellow-rump warbler,"* "golden-crowned warbler,"† and "belted-warbler,"‡ I am persuaded that the whole three have been drawn from the present species.

* Arctic Zoology, p. 400, No. 188.
† *Ibid.* No. 294. ‡ *Ibid.* No. 306.

CERULEAN WARBLER. (*Sylvia cœrulea.*)

PLATE XVII.—Fig. 5.

Peale's Museum, No. 7309.

SYLVICOLA CŒRULEA.—Swainson.—Male.*

Sylvia azurea, *Bonap. Synop.* p. 85.—Sylvia azurea, Azure Warbler, *Steph. Sh. Zool.* x. p. 653.—Sylvia cœrulea, Cerulean Warbler, *Steph. Sh. Zool.* x. p. 652.—Sylvia bifasciata, *Say*, *Journ. to Rocky Mount.* i. p. 170.—The Azure Warbler, Sylvia azurea, *Aud.* pl. 48, male and female, *Orn. Biog.* i. p. 255.

This delicate little species is now, for the first time, introduced to public notice. Except my friend Mr Peale, I know of no other naturalist who seems to have hitherto known of its existence. At what time it arrives from the south I cannot positively say, as I never met with it in spring, but have several times found it during summer. On the borders of streams and marshes, among the branches of the poplar, it is sometimes to be found. It has many of the habits of the flycatcher; though, like the preceding, from the formation of its bill, we must arrange it with the warblers. It is one of our scarce birds in Pennsylvania, and its nest has hitherto eluded my search. I have never observed it after the 20th of August, and therefore suppose it retires early to the south.

This bird is four inches and a half long, and seven and a half broad; the front and upper part of the head is of a fine verditer blue; the hind head and back of the same colour, but not quite so brilliant; a few lateral streaks of black mark the upper part of the back; wings and tail, black edged with sky-blue; the three secondaries next the body edged with white, and the first and second row of coverts also tipt with white; tail-coverts, large, black, and broadly tipt with blue;

* Female figured Vol. III. Pl. XI.—the continuation by Bonaparte.

lesser wing-coverts, black, also broadly tipt with blue, so as to appear nearly wholly of that tint; sides of the breast, spotted or streaked with blue; belly, chin, and throat, pure white; the tail is forked, the five lateral feathers on each side with each a spot of white, the two middle more slightly marked with the same; from the eye backwards extends a line of dusky blue; before and behind the eye, a line of white; bill dusky above, light blue below; legs and feet, light blue.

SOLITARY FLYCATCHER. (*Muscicapa solitaria.*)

PLATE XVII.—FIG. 6.

VIREO SOLITARIUS.—VIEILLOT.

Vireo solitarius, *Bonap. Synop.* p. 70.

THIS rare species I can find nowhere described. I have myself never seen more than three of them, all of whom corresponded in their markings, and, on dissection, were found to be males. It is a silent, solitary bird. It is also occasionally found in the State of Georgia, where I saw a drawing of it in the possession of Mr Abbot, who considered it a very scarce species. He could give me no information of the female. The one from which the figure in the plate was taken was shot in Mr Bartram's woods, near Philadelphia, among the branches of dogwood, in the month of October. It appears to belong to a particular family or subdivision of the *Muscicapa* genus, among which are the white-eyed, the yellow-throated, and several others already described in the present work. Why one species should be so rare, while another, much resembling it, is so numerous, at least a thousand for one, is a question I am unable to answer, unless by supposing the few we meet with here to be accidental stragglers from the great body, which may have their residence in some other parts of our extensive continent.

Drawn from Nature by A.Wilson

Engraved by W.H.Lizars

1. Cow Bunting? 2. Female. 3. Young. 4. Maryland Yellow throat. 5. Blue grey Flycatcher. 6. White eyed F.

18.

The solitary flycatcher is five inches long, and eight inches in breadth; cheeks and upper part of the head and neck, a fine bluish gray; breast, pale cinereous; flanks and sides of the breast, yellow; whole back and tail-coverts, green olive; wings, nearly black; the first and second row of coverts tipt with white; the three secondaries next the body edged with pale yellowish white; the rest of the quills bordered with light green; tail, slightly forked, of the same tint as the wings, and edged with light green; from the nostrils a line of white proceeds to and encircles the eye; lores, black; belly and vent, white; upper mandible, black; lower, light blue; legs and feet, light blue; eyes, hazel.

COW BUNTING.* (*Emberiza pecoris.*)

PLATE XVIII.—FIGS. 1, 2, AND 3.

Le Brunet, *Buff.* iv. 138.—Le Pinçon de Virginie, *Briss.* iii. 165.—Cowpen Bird, *Catesb.* i. 34.—*Lath.* ii. 269.—*Arct. Zool.* p. 371, No. 241.—Sturnus stercorarius, *Bartram*, p. 291.—*Peale's Museum*, No. 6378, male; 6379, female.

MOLOTHRUS PECORIS.—SWAINSON.

Fringilla pecoris, *Sab. Frank. Journ.* p. 676.—Sturnus junceti, *Lath. Ind. Orn.*—Emberiza pecoris, *Bonap. Nomencl.* No. 89.—Icterus pecoris, *Bonap. Synop.* p. 53.—Aglaius pecoris, *Sw. Synop. Birds of Mex. Phil. Mag.* June 1827, p. 436.—The Cowpen Bird, *Aud.* pl. 99, *Orn. Biog.* i. p. 493.—Molothrus pecoris, *North. Zool.* ii. p. 277.

THERE is one striking peculiarity in the works of the great Creator, which becomes more amazing the more we reflect on it; namely, that He has formed no species of animals so minute or obscure that are not invested with certain powers and peculiarities, both of outward conformation and internal

* The American cuckoo (*Cuculus Carolinensis*) is by many people called the cow bird, from the sound of its notes resembling the words *cow, cow.* This bird builds its own nest very artlessly in a cedar or an apple tree, and lays four greenish blue eggs, which it hatches, and rears its young with great tenderness.

faculties, exactly suited to their pursuits, sufficient to distinguish them from all others, and forming for them a character solely and exclusively their own. This is particularly so among the feathered race. If there be any case where these characteristic features are not evident, it is owing to our want of observation, to our little intercourse with that particular tribe, or to that contempt for inferior animals, and all their habitudes, which is but too general, and which bespeaks a morose, unfeeling, and unreflecting mind. These peculiarities are often surprising, always instructive where understood, and (as in the subject of our present chapter) at least amusing, and worthy of being further investigated.*

The most remarkable trait in the character of this species is the unaccountable practice it has of dropping its eggs into the nests of other birds, instead of building and hatching for itself, and thus entirely abandoning its progeny to the care

* In this curious species we have another instance of those wonderful provisions of nature, which have hitherto baffled the knowledge and perseverance of man to discover for what uses they were intended. The only authenticated instance of a like circumstance that we are aware of is in the economy of the common cuckoo of Europe. Some foreign species, which rank as true *cuculi*, are said to deposit their eggs in the nests of other birds; but I am not sure that the fact is confirmed. With regard to the birds in question, there is little common between them, except that both are migratory, and both deposit their eggs in the nest of an alien. The cow bunting is polygamous; and I strongly suspect that our cuckoo is the same. In the deposition of the egg, the mode of procedure is nearly similar; great uneasiness, and a sort of fretting, previously, with a calm of quiet satisfaction afterwards. In both species we have beautiful provisions to ensure the non-disturbance of the intruder by its foster progeny: in the one, by a greater strength easily overcoming and driving out the natural but more tender young; in all love of the natural offspring being destroyed in the parents, and succeeded by a powerful desire to preserve and rear to maturity the usurper of their rights: in the other, where the young would, in some instances, be of a like size and strength, and where a combat might prove fatal in an opposite direction to the intentions of Providence, all necessity of contest is at once avoided by the eggs of the cow bunting requiring a shorter period to hatch than any of the birds chosen as foster parents.—Ed.

and mercy of strangers. More than two thousand years ago, it was well known, in those countries where the bird inhabits, that the cuckoo of Europe (*Cuculus canorus*) never built herself a nest, but dropped her eggs in the nests of other birds; but, among the thousands of different species that spread over that and other parts of the globe, no other instance of the same uniform habit has been found to exist, until discovered in the bird now before us. Of the reality of the former there is no doubt; it is known to every schoolboy in Britain; of the truth of the latter I can myself speak with confidence, from personal observation, and from the testimony of gentlemen, unknown to each other, residing in different and distant parts of the United States. The circumstances by which I became first acquainted with this peculiar habit of the bird are as follows:—

I had, in numerous instances, found in the nests of three or four particular species of birds, one egg, much larger, and differently marked from those beside it. I had remarked that these odd-looking eggs were all of the same colour, and marked nearly in the same manner, in whatever nest they lay, though frequently the eggs beside them were of a quite different tint; and I had also been told, in a vague way, that the cow bird laid in other birds' nests. At length I detected the female of this very bird in the nest of the red-eyed flycatcher, which nest is very small, and very singularly constructed. Suspecting her purpose, I cautiously withdrew without disturbing her; and had the satisfaction to find, on my return, that the egg which she had just dropped corresponded, as nearly as eggs of the same species usually do, in its size, tint, and markings, to those formerly taken notice of. Since that time, I have found the young cow bunting in many instances in the nests of one or other of these small birds. I have seen these last followed by the young cow bird calling out clamorously for food, and often engaged in feeding it; and I have now, in a cage before me, a very fine one, which six months ago I took from the nest of the Maryland yellow-throat, and from which the figures of the young bird and male cow bird in the plate were taken. The figure in the act of feeding it is the female Maryland yellow-throat in whose nest it was found. I claim, however, no merit for a discovery not originally my own, these singular habits having long been known to people of observation resident in the country, whose information, in this case, has preceded that of all our school philosophers and closet naturalists, to whom the matter has, till now, been totally unknown.

About the 25th of March, or early in April, the cowpen bird makes his first appearance in Pennsylvania from the south, sometimes in company with the red-winged blackbird, more frequently in detached parties, resting early in the morning, an hour at a time, on the tops of trees near streams of water, appearing solitary, silent, and fatigued. They continue to be occasionally seen in small solitary parties, particularly along creeks and banks of rivers, so late as the middle of June; after which, we see no more of them until about the beginning or middle of October, when they reappear in much larger flocks, generally accompanied by numbers of the redwings, between whom and the present species there is a considerable similarity of manners, dialect, and personal resemblance. In these aërial voyages, like other experienced navigators, they take advantage of the direction of the wind, and always set out with a favourable gale. My venerable and observing friend, Mr Bartram, writes me on the 13th of October as follows:—"The day before yesterday, at the height of the northeast storm, prodigious numbers of the cowpen birds came by us, in several flights of some thousands in a flock. Many of them settled on trees in the garden to rest themselves, and then resumed their voyage southwards. There were a few of their *cousins*, the redwings, with them. We shot three, a male and two females."

From the early period at which these birds pass in the spring, it is highly probable that their migrations extend very far north. Those which pass in the months of March and April can have no opportunity of depositing their eggs here, there being not more than one or two of our small birds which build so early. Those that pass in May and June are frequently observed loitering singly about solitary thickets, reconnoitring, no doubt, for proper nurses, to whose care they may commit the hatching of their eggs and the rearing of their helpless orphans. Among the birds selected for this duty are the following, all of which are figured and described in this volume:—The blue bird, which builds in a hollow tree; the chipping sparrow, in a cedar bush; the golden-crowned thrush, on the ground, in the shape of an oven; the red-eyed flycatcher, a neat pensile nest, hung by the two upper edges on a small sapling or drooping branch; the yellow bird, in the fork of an alder; the Maryland yellow-throat, on the ground, at the roots of brier bushes; the white-eyed flycatcher, a pensile nest on the bending of a smilax vine; and the small blue-gray flycatcher, also a pensile nest, fastened to the slender twigs of a tree, sometimes at the height of fifty or sixty feet from the ground. The three last-mentioned nurses are represented on the same plate with the bird now under consideration. There are, no doubt, others to whom the same charge is committed; but all these I have myself met with acting in that capacity.

Among these, the yellow-throat and the red-eyed flycatcher appear to be particular favourites; and the kindness and affectionate attention which these two little birds seem to pay to their nurslings, fully justify the partiality of the parents.

It is well known to those who have paid attention to the manners of birds, that, after their nest is fully finished, a day or two generally elapses before the female begins to lay. This delay is in most cases necessary to give firmness to the yet damp materials, and allow them time to dry. In this state it is sometimes met with, and laid in by the cow bunting; the result of which I have invariably found to be the desertion of the nest by its rightful owner, and the consequent loss of the egg thus dropped in it by the intruder. But when the owner herself has begun to lay, and there are one or more eggs in the nest before the cow bunting deposits hers, the attachment of the proprietor is secured, and remains unshaken until incubation is fully performed, and the little stranger is able to provide for itself.

The well-known practice of the young cuckoo of Europe in turning out all the eggs and young which it feels around it, almost as soon as it is hatched, has been detailed in a very satisfactory and amusing manner by the amiable Dr Jenner,* who has since risen to immortal celebrity in a much nobler pursuit, and to whose genius and humanity the whole human race are under everlasting obligations. In our cow bunting, though no such habit has been observed, yet still there is something mysterious in the disappearance of the nurse's own eggs soon after the foundling is hatched, which happens regularly before all the rest. From twelve to fourteen days is the usual time of incubation with our small birds; but although I cannot exactly fix the precise period requisite for the egg of the cow bunting, I think I can say almost positively, that it is a day or two less than the shortest of the above-mentioned spaces! In this singular circumstance we see a striking provision of the Deity; for did this egg require a day or two more, instead of so much less, than those among which it has been dropped, the young it contained would in every instance most inevitably perish, and thus, in a few years, the whole species must become extinct. On the first appearance of the young cow bunting, the parent being frequently obliged to leave the nest to provide sustenance for the foundling, the business of incubation is thus necessarily interrupted; the disposition to continue it abates; nature has now given a new direction to the zeal of the parent; and the remaining eggs, within a day or two at most, generally disappear. In some instances, indeed, they have been found on the ground near, or below, the nest; but this is rarely the case.

I have never known more than one egg of the cow bunting

* See Philosophical Transactions for 1788, part ii.

dropped in the same nest. This egg is somewhat larger than that of the blue bird, thickly sprinkled with grains of pale brown on a dirty white ground. It is of a size proportionable to that of the bird.

So extraordinary and unaccountable is this habit, that I have sometimes thought it might not be general among the whole of this species in every situation; that the extreme heat of our summers, though suitable enough for their young, might be too much for the comfortable residence of the parents; that, therefore, in their way to the north through our climate, they were induced to secure suitable places for their progeny; and that in the regions where they more generally pass the summer, they might perhaps build nests for themselves, and rear their own young, like every other species around them. On the other hand, when I consider that many of them tarry here so late as the middle of June, dropping their eggs, from time to time, into every convenient receptacle—that in the States of Virginia, Maryland, Delaware, New Jersey, and Pennsylvania, they uniformly retain the same habits—and, in short, that in all these places I have never yet seen or heard of their nest,—reasoning from these facts, I think I may safely conclude that they never build one, and that in those remote northern regions their manners are the same as we find them here.

What reason Nature may have for this extraordinary deviation from her general practice, is, I confess, altogether beyond my comprehension. There is nothing singular to be observed in the anatomical structure of the bird that would seem to prevent or render it incapable of incubation. The extreme heat of our climate is probably one reason why, in the months of July and August, they are rarely to be seen here. Yet we have many other migratory birds that regularly pass through Pennsylvania to the north, leaving a few residents behind them, who, without exception, build their own nests and rear their own young. This part of the country also abounds with suitable food, such as they usually subsist on. Many conjectures indeed might be formed as to the probable cause; but all of them that have occurred to me are unsatisfactory and inconsistent. Future and more numerous observations, made with care, particularly in those countries where they most usually pass the summer, may throw more light on this matter; till then, we can only rest satisfied with the reality of the fact.

This species winters regularly in the lower parts of North and South Carolina and Georgia; I have also met with them near Williamsburg, and in several other parts of Virginia. In January 1809, I observed strings of them for sale in the market of Charleston, South Carolina. They often frequent corn and rice fields, in company with their cousins, as Mr Bartram calls them, the red-winged blackbirds; but are more commonly found accompanying the cattle, feeding on the seeds, worms, &c., which they pick up amongst the fodder and from the excrements of the cattle, which they scratch up for this purpose. Hence they have pretty generally obtained the name of *cowpen birds*, *cow birds*, or *cow blackbirds*. By the naturalists of Europe they have hitherto been classed with the finches; though improperly, as they have no family resemblance to that tribe sufficient to justify that arrangement. If we are to be directed by the conformation of their bill, nostrils, tongue, and claws, we cannot hesitate a moment in classing them with the red-winged blackbirds, *oriolus phœniceus;* not, however, as orioles, but as buntings, or some new intermediate genus; the notes or dialect of the cow bunting and those of the redwings, as well as some other peculiarities of voice and gesticulation, being strikingly similar.

Respecting this extraordinary bird, I have received communications from various quarters, all corroborative of the foregoing particulars. Among these is a letter from Dr Potter of Baltimore, which, as it contains some new and interesting facts, and several amusing incidents, illustrative of the character of the bird, I shall with pleasure lay before the reader, apologising to the obliging writer for a few unimportant omissions which have been anticipated in the preceding pages :—

"I regret exceedingly that professional avocations have put it out of my power to have replied earlier to your favour of the 19th of September ; and although I shall not now reflect all the light you desire, a faithful transcript from memoranda, noted at the moment of observation, may not be altogether uninteresting.

"The *Fringilla pecoris* is generally known in Maryland by the name of the cow blackbird ; and none but the naturalist view it as a distinct species. It appears about the last of March, or first week in April, though sometimes a little earlier when the spring is unusually forward. It is less punctual in its appearance than many other of our migratory birds.

"It commonly remains with us till about the last of October, though unusually cold weather sometimes banishes it much earlier. It, however, sometimes happens that a few of them remain with us all winter, and are seen hovering about our barns and farmyards when straitened for sustenance by snow or hard frost. It is remarkable that in some years I have not been able to discover one of them during the months of July and August, when they have suddenly appeared in September in great numbers. I have noticed this fact always immediately after a series of very hot weather, and then only. The general opinion is, that they then retire to the deep recesses of the shady forest ; but, if this had been the fact, I should probably have discovered them in my rambles in every part of the woods. I think it more likely that they migrate farther north, till they find a temperature more congenial to their feelings, or find a richer repast in following the cattle in a better pasture.*

* "It may not be improper to remark here, that the appearance of this bird in spring is sometimes looked for with anxiety by the farmers. If the horned cattle happen to be diseased in spring, they ascribe it to worms, and consider the pursuit of the birds as an unerring indication of the necessity of medicine. Although this hypothesis of the worms infesting the cattle so as to produce much disease is problematical,

"In autumn, we often find them congregated with the marsh blackbirds, committing their common depredations upon the ears of the Indian-corn; and at other seasons, the similarity of their pursuits in feeding introduces them into the same company. I could never observe that they would keep the company of any other bird.

"The cowpen finch differs, moreover, in another respect from all the birds with which I am acquainted. After an observance of many years, I could never discover anything like *pairing*, or a mutual attachment between the sexes. Even in the season of love, when other birds are separated into pairs, and occupied in the endearing office of providing a receptacle for their offspring, the *Fringillæ* are seen feeding in odd as well as even numbers, from one to twenty, and discovering no more disposition towards perpetuating their species than birds of any other species at other seasons, excepting a promiscuous concubinage, which pervades the whole tribe. When the female separates from the company, her departure is not noticed; no gallant partner accompanies her, nor manifests any solicitude in her absence; nor is her return greeted by that gratulatory tenderness that so eminently characterises the males of other birds. The male proffers the same civilities to any female indiscriminately, and they are reciprocated accordingly, without exciting either resentment or jealousy in any of the party. This want of sexual attachment is not inconsistent with the general economy of this singular bird; for, as they are neither their own architect, nor nurse of their own young, the degree of attachment that governs others would be superfluous.

their superabundance at this season cannot be denied. The larvæ of several species are deposited in the vegetables when green, and the cattle are fed on them as fodder in winter. This furnishes the principal inducement for the bird to follow the cattle in spring, when the aperient effects of the green grasses evacuate great numbers of worms. At this season the *Pecoris* often stuffs its crop with them till it can contain no more. There are several species, but the most numerous is a small white one similar to, if not the same as, the *Ascaris* of the human species."

"That the *Fringilla* never builds a nest for itself, you may assert without the hazard of a refutation. I once offered a premium for the nest, and the negroes in the neighbourhood brought me a variety of nests; but they were always traced to some other bird. The time of depositing their eggs is from the middle of April to the last of May, or nearly so, corresponding with the season of laying observed by the small birds on whose property it encroaches. It never deposits but one egg in the same nest, and this is generally after the rightful tenant begins to deposit hers, but never, I believe, after she has commenced the process of incubation. It is impossible to say how many they lay in a season, unless they could be watched when confined in an aviary.

"By a minute attention to a number of these birds when they feed in a particular field in the laying season, the deportment of the female, when the time of laying draws near, becomes particularly interesting. She deserts her associates, assumes a drooping, sickly aspect, and perches upon some eminence where she can reconnoitre the operations of other birds in the process of nidification. If a discovery suitable to her purpose cannot be made from her stand, she becomes more restless, and is seen flitting from tree to tree, till a place of deposit can be found. I once had an opportunity of witnessing a scene of this sort which I cannot forbear to relate. Seeing a female prying into a bunch of bushes in search of a nest, I determined to see the result, if practicable; and knowing how easily they are disconcerted by the near approach of man, I mounted my horse, and proceeded slowly, sometimes seeing and sometimes losing sight of her, till I had travelled nearly two miles along the margin of a creek. She entered every thick place, prying with the strictest scrutiny into places where the small birds usually build, and at last darted suddenly into a thick copse of alders and briers, where she remained five or six minutes, when she returned, soaring above the underwood, and returned to the company she had left feeding in the field. Upon entering the covert, I found the nest of a yellow-throat, with an egg of each. Knowing the precise time of deposit, I noted the spot and date with a view of determining a question of importance,—the time required to hatch the egg of the cow bird, which I supposed to commence from the time of the yellow-throat's laying the last egg. A few days after, the nest was removed, I knew not how, and I was disappointed. In the progress of the cow bird along the creek's side, she entered the thick boughs of a small cedar, and returned several times before she could prevail on herself to quit the place; and upon examination, I found a sparrow sitting on its nest, on which she no doubt would have stolen in the absence of the owner. It is, I believe, certain, that the cowpen finch never makes a forcible entry upon the premises, by attacking other birds and ejecting them from their rightful tenements, although they are all perhaps inferior in strength, except the blue bird, which, although of a mild as well as affectionate disposition, makes a vigorous resistance when assaulted. Like most other tyrants and thieves, they are cowardly, and accomplish by stealth what they cannot obtain by force.

"The deportment of the yellow-throat on this occasion is not to be omitted. She returned while I waited near the spot, and darted into her nest, but returned immediately and perched upon a bough near the place, remained a minute or two, and entered it again, returned, and disappeared. In ten minutes she returned with the male. They chattered with great agitation for half an hour, seeming to participate in the affront, and then left the place. I believe all the birds thus intruded on manifest more or less concern at finding the egg of a stranger in their own nests. Among these, the sparrow is particularly punctilious; for she sometimes chirps her complaints for a day or two, and often deserts the premises altogether, even after she has deposited one or more eggs. The following anecdote will show, not only that the cowpen finch insinuates herself slyly into the nests of other birds, but that even the most pacific of them will resent the insult. A blue bird had built for three successive seasons in the cavity of a mulberry tree near my dwelling. One day, when the nest was nearly finished, I discovered a female cow bird perched upon a fence stake near it, with her eyes apparently fixed upon the spot, while the builder was busy in adjusting her nest. The moment she left it, the intruder darted into it, and in five minutes returned, and sailed off to her companions with seeming delight, which she expressed by her gestures and notes. The blue bird soon returned and entered the nest, but instantaneously fluttered back with much apparent hesitation, and perched upon the highest branch of the tree, uttering a rapidly repeated note of complaint and resentment, which soon brought the male, who reciprocated her feelings by every demonstration of the most vindictive resentment. They entered the nest together and returned several times, uttering their uninterrupted complaints for ten or fifteen minutes. The male then darted away to the neighbouring trees as if in quest of the offender, and fell upon a cat bird, which he chastised severely, and then turned to an innocent sparrow that was chanting its ditty in a peach tree. Notwithstanding the affront was so passionately resented, I found the blue bird had laid an egg the next day. Perhaps a tenant less attached to a favourite spot would have acted more fastidiously, by deserting the premises altogether. In this instance, also, I determined to watch the occurrences that were to follow; but on one of my morning visits, I found the common enemy of the eggs and young of all the small birds had despoiled the nest,—a coluber was found coiled in the hollow, and the eggs sucked.

"Agreeably to my observation, all the young birds destined to cherish the young cow bird are of a mild and affectionate disposition; and it is not less remarkable that they are all smaller than the intruder; the blue bird is the only one nearly as large. This is a good-natured mild creature, although it makes a vigorous defence when assaulted. The yellow-throat, the sparrow, the goldfinch, the indigo bird, and the blue bird, are the only birds in whose nests I have found the eggs or the young of the cowpen finch, though doubtless there are some others.

"What becomes of the eggs or young of the proprietor? This is the most interesting question that appertains to this subject. There must be some special law of nature which determines that the young of the proprietors are never to be found tenants in common with the young cow bird. I shall offer the result of my own experience on this point, and leave it to you and others better versed in the mysteries of nature than I am to draw your own conclusions. Whatever theory may be adopted, the facts must remain the same. Having discovered a sparrow's nest with five eggs, four and one, and the sparrow sitting, I watched the nest daily. The egg of the cow bird occupied the centre, and those of the sparrow were pushed a little up the sides of the nest. Five days after the discovery, I perceived the shell of the finch's egg broken, and the next the bird was hatched. The sparrow returned while I was near the nest, with her mouth full of food, with which she fed the young cow bird with every possible mark of affection, and discovered the usual concern at my approach. On the succeeding day, only two of the sparrow's eggs remained, and the next day there were none. I sought in vain for them on the ground, and in every direction.

"Having found the egg of the cow bird in the nest of a yellow-throat, I repeated my observations. The process of incubation had commenced, and on the seventh day from the discovery, I found a young cow bird that had been hatched during my absence of twenty-four hours, all the eggs of the proprietor remaining. I had not an opportunity of visiting the nest for three days, and, on my return, there was only one egg remaining, and that rotten. The yellow-throat attended the young interloper with the same apparent care and affection as if it had been its own offspring.

"The next year my first discovery was in a blue bird's nest built in a hollow stump. The nest contained six eggs, and the process of incubation was going on. Three or four days after my first visit, I found a young cow bird and three eggs remaining. I took the eggs out; two contained young birds, apparently come to their full time, and the other was rotten. I found one of the other eggs on the ground at the foot of the stump, differing in no respect from those in the nest, no signs of life being discoverable in either.

"Soon after this I found a goldfinch's nest with one egg of each only, and I attended it carefully till the usual complement of the owner were laid. Being obliged to leave home, I could not ascertain precisely when the process of incubation commenced; but from my reckoning, I think the egg of the cow bird must have been hatched in nine or ten days from the commencement of incubation. On my return, I found the young cow bird occupying nearly the whole nest, and the foster-mother as attentive to it as she could have been to her own. I ought to acknowledge here, that in none of these instances could I ascertain exactly the time required to hatch the cow bird's eggs, and that of course none of them are decisive; but is it not strange that the egg of the intruder should be so uniformly the first hatched? The idea of the egg being larger, and therefore from its own gravity finding the centre of the nest, is not sufficient to explain the phenomenon; for in this situation the other eggs would be proportionably elevated at the sides, and therefore receive as much or more warmth from the body of the incumbent than the other.* This principle would scarcely apply to the eggs of the blue bird, for they are nearly of the same size; if there be any difference, it would be in favour of the eggs of the builder of the nest. How do the eggs get out of the nest? Is it by the size and nestling of the young cow bird? This cannot always be the case; because, in the instance of the

* The ingenious writer seems not to be aware, that almost all birds are in the habit, while sitting, of changing the eggs from the centre to the circumference, and *vice versa*, that all of them may receive an equal share of warmth.

blue bird's nest in the hollow stump, the cavity was a foot deep, the nest at the bottom, and the ascent perpendicular; nevertheless the eggs were removed, although filled with young ones; moreover, a young cowpen finch is as helpless as any other young bird, and, so far from having the power of ejecting others from the nest, or even the eggs, that they are sometimes found on the ground under the nest, especially when the nest happens to be very small. I will not assert that the eggs of the builder of the nest are never hatched; but I can assert that I have never been able to find one instance to prove the affirmative. If all the eggs of both birds were to be hatched, in some cases the nest would not hold half of them; for instance, those of the sparrow or yellow bird. I will not assert that the supposititious egg is brought to perfection in less time than those of the bird to which the nest belongs; but, from the facts stated, I am inclined to adopt such an opinion. How are the eggs removed after the accouchement of the spurious occupant? By the proprietor of the nest unquestionably; for this is consistent with the rest of her economy. After the power of hatching them is taken away by her attention to the young stranger, the eggs would be only an encumbrance, and therefore instinct prompts her to remove them. I might add, that I have sometimes found the eggs of the sparrow, in which were unmatured young ones, lying near the nest containing a cow bird, and therefore I cannot resist this conclusion. Would the foster parent feed two species of young at the same time? I believe not. I have never seen an instance of any bird feeding the young of another, unless immediately after losing her own. I should think the sooty looking stranger would scarcely interest a mother while the cries of her own offspring, always intelligible, were to be heard. Should such a competition ever take place, I judge the stranger would be the sufferer, and probably the species soon become extinct. Why the *lex naturæ conservatrix* should decide in favour of the surreptitious progeny is not for me to determine.

"As to the vocal powers of this bird, I believe its pretensions are very humble, none of its notes deserving the epithet musical. The sort of simple crackling complaint it utters at being disturbed, constitutes also the expression of its pleasure at finding its companions, varying only in a more rapidly repeated monotony. The deportment of the male, during his promiscuous intercourse with the other sex, resembles much that of a pigeon in the same situation. He uses nearly the same gestures; and, by attentively listening, you will hear a low, guttural sort of muttering, which is the most agreeable of his notes, and not unlike the cooing of a pigeon.

"This, sir, is the amount of my information on this subject, and is no more than a transcript from my notes made several years ago. For ten years past, since I have lived in this city, many of the impressions of nature have been effaced, and artificial ideas have occupied their places. The pleasure I formerly received in viewing and examining the objects of nature, are, however, not entirely forgotten; and those which remain, if they can interest you, are entirely at your service. With the sincerest wishes for the success of your useful and arduous undertaking, I am, dear sir, yours very respectfully,

"NATHANIEL POTTER."

To the above very interesting detail I shall add the following recent fact, which fell under my own observation, and conclude my account of this singular species.

In the month of July last, I took from the nest of the Maryland yellow-throat, which was built among the dry leaves at the root of a brier bush, a young male cow bunting, which filled and occupied the whole nest. I had previously watched the motions of the foster parents for more than an hour, in order to ascertain whether any more of their young were lurking about or not; and was fully satisfied that there were none. They had, in all probability, perished in the manner before mentioned. I took this bird home with me, and placed it in the same cage with a red bird (*Loxia cardinalis*), who, at first, and for several minutes after, examined it closely, and seemingly with great curiosity. It soon became clamorous for food, and from that moment the red bird seemed to adopt it as his own, feeding it with all the assiduity and tenderness of the most affectionate nurse. When he found that the grasshopper which he had brought it was too large for it to swallow, he took the insect from it, broke it in small portions, chewed them a little to soften them, and with all the gentleness and delicacy imaginable, put them separately into its mouth. He often spent several minutes in looking at and examining it all over, and picking off any particles of dirt that he observed on its plumage. In teaching and encouraging it to learn to eat of itself, he often reminded me of the lines of Goldsmith—

> He tried each art, reproved each dull delay,
> Allured to "*fav'rite food*," and led the way.

This cow bird is now six months old, is in complete plumage, and repays the affectionate services of his foster parent with a frequent display of all the musical talents with which nature has gifted him. These, it must be confessed, are far from being ravishing, yet, for their singularity, are worthy of notice. He spreads his wings, swells his body into a globular form, bristling every feather in the manner of a turkey cock, and, with great seeming difficulty, utters a few low, spluttering notes, as if proceeding from his belly; always, on these occasions, strutting in front of the spectator with great consequential affectation.

To see the red bird, who is himself so excellent a performer, silently listening to all this guttural splutter, reminds me of the great Handel contemplating a wretched catgut scraper. Perhaps, however, these may be meant for the notes of *love* and *gratitude*, which are sweeter to the ear and dearer to the heart than all the artificial solos or concertos on this side heaven.

The length of this species is seven inches; breadth, eleven

inches; the head and neck are of a very deep silky drab; the upper part of the breast a dark changeable violet; the rest of the bird is black, with a considerable gloss of green when exposed to a good light; the form of the bill is faithfully represented in the plate—it is evidently that of an *Emberiza;* the tail is slightly forked; legs and claws, glossy black, strong and muscular; iris of the eye, dark hazel. Catesby says of this bird, "It is all over of a brown colour, and something lighter below;" a description that applies only to the female, and has been repeated, in nearly the same words, by almost all succeeding ornithologists. The young male birds are at first altogether brown, and for a month or more are naked of feathers round the eye and mouth; the breast is also spotted like that of a thrush with light drab and darker streaks. In about two months after they leave the nest, the black commences at the shoulders of the wings, and gradually increases along each side, as the young feathers come out, until the bird appears mottled on the back and breast with deep black and light drab. At three months, the colours of the plumage are complete, and, except in moulting, are subject to no periodical change.

MARYLAND YELLOW-THROAT. (*Sylvia Marilandica.*)

PLATE XVIII.—FIG. 4, FEMALE.

TRICHAS PERSONATUS.—SWAINSON.—FEMALE.

THE male of this species having already been represented,* accompanied by a particular detail of its manners, I have little further to add here relative to this bird. I found several of them round Wilmington, North Carolina, in the month of January, along the margin of the river, and by the Cypress Swamp, on the opposite side. The individual from which the

* See Plate VI. fig. 1.

figure in the plate was taken was the actual nurse of the young cowpen bunting which it is represented in the act of feeding.

It is five inches long, and seven in extent; the whole upper parts, green olive; something brownish on the neck, tips of the wings, and head; the lower parts, yellow, brightest on the throat and vent; legs, flesh coloured. The chief difference between this and the male, in the markings of their plumage, is, that the female is destitute of the black bar through the eyes, and the bordering one of pale bluish white.

SMALL BLUE-GRAY FLYCATCHER. (*Muscicapa cœrulea.*)

PLATE XVIII.—Fig. 5.

Motacilla cœrulea, *Turton, Syst.* i. p. 612.—Blue Flycatcher, *Edw.* pl. 302.—Regulus griseus, the Little Bluish-grey Wren, *Bartram*, p. 291.—Le Figuier gris de fer, *Buff.* v. p. 309.—Cerulean Warbler, *Arct. Zool.* ii. No. 299.—*Lath. Syn.* iv. p. 490, No. 127.—*Peale's Museum*, No. 6829.

CULICIVORA CŒRULEA.—Swainson.*

Culicivora, *Sw. New Groups in Orn. Zool. Journ.* No. 11, p. 359.—Sylvia cœrulea, *Bonap. Synop.* p. 85.—The Blue-Gray Flycatcher, *Aud.* pl. 84, male and female, *Orn. Biog.* i. p. 431.

This diminutive species, but for the length of the tail, would rank next to our humming bird in magnitude. It is a very dexterous flycatcher, and has also something of the manners of the titmouse, with whom, in early spring and fall, it frequently associates. It arrives in Pennsylvania from the south about the middle of April; and about the beginning of May builds its nest, which it generally fixes among the twigs of a tree, sometimes at the height of ten feet from the ground, sometimes fifty feet high, on the extremities of the tops of a

* This species will represent another lately formed genus, of which the *Muscicapa stenura* of Temminck's *Pl. coloriées* forms the type. It is a curious group, connecting *Tyrannula, Setophaga*, the Flycatchers, and the *Sylviadæ.*—Ed.

high tree in the woods. This nest is formed of very slight and perishable materials,—the husks of buds, stems of old leaves, withered blossoms of weeds, down from the stalks of fern, coated on the outside with gray lichen, and lined with a few horse hairs. Yet in this frail receptacle, which one would think scarcely sufficient to admit the body of the owner, and sustain even its weight, does the female cow bird venture to deposit her egg; and to the management of these pigmy nurses leaves the fate of her helpless young. The motions of this little bird are quick; he seems always on the look-out for insects; darts about from one part of the tree to another, with hanging wings and erected tail, making a feeble chirping, *tsee, tsee,* no louder than a mouse. Though so small in itself, it is ambitious of hunting on the highest branches, and is seldom seen among the humbler thickets. It remains with us until the 20th or 28th of September, after which we see no more of it until the succeeding spring. I observed this bird near Savannah, in Georgia, early in March; but it does not winter even in the southern parts of that State.

The length of this species is four inches and a half; extent, six and a half; front, and line over the eye, black; bill, black, very slender, overhanging at the tip, notched, broad, and furnished with bristles at the base; the colour of the plumage above is a light bluish gray, bluest on the head, below bluish white; tail, longer than the body, a little rounded, and black, except the exterior feathers, which are almost all white, and the next two also tipt with white; tail-coverts, black; wings, brownish black, some of the secondaries next the body edged with white; legs, extremely slender, about three-fourths of an inch long, and of a bluish black colour. The female is distinguished by wanting the black line round the front.

The food of this bird is small winged insects and their larvæ, but particularly the former, which it seems almost always in pursuit of.

WHITE-EYED FLYCATCHER. (*Muscicapa cantatrix.*)

PLATE XVIII.—Fig. 6.

Muscicapa noveboracensis, *Gmel. Syst.* i. p. 947.—Hanging Flycatcher, *Lath. Syn. Supp.* p. 174.—*Arct. Zool.* p. 389, No. 274.—Muscicapa cantatrix, the Little Domestic Flycatcher, or Green Wren, *Bartram*, p. 290.—*Peale's Museum*, No. 6778.

VIREO NOVEBORACENSIS.—Bonaparte.

Vireo noveboracensis, *Bonap. Synop.* p. 70.—The White-Eyed Flycatcher, or Vireo, *Aud.* pl. 63, male, *Orn. Biog.* i. p. 328.

This is another of the cow bird's adopted nurses, a lively, active, and sociable little bird, possessing a strong voice for its size, and a great variety of notes, and singing, with little intermission, from its first arrival, about the middle of April, till a little before its departure in September. On the 27th of February, I heard this bird in the southern parts of the State of Georgia, in considerable numbers, singing with great vivacity. They had only arrived a few days before. Its arrival in Pennsylvania, after an interval of seven weeks, is a proof that our birds of passage, particularly the smaller species, do not migrate at once from south to north; but progress daily, keeping company, as it were, with the advances of spring. It has been observed in the neighbourhood of Savannah so late as the middle of November; and probably winters in Mexico and the West Indies.

This bird builds a very neat little nest, often in the figure of an inverted cone; it is suspended by the upper edge of the two sides, on the circular bend of a prickly vine,—a species of smilax that generally grows in low thickets. Outwardly, it is constructed of various light materials, bits of rotten wood, fibres of dry stalks of weeds, pieces of paper, commonly newspapers, an article almost always found about its nest, so that some of my friends have given it the name of the *Politician ;* all these substances are interwoven with the silk of caterpillars,

Drawn from Nature by A. Wilson *Engraved by W.H. Lizars*

1. Mottled Owl. 2. Meadow Lark. 3. Black and white Creeper. 4. Pine-creeping Warbler.

19.

and the inside is lined with fine dry grass and hair. The female lays five eggs, pure white, marked near the great end with a very few small dots of deep black or purple. They generally raise two broods in a season. They seem particularly attached to thickets of this species of smilax, and make a great ado when any one comes near their nest, approaching within a few feet, looking down, and scolding with great vehemence. In Pennsylvania they are a numerous species.

The white-eyed flycatcher is five inches and a quarter long, and seven in extent; the upper parts are a fine yellow olive, those below, white, except the sides of the breast and under the wings, which are yellow; line round the eye, and spot near the nostril, also rich yellow; wings, deep dusky black, edged with olive green, and crossed with two bars of pale yellow; tail, forked, brownish black, edged with green olive; bill, legs, and feet, light blue; the sides of the neck incline to a grayish ash. The female and young of the first season are scarcely distinguishable in plumage from the male.

MOTTLED OWL. (*Strix nœvia.*)

PLATE XIX.—FIG. 1.

Arct. Zool. 231, No. 118.—*Lath.* i. 126.—*Turton*, i. 167.—*Peale's Museum*, No. 444.

STRIX ASIO.—LINNÆUS.*

Strix asio, *Bonap. Synop.* p. 36.—Hibou asio, *Temm. Pl. col.* pl. 80.—The Little Screech Owl, *Aud.* pl. 97, adult and young, *Orn. Biog.* i. p. 486.

ON contemplating the grave and antiquated figure of this *night wanderer*, so destitute of everything like gracefulness of shape, I can scarcely refrain from smiling at the conceit of

* The difference in the plumage of the young and old has caused Wilson to fall into a mistake, and multiply species by introducing the different states under distinct specific appellations. On Plate XLII. is represented the young plumage of the bird, under the name which must

the ludicrous appearance this bird must have made had Nature bestowed on it the powers of song, and given it the faculty of warbling out sprightly airs, while robed in such a solemn exterior. But the great God of nature hath, in His wisdom, assigned to this class of birds a more unsocial and less noble, though, perhaps, not less useful, disposition, by assimilating them, not only in form of countenance, but in voice, manners, and appetite to some particular beasts of prey, secluding them from the enjoyment of the gay sunshine of day, and giving them little more than the few solitary hours of morning and evening twilight to procure their food and pursue their amours, while all the tuneful tribes, a few excepted, are wrapt in silence and repose. That their true character, however, should not be concealed from those weaker animals on whom they feed (for Heaven abhors deceit and hypocrisy), He has stamped their countenance with strong traits of their murderer, the cat; and birds in this respect are, perhaps, better physiognomists than men.

be adopted for it, as the original one of Linnæus. The tawny owls of this country present similar changes, and were long held as distinct, until accurate observers proved their difference. C. L. Bonaparte appears to have been the first who made public mention of the confusion which existed; and Mr Audubon has illustrated the sexes and young in one of his best plates. The species appears peculiar to America. They are scarce in the southern districts; but above the Falls of the Ohio they increase in number, and are plentiful in Virginia, Maryland, and all the eastern districts. Its range to the northward perhaps is not very extensive; it does not appear to have been met with in the last Overland Expedition, no mention being made of it in the "Northern Zoology." The flight of this owl, like its congeners, is smooth and noiseless. By Audubon, it is said sometimes to rise above the top branches of the highest forest trees, while in pursuit of large beetles, and at other times to sail low and swiftly over the fields or through the woods, in search of small birds, field mice, moles, or wood rats, from which it chiefly derives its subsistence. According to some gentlemen, the nest is placed at the bottom of the hollow trunk of a tree, often not at a greater height than six or seven feet from the ground, at other times so high as from thirty to forty. It is composed of a few grasses and feathers. The eggs are four or five, of a nearly globular form, and pure white colour.—Ed.

The owl now before us is chiefly a native of the northern regions, arriving here with several others about the commencement of cold weather; frequenting the uplands and mountainous districts in preference to the lower parts of the country, and feeding on mice, small birds, beetles, and crickets. It is rather a scarce species in Pennsylvania; flies usually in the early part of night and morning; and is sometimes observed sitting on the fences during day, when it is easily caught, its vision at that time being very imperfect.

The bird represented in the plate was taken in this situation, and presented to me by a friend. I kept it in the room beside me for some time, during which its usual position was such as I have given it. Its eyelids were either half shut, or slowly and alternately opening and shutting, as if suffering from the glare of day; but no sooner was the sun set than its whole appearance became lively and animated; its full and globular eyes shone like those of a cat; and it often lowered its head in the manner of a cock when preparing to fight, moving it from side to side, and also vertically, as if reconnoitring you with great sharpness. In flying through the room, it shifted from place to place with the silence of a spirit (if I may be allowed the expression), the plumage of its wings being so extremely fine and soft as to occasion little or no friction with the air,—a wise provision of Nature, bestowed on the whole genus, to enable them, without giving alarm, to seize their prey in the night. For an hour or two in the evening, and about break of day, it flew about with great activity. When angry, it snapped its bill repeatedly with violence, and so loud as to be heard in the adjoining room, swelling out its eyes to their full dimensions, and lowering its head as before described. It swallowed its food hastily, in large mouthfuls; and never was observed to drink. Of the eggs and nest of this species, I am unable to speak.

The mottled owl is ten inches long, and twenty-two in

extent; the upper part of the head, the back, ears, and lesser wing-coverts, are dark brown, streaked and variegated with black, pale brown, and ash; wings, lighter; the greater coverts and primaries, spotted with white; tail, short, even, and mottled with black, pale brown, and whitish, on a dark brown ground; its lower side, grey; horns (as they are usually called), very prominent, each composed of ten feathers, increasing in length from the front backwards, and lightest on the inside; face, whitish, marked with small touches of dusky, and bounded on each side with a circlet of black; breast and belly, white, beautifully variegated with ragged streaks of black, and small transverse touches of brown; legs, feathered nearly to the claws, with a kind of hairy down, of a pale brown colour; vent and under tail-coverts, white, the latter slightly marked with brown; iris of the eye, a brilliant golden yellow; bill and claws, bluish horn colour.

This was a female. The male is considerably less in size; the general colours darker; and the white on the wing-coverts not so observable.

Hollow trees, either in the woods or orchard, or close evergreens in retired situations, are the usual roosting places of this and most of our other species. These retreats, however, are frequently discovered by the nuthatch, titmouse, or blue jay, who instantly raise the alarm; a promiscuous group of feathered neighbours soon collect round the spot, like crowds in the streets of a large city when a thief or murderer is detected; and, by their insults and vociferation, oblige the recluse to seek for another lodging elsewhere. This may account for the circumstance of sometimes finding them abroad during the day, on fences and other exposed situations.

MEADOW LARK. (*Alauda magna.*)

PLATE XIX.—FIG. 2.

Linn. Syst. 289.—Crescent Stare, *Arct. Zool.* 330, No. 192.—*Lath.* iii. 6, var. A.—Le Fer-à-cheval, ou Merle à collier d'Amerique, *Buff.* iii. p. 371.—*Catesb. Car.* i. pl. 33.—*Bartram*, p. 290.—*Peale's Museum*, No. 5212.

STURNELLA LUDOVICIANA.—SWAINSON.*

Sturnus Ludovicianus (subgenus Sturnella), *Bonap. Synop.* p. 49.—Sturnella collaris, *Vieill. Gal. des Ois.* pl. 80.—Sturnella Ludoviciana, *North. Zool.* ii. p. 282.

THOUGH this well-known species cannot boast of the powers of song which distinguish that "harbinger of day," the sky-lark of Europe, yet in richness of plumage, as well as in sweetness of voice (as far as his few notes extend), he stands eminently its superior. He differs from the greater part of his tribe in wanting the long, straight hind claw, which is probably the reason why he has been classed, by some late naturalists, with the starlings. But in the particular form of

* In changing the specific name of this species, C. L. Bonaparte thinks that Wilson must have been misled by some European author, as he was acquainted with the works wherein it was previously described. It ought to remain under the appellation bestowed on it by Linnæus, Brisson, &c. With regard to the generic term, this curious form has been chosen by Vieillot as the type of his genus *Sturnella*, containing yet only two species,—that of Wilson, and another from the southern continent. The form is peculiar to the New World, and seems to have been a subject of uncertainty to most ornithologists, as we find it placed in the genera *Turdus*, *Sturnus*, *Alauda*, and *Cassicus*, to all of which it is somewhat allied, but to none can it rank as a congener. In the bill, head, and wings, with some modification, we have the forms of the two first and last; in the colours of the plumage, the elongation of the scapularies, and tail-coverts, in the legs, feet, and hinder claw, that of the *Alaudæ*. The tarsi and feet are decidedly ambulatorial, as is confirmed by the habits of the species, though the tail indicates that of a scansorial bird; but as far as we yet know, it is the only indication of this power. In the structure of the nest, we have the weaving of the *Icteri*, the situation of many of the warblers, and the form of the true wrens.—ED.

his bill, in his manners, plumage, mode and place of building his nest, Nature has clearly pointed out his proper family.

This species has a very extensive range, having myself found them in Upper Canada, and in each of the States, from New Hampshire to New Orleans. Mr Bartram also informs me, that they are equally abundant in East Florida. Their favourite places of retreat are pasture fields and meadows, particularly the latter, which have conferred on them their specific name, and no doubt supplies them abundantly with the particular seeds and insects on which they feed. They are rarely or never seen in the depth of the woods, unless where, instead of underwood, the ground is covered with rich grass, as in the Chactaw and Chickasaw countries, where I met with them in considerable numbers in the months of May and June. The extensive and luxuriant prairies between Vincennes and St Louis also abound with them.

It is probable that, in the more rigorous regions of the north, they may be birds of passage, as they are partially so here; though I have seen them among the meadows of New Jersey, and those that border the rivers Delaware and Schuylkill, in all seasons, even when the ground was deeply covered with snow. There is scarcely a market-day in Philadelphia, from September to March, but they may be found exposed to sale. They are generally considered, for size and delicacy, little inferior to the quail, or what is here usually called the partridge, and valued accordingly. I once met with a few of these birds in the month of February, during a deep snow, among the heights of the Alleghany, between Shippensburgh and Somerset, gleaning on the road, in company with the small snow birds. In the State of South Carolina and Georgia, at the same season of the year, they swarm among the rice plantations, running about the yards and outhouses, accompanied by the killdeers, with little appearance of fear, as if quite domesticated.

These birds, after the building season is over, collect in flocks; but seldom fly in a close, compact body; their flight is something in the manner of the grouse and partridge, laborious and steady, sailing and renewing the rapid action of the wings alternately. When they alight on trees or bushes, it is generally on the tops of the highest branches, whence they send forth a long, clear, and somewhat melancholy note, that, in sweetness and tenderness of expression, is not surpassed by any of our numerous warblers. This is sometimes followed by a kind of low, rapid chattering, the particular call of the female; and again the clear and plaintive strain is repeated as before. They afford tolerably good amusement to the sportsman, being most easily shot while on wing; as they frequently squat among the long grass, and spring within gunshot. The nest of this species is built generally in, or below, a thick tuft or tussock of grass; it is composed of dry grass and fine bent, laid at the bottom, and wound all around, leaving an arched entrance level with the ground; the inside is lined with fine stalks of the same materials, disposed with great regularity. The eggs are four, sometimes five, white, marked with specks, and several large blotches of reddish brown, chiefly at the thick end. Their food consists of caterpillars, grub worms, beetles, and grass seeds, with a considerable proportion of gravel. Their general name is the meadow lark; among the Virginians, they are usually called the old field-lark.

The length of this bird is ten inches and a half; extent, sixteen and a half; throat, breast, belly, and line from the eye to the nostrils, rich yellow; inside lining and edge of the wing, the same; an oblong crescent of deep velvety black ornaments the lower part of the throat; lesser wing-coverts, black, broadly bordered with pale ash; rest of the wing-feathers, light brown, handsomely serrated with black; a line of yellowish white divides the crown, bounded on each side by a stripe of black, intermixed with bay, and another line of yellowish white passes over each eye, backwards; cheeks, bluish white; back, and rest of the upper parts, beautifully variegated with black, bright bay, and pale ocher; tail, wedged, the feathers neatly pointed, the four outer ones on each side nearly all white; sides, thighs, and vent, pale yellow ochre, streaked with black; upper mandible, brown; lower, bluish white; eyelids furnished with strong black hairs; legs and feet, very large, and of a pale flesh colour.

The female has the black crescent more skirted with gray, and not of so deep a black. In the rest of her markings, the plumage differs little from that of the male. I must here take notice of a mistake committed by Mr Edwards in his "History of Birds," vol. vi. p. 123, where, on the authority of a bird dealer of London, he describes the calandre lark (a native of Italy and Russia) as belonging also to North America, and having been brought from Carolina. I can say with confidence, that, in all my excursions through that and the rest of the southern States, I never met such a bird, nor any person who had ever seen it. I have no hesitation in believing that the calandre is not a native of the United States.

BLACK AND WHITE CREEPER. (*Certhia maculata.*)

PLATE XIX.—FIG. 3.

Edw. pl. 300.—White Poll Warbler, *Arct. Zool.* 402, No. 293.—Le Figuier varié, *Buff.* v. 305.—*Lath.* ii. 488.—*Turton*, i. p. 803.—*Peale's Museum*, No. 7092.

SYLVICOLA VARIA.—JARDINE.*

Sylvia varia, *Bonap. Synop.* p. 81.—Le Mniotilla varié, Mniotilla varia, *Vieill. Gall. des Ois.* pl. 169.

THIS nimble and expert little species seldom perches on the small twigs, but circumambulates the trunk and larger branches, in quest of ants and other insects, with admirable dexterity. It arrives in Pennsylvania from the south about the 20th of April; the young begin to fly early in July; and the whole tribe abandon the country about the beginning of

* This forms the type of Vieillot's *Mniotilla*, and will, perhaps, show the scansorial form in *Sylvicola*.—ED.

October. Sloane describes this bird as an inhabitant of the West India islands, where it probably winters. It was first figured by Edwards from a dried skin sent him by Mr William Bartram, who gave it its present name. Succeeding naturalists have classed it with the warblers,—a mistake which I have endeavoured to rectify.

The genus of creepers comprehends about thirty different species, many of which are richly adorned with gorgeous plumage; but, like their congenial tribe, the woodpeckers, few of them excel in song; their tongues seem better calculated for extracting noxious insects from the bark of trees than for trilling out sprightly airs; as the hardened hands of the husbandman are better suited for clearing the forest or guiding the plough than dancing among the keys of a forte-piano. Which of the two is the more honourable and useful employment is not difficult to determine. Let the farmer, therefore, respect this little bird for its useful qualities, in clearing his fruit and forest trees from destructive insects, though it cannot serenade him with its song.

The length of this species is five inches and a half; extent, seven and a half; crown, white, bordered on each side with a band of black, which is again bounded by a line of white passing over each eye; below this is a large spot of black covering the ear-feathers; chin and throat, black; wings, the same, crossed transversely by two bars of white; breast and back, streaked with black and white; tail, upper and also under coverts, black, edged and bordered with white; belly, white; legs and feet, dirty yellow; hind claw the longest, and all very sharp pointed; bill, a little compressed sidewise, slightly curved, black above, paler below; tongue, long, fine pointed, and horny at the extremity. These last circumstances, joined to its manners, characterise it decisively as a creeper.

The female and young birds of the first year want the black on the throat, having that part of a grayish white.

PINE-CREEPING WARBLER. (*Sylvia pinus.*)

PLATE XIX.—Fig. 4.

Pine Creeper, *Catesby*, i. 61.—*Peale's Museum*, No. 7312.

SYLVICOLA PINUS.—Jardine.

Sylvia pinus, *Bonap. Synop.* p. 81.

This species inhabits the pine woods of the southern States, where it is resident, and where I first observed it, running along the bark of the pines, sometimes alighting and feeding on the ground, and almost always, when disturbed, flying up, and clinging to the trunks of the trees. As I advanced towards the south, it became more numerous. Its note is a simple reiterated chirrup, continued for four or five seconds.

Catesby first figured and described this bird, but so imperfectly, as to produce among succeeding writers great confusion, and many mistakes as to what particular bird was intended. Edwards has supposed it to be the blue-winged yellow warbler! Latham has supposed another species to be meant; and the worthy Mr Pennant has been led into the same mistakes, describing the male of one species and the female of another as the male and female pine creeper. Having shot and examined great numbers of these birds, I am enabled to clear up these difficulties by the following descriptions, which will be found to be correct:—

The pine-creeping warbler is five and a half inches long, and nine inches in extent; the whole upper parts are of a rich green olive, with a considerable tinge of yellow; throat, sides, and breast, yellow; wings and tail, brown, with a slight cast of bluish, the former marked with two bars of white, slightly tinged with yellow; tail, forked, and edged with ash; the three exterior feathers, marked near the tip with a broad spot of white; middle of the belly and vent-feathers, white. The

Drawn from Nature by A.Wilson

Engraved by W.H. Lizars

1. Louisiana Tanager. 2. Clarks Crow. 3. Lewis's Woodpecker.

20.

female is brown, tinged with olive green on the back ; breast, dirty white, or slightly yellowish. The bill in both is truly that of a warbler ; and the tongue, slender, as in the *Motacilla* genus, notwithstanding the habits of the bird.

The food of these birds is the seeds of the pitch pine, and various kinds of bugs. The nest, according to Mr Abbot, is suspended from the horizontal fork of a branch, and formed outwardly of slips of grape-vine bark, rotten wood, and caterpillars' webs, with sometimes pieces of hornets' nests interwoven, and is lined with dry pine leaves, and fine roots of plants. The eggs are four, white, with a few dark brown spots at the great end.

These birds, associating in flocks of twenty or thirty individuals, are found in the depth of the pine barrens ; and are easily known by their manner of rising from the ground and alighting on the body of the tree. They also often glean among the topmost boughs of the pine tree, hanging head downwards, like the titmouse.

LOUISIANA TANAGER. (*Tanagra Columbianus.*)

PLATE XIX.—Fig. 2.

Peale's Museum, No. 6236.

PYRANGA LUDOVICIANA.—Jardine.*

Tanagra Ludoviciana, *Bonap. Synop.* p. 105.—Pyranga erythropis, *Vieill.* auct. *Bonap.*

This bird, and the two others that occupy the same plate, were discovered in the remote regions of Louisiana, by an exploring party under the command of Captain George Merriwether Lewis, and Lieutenant, now General, William Clark, in their memorable expedition across the continent to the

* It is impossible to decide the generic station of this bird. It appears very rare ; and it is probable that the British collections do not possess any specimen.—Ed.

Pacific Ocean. They are entitled to a distinguished place in the pages of AMERICAN ORNITHOLOGY, both as being, till now, altogether unknown to naturalists, and as natives of what *is* or at least *will be*, and that at no distant period, part of the western territory of the United States.

The frail remains of the bird now under consideration, as well as of the other two, have been set up by Mr Peale in his museum, with as much neatness as the state of the skins would permit. Of three of these, which were put into my hands for examination, the most perfect was selected for the drawing. Its size and markings were as follows :—Length, six inches and a half ; back, tail, and wings, black ; the greater wing-coverts, tipt with yellow ; the next superior row, wholly yellow ; neck, rump, tail-coverts, and the whole lower parts, greenish yellow ; forepart of the head, to and beyond the eyes, light scarlet ; bill, yellowish horn colour ; edges of the upper mandible, ragged, as in the rest of its tribe ; legs, light blue ; tail, slightly forked, and edged with dull whitish ; the whole figure about the size, and much resembling in shape, the scarlet tanager (Plate XI. fig. 3), but evidently a different species, from the black back and yellow coverts. Some of the feathers on the upper part of the back were also skirted with yellow. A skin of what I supposed to be the female, or a young bird, differed in having the wings and back brownish, and in being rather less.

The family or genus to which this bird belongs is particularly subject to changes of colour, both progressively, during the first and second seasons, and also periodically afterwards. Some of those that inhabit Pennsylvania change from an olive green to a greenish yellow, and lastly to a brilliant scarlet ; and, I confess, when the preserved specimen of the present species was first shown me, I suspected it to have been passing through a similar change at the time it was taken. But having examined two more skins of the same species, and finding them all marked very nearly alike, which is seldom the case with those birds that change while moulting, I began to think that this might be its most permanent, or, at least, its summer or winter dress.

The little information I have been able to procure of the species generally, or at what particular season these were shot, prevents me from being able to determine this matter to my wish.

I can only learn that they inhabit the extensive plains or prairies of the Missouri, between the Osage and Mandan nations ; building their nests in low bushes, and often among the grass. With us, the tanagers usually build on the branches of a hickory or white oak sapling. These birds delight in various kinds of berries, with which those rich prairies are said to abound.

CLARK'S CROW. (*Corvus Columbianus.*)

PLATE XX.—Fig. 2.

Peale's Museum, No. 1371.

CORVUS COLUMBIANUS.—Swainson.

Corvus Columbianus, *Bonap. Synop.* p. 56.

This species resembles a little the jackdaw of Europe (*Corvus monedula*) ; but is remarkable for its formidable claws, which approach to those of the *Falco* genus, and would seem to intimate that its food consists of living animals, for whose destruction these weapons must be necessary. In conversation with different individuals of the party, I understood that this bird inhabits the shores of the Columbia and the adjacent country in great numbers, frequenting the rivers and sea-shore, probably feeding on fish ; and that it has all the gregarious and noisy habits of the European species, several of the party supposing it to be the same. The figure in the plate was drawn with particular care, after a minute examination and measurement of the only preserved skin that was saved, and which is now deposited in Mr Peale's museum.

This bird measures thirteen inches in length; the wings, the two middle tail-feathers, and the interior vanes of the next (except at the tip), are black, glossed with steel blue; all the secondaries, except the three next the body, are white for an inch at their extremities, forming a large spot of white on that part when the wing is shut; the tail is rounded; yet the two middle feathers are somewhat shorter than those adjoining; all the rest are pure white, except as already described; the general colour of the head, neck, and body, above and below, is a light silky drab, darkening almost to a dove colour on the breast and belly; vent, white; claws, black, large, and hooked, particularly the middle and hind claw; legs also black; bill, a dark horn colour; iris of the eye, unknown.

In the State of Georgia, and several parts of West Florida, I discovered a crow not hitherto taken notice of by naturalists, rather larger than the present species, but much resembling it in the form and length of its wings, in its tail, and particularly its claws. This bird is a constant attendant along the borders of streams and stagnating ponds, feeding on small fish and lizards, which I have many times seen him seize as he swept along the surface. A well-preserved specimen of this bird was presented to Mr Peale, and is now in his museum. It is highly probable that, with these external resemblances, the habits of both may be nearly alike.

LEWIS'S WOODPECKER. (*Picus torquatus.*)

PLATE XX.—Fig. 3.

Peale's Museum, No. 2020.

MELANERPES? TORQUATUS.—Jardine.*

Picus torquatus, *Bonap. Synop.* p. 46.

Of this very beautiful and singularly marked species I am unable to give any further account than as relates to its external appearance. Several skins of this species were preserved, all of which I examined with care, and found little or no difference among them, either in the tints or disposition of the colours.

The length of this was eleven inches and a half; the back, wings, and tail were black, with a strong gloss of green; upper part of the head, the same; front, chin, and cheeks, beyond the eyes, a dark rich red; round the neck passes a broad collar of white, which spreads over the breast, and looks as if the fibres of the feathers had been silvered: these feathers are also of a particular structure, the fibres being separate, and of a hair-like texture; belly, deep vermilion, and of the same strong hair-like feathers, intermixed with silvery ones; vent, black; legs and feet, dusky, inclining to greenish blue; bill, dark horn colour.

For a more particular, and doubtless a more correct, account of this and the two preceding species, the reader is referred to General Clark's History of the Expedition. The three birds

* Having no authority from the founder of the genus, and not having seen the bird, I place it with the red-headed woodpecker provisionally. The lengthened wings, proportion of toes, and distribution of the colours, seem however to warrant it.

The female is said by Bonaparte, on the authority of Mr Peale, who shot them breeding on the Rocky Mountains, to resemble the male closely.—Ed.

I have here introduced are but a small part of the valuable collection of new subjects in natural history discovered and preserved, amidst a thousand dangers and difficulties, by those two enterprising travellers, whose intrepidity was only equalled by their discretion, and by their active and laborious pursuit of whatever might tend to render their journey useful to science and to their country. It was the request and particular wish of Captain Lewis, made to me in person, that I should make drawings of such of the feathered tribes as had been preserved, and were new. That brave soldier, that amiable and excellent man, over whose solitary grave in the wilderness I have since shed tears of affliction, having been cut off in the prime of his life, I hope I shall be pardoned for consecrating this humble note to his memory, until a more able pen shall do better justice to the subject.

CANADA JAY. (*Corvus Canadensis.*)

PLATE XXI.—Fig. 1.

Linn. Syst. 158.—Cinereous Crow, *Arct. Zool.* p. 248, No. 137.—*Lath.* i. 389.—Le Geay brun de Canada, *Briss.* ii. 54.—*Buff.* iii. 117.

GARRULUS CANADENSIS.—Swainson.

Corvus Canadensis, *Bonap. Synop.* p. 58.—Garrulus Canadensis, *North. Zool.* ii. p. 295.

Were I to adopt the theoretical reasoning of a celebrated French naturalist, I might pronounce this bird to be a debased descendant from the common blue jay of the United States, degenerated by the influence of the bleak and chilling regions of Canada; or perhaps a *spurious* production between the blue jay and the cat bird: or, what would be more congenial to the Count's ideas, trace its degradation to the circumstance of migrating, some thousand years ago, from the genial shores of Europe, where nothing like degeneracy or degradation ever takes place among any of God's creatures. I shall, however, on the present occasion, content myself with stating

Drawn from Nature by A. Wilson *Engraved by W.H. Lizars*

1. Canada Jay. 2. Snow Bunting. 3. Rusty Grakle. 4. Purple Grakle.

21.

a few particulars better supported by facts, and more consonant to the plain homespun of common sense.

This species inhabits the country extending from Hudson's Bay, and probably farther north, to the river St Lawrence; also, in winter, the inland parts of the district of Maine, and northern tracts of the States of Vermont and New York. When the season is very severe, with deep snow, they sometimes advance farther south; but generally return northward as the weather becomes more mild.

The character given of this bird by the people of those parts of the country where it inhabits is, that it feeds on black moss, worms, and even flesh; when near habitations or tents, pilfers everything it can come at; is bold, and comes even into the tent, to eat meat out of the dishes; watches the hunters while baiting their traps for martens, and devours the bait as soon as their backs are turned; that they breed early in spring, building their nests on pine trees, forming them of sticks and grass, and lay blue eggs; that they have two, rarely three, young at a time, which are at first quite black, and continue so for some time; that they fly in pairs; lay up hoards of berries in hollow trees; are seldom seen in January, unless near houses; are a kind of mock bird; and, when caught, pine away, though their appetite never fails them: notwithstanding all which ingenuity and good qualities, they are, as we are informed, detested by the natives.*

The only individuals of this species that I ever met with in the United States were on the shores of the Mohawk, a short way above the Little Falls. It was about the last of November, when the ground was deeply covered with snow. There were three or four in company, or within a small distance of each other, flitting leisurely along the roadside, keeping up a kind of low chattering with one another, and seemed nowise apprehensive at my approach. I soon secured the whole, from the best of which the drawing in the plate was carefully made.

* Hearne's Journey, p. 405.

On dissection, I found their stomachs occupied by a few spiders, and the aureliæ of some insects. I could perceive no difference between the plumage of the male and female.

The Canada jay is eleven inches long, and fifteen in extent; back, wings, and tail, a dull leaden gray, the latter long, cuneiform, and tipt with dirty white; interior vanes of the wings, brown, and also partly tipt with white; plumage of the head, loose and prominent; the forehead, and feathers covering the nostril, as well as the whole lower parts, a dirty brownish white, which also passes round the bottom of the neck like a collar; part of the crown and hind head, black; bill and legs, also black; eye, dark hazel. The whole plumage on the back is long, loose, unwebbed, and in great abundance, as if to protect it from the rigours of the regions it inhabits.

A gentleman of observation, who resided for many years near the North River, not far from Hudson, in the State of New York, informs me that he has particularly observed this bird to arrive there at the commencement of cold weather; he has often remarked its solitary habits; it seemed to seek the most unfrequented, shaded retreats, keeping almost constantly on the ground, yet would sometimes, towards evening, mount to the top of a small tree, and repeat its notes (which a little resemble those of the baltimore) for a quarter of an hour together; and this it generally did immediately before snow or falling weather.

SNOW BUNTING. (*Emberiza nivalis.*)

PLATE XXI.—Fig. 2.

Linn. Syst. 308.—*Arct. Zool.* p. 355, No. 222.—Tawny Bunting, *Br. Zool.* No. 121.—L'Ortolan de Neige, *Buff.* iv. 329, *Pl. enl.* 497.—*Peale's Museum*, No. 5900.

PLECTROPHANES NIVALIS.—Meyer.*

Emberiza nivalis, *Flem. Br. Anim.* p. 79.—Snow Bunting, *Mont. Orn. Dict.* i.—*Bew. Br. Birds*, i. p. 148.—*Selb. Ill. Orn.* i. 247, pl. 52.—Tawny Bunting, *Mont. Orn. Dict. Bew. Br. Birds*, i. 150.—Bruent de neize, *Temm. Man. d'Orn.* i. p. 319.—Emberiza nivalis, *Bonap. Synop.* p. 103.—Emberiza (plectrophanes) nivalis, *North. Zool.* ii. p. 246.

This being one of those birds common to both continents, its migrations extending almost from the very pole to a distance of forty or fifty degrees around, and its manners

* This species, from its various changes of plumage, has been multiplied into several; and in form being allied to many genera, it has been variously placed by different ornithologists. Meyer was the first to institute a place for itself, and, with a second, the *Fringilla Lapponica*, it will constitute his genus *Plectrophanes*, which is generally adopted into our modern systems. The discrepancies of form were also seen by Vieillot, who, without attending to his predecessor, made the genus *Passerina* of the Lapland finch. They are both natives of America; the latter has been added by the Prince of Musignano, and figured in Volume III. It has also been lately discovered to be an occasional visitant in this country, being taken by the bird-catchers about London. The following very proper observations occur in Mr Selby's account of the Lapland finch :—

"The appropriate station for this genus, I conceive to be intermediate between *Alauda* and *Emberiza*, forming, as it were, the medium of connection or passage from one genus to the other. In *Alauda*, it is met with that section of the genus which, in the increasing thickness and form of the bill, shows a deviation from the more typical species, and a nearer approach to the thick-billed *Fringillidæ;* to this section *Alauda calandra* and *brachydactyla* belong. Its affinity to the larks is also shown by the form of the feet and production of the hinder claw; this in *Lapponica* is nearly straight, and longer than the toe, resembling in every respect that of many of the true larks. The habits and manners of the two known species also bear a much greater resemblance to those of the larks than the buntings. Like the members of the first genus, they live entirely upon the ground, and never perch. Their and peculiarities having been long familiarly known to the naturalists of Europe, I shall in this place avail myself of the most interesting parts of their accounts, subjoining such particulars as have fallen under my own observation.

"These birds," says Mr Pennant, "inhabit not only Greenland,* but even the dreadful climate of Spitzbergen, where vegetation is nearly extinct, and scarcely any but *cryptogamous* plants are found. It therefore excites wonder how birds which are graminivorous in every other than those frost-bound regions subsist, yet are there found in great flocks, both on the land and ice of Spitzbergen.† They annually pass to this country by way of Norway; for, in the spring, flocks innumerable appear, especially on the Norwegian isles, continue only three weeks, and then at once disappear.‡ As they do

mode of progression is also the same, being by successive steps, and not the hopping motion used by all the true *Emberizæ*. A power of flight superior to that possessed by the true buntings is also indicated by the greater length of the wings and form of the tail-feathers. In *Plectrophanes*, the first and second quills are nearly equal in length, and the longest in the wing; in *Emberiza*, on the contrary, the second and third are equal, and longer than the first. The affinity of our genus to *Emberiza* is shown in the form of the bill, which, with the exception of being shorter and more rounded on the back, possesses the characteristic distinctions of that genus."

During the spring and breeding season, the plumage assumes a pure white on the under parts, and deep black on all the brown markings of the upper. The feathers are at first edged with brown, which gradually drop off as the summer advances. A third species is figured in the "Northern Zoology" (*Plectrophanes picta*, Sw.) Only one specimen was obtained, associating with the Lapland buntings, on the banks of the Saskatchewan. The description of the bird in the summer plumage is nearly thus given:—"Head and sides velvet black; three distinct spots of pure white on the sides of the head, one bordering on the chin, another on the ear, a third above the eye, a less distinct spot on the middle of the nape; the neck above, wood brown, the dorsal plumage and lowest rows of wing-coverts, blackish brown; the under plumage, entirely of a colour intermediate between wood brown and buff orange."—Ed.

* Crantz, i. 77.

† Lord Mulgrave's Voyage, 188; Martin's Voyage, 73.

‡ Leems, 256.

not breed in Hudson's Bay, it is certain that many retreat to this last of lands, and totally uninhabited, to perform, in full security, the duties of love, incubation, and nutrition. That they breed in Spitzbergen is very probable; but we are assured that they do so in Greenland. They arrive there in April, and make their nests in the fissures of the rocks on the mountains in May. The outside of their nest is grass, the middle of feathers, and the lining the down of the arctic fox. They lay five eggs, white, spotted with brown: they sing finely near their nest.

"They are caught by the boys in autumn when they collect near the shores in great flocks in order to migrate, and are eaten dried.*

"In Europe, they inhabit, during summer, the most naked Lapland alps, and descend in rigorous seasons into Sweden, and fill the roads and fields; on which account the Dalecarlians call them *illwarsfogel,* or bad-weather birds—the Uplanders, *hardwarsfogel,* expressive of the same. The Laplanders style them *alaipg.* Leems † remarks, I know not with what foundation, that they fatten on the flowing of the tides in Finmark, and grow lean on the ebb. The Laplanders take them in great numbers in hair springs for the table, their flesh being very delicate.

"They seem to make the countries within the whole arctic circle their summer residence, from whence they overflow the more southern countries in amazing multitudes at the setting in of winter in the frigid zone. In the winter of 1778–79, they came in such multitudes into Birsa, one of the Orkney Islands, as to cover the whole barony; yet of all the numbers, hardly two agreed in colours.

"Lapland, and perhaps Iceland, furnishes the north of Britain with the swarms that frequent these parts during winter, as low as the Cheviot Hills, in lat. 52° 32′. Their resting-places are the Feroe Isles, Shetland, and the Orkneys. The Highlands of Scotland, in particular, abound with them.

* Faun. Greenl. 118. † Finmark, 255.

Their flights are immense, and they mingle so closely together in form of a ball, that the fowlers make great havoc among them. They arrive lean, soon become very fat, and are delicious food. They either arrive in the Highlands very early, or a few breed there, for I had one shot for me at Invercauld, the 4th of August. But there is a certainty of their migration; for multitudes of them fall, wearied with their passage, on the vessels that are sailing through the Pentland Firth.*

"In their summer dress, they are sometimes seen in the south of England,† the climate not having severity sufficient to affect the colours; yet now and then a milk-white one appears, which is usually mistaken for a white lark.

"Russia and Siberia receive them in their severe seasons annually, in amazing flocks, overflowing almost all Russia. They frequent the villages, and yield a most luxurious repast. They vary there infinitely in their winter colours; are pure white, speckled, and even quite brown.‡ This seems to be the influence of difference of age more than of season. Germany has also its share of them. In Austria, they are caught and fed with millet, and afford the epicure a treat equal to that of the ortolan." §

These birds appear in the northern districts of the United States early in December, or with the first heavy snow, particularly if drifted by high winds. They are usually called the *white* snow bird, to distinguish them from the small dark bluish snow bird already described. Their numbers increase with the increasing severity of weather and depth of snow. Flocks of them sometimes reach as far south as the borders of Maryland; and the whiteness of their plumage is observed to be greatest towards the depth of winter. They spread over the Gennesee country and the interior of the district of Maine, flying in close compact bodies, driving about most in a high wind; sometimes alighting near the doors, but seldom sitting

* Bishop Pocock's Journal, MS.

† Morton's Northamp. p. 427.

‡ Bell's Travels, i. 198.

§ Kramer, Anim. Austr. 372.

long, being a roving, restless bird. In these plentiful regions, where more valuable game is abundant, they hold out no temptation to the sportsman or hunter; and except the few caught by boys in snares, no other attention is paid to them. They are, however, universally considered as the harbingers of severe cold weather. How far westward they extend I am unable to say. One of the most intelligent and expert hunters who accompanied Captains Lewis and Clark on their expedition to the Pacific Ocean, informs me that he has no recollection of seeing these birds in any part of their tour, not even among the bleak and snowy regions of the Stony Mountains, though the little blue one was in abundance.

The snow bunting derives a considerable part of its food from the seeds of certain aquatic plants, which may be one reason for its preferring these remote northern countries, so generally intersected with streams, ponds, lakes, and shallow arms of the sea, that probably abound with such plants. In passing down the Seneca river towards Lake Ontario, late in the month of October, I was surprised by the appearance of a large flock of these birds feeding on the surface of the water, supported on the tops of a growth of weeds that rose from the bottom, growing so close together that our boat could with great difficulty make its way through them. They were running about with great activity; and those I shot and examined were filled, not only with the seeds of this plant, but with a minute kind of shell-fish that adheres to the leaves. In these kind of aquatic excursions they are doubtless greatly assisted by the length of their hind heel and claws. I also observed a few on Table Rock, above the Falls of Niagara, seemingly in search of the same kind of food.

According to the statements of those traders who have resided near Hudson's Bay, the snow buntings are the earliest of their migratory birds, appearing there about the 11th of April, staying about a month or five weeks, and proceeding farther north to breed. They return again in September, stay till November, when the severe frosts drive them southward.*

The summer dress of the snow bunting is a tawny brown, interspersed with white, covering the head, neck, and lower parts; the back is black, each feather being skirted with brown; wings and tail, also black, marked in the following manner:—The three secondaries next the body are bordered with bay, the next with white, and all the rest of the secondaries, as well as their coverts and shoulder of the wing, pure white; the first six primaries are black from their coverts downwards to their extremities; tail, forked, the three exterior feathers on each side white, marked on the outer edge near the tip with black, the rest nearly all black; tail-coverts, reddish brown, fading into white; bill, pale brown; legs and feet, black; hind claw, long, like that of the lark, though more curved. In winter, they become white on the head, neck, and whole under side, as well as great part of the wings and rump; the back continues black, skirted with brown. Some are even found pure white. Indeed, so much does their plumage vary according to age and season, that no two are found at any time alike.

RUSTY GRAKLE. (*Gracula ferruginea.*)

PLATE XXI.—Fig. 3.

Black Oriole, *Arct. Zool.* p. 259, No. 144.—Rusty Oriole, *Ibid.* p. 260, No. 146.—New York Thrush, *Ibid.* p. 339, No. 205.—Hudsonian Thrush, *Ibid.* No. 234, female.—Labrador Thrush, *Ibid.* p. 340, No. 206.—*Peale's Museum*, No. 5514.

SCOLEPHAGUS FERRUGINEUS.—Swainson.

Quiscalus ferrugineus, *Bonap. Synop.* p. 55.—Scolephagus ferrugineus, *North. Zool.* ii. p. 286.

Here is a single species described by one of the most judicious naturalists of Great Britain no less than five different times!—the greater part of these descriptions is copied by succeeding

* London Philosophical Transactions, lxii. 403.

naturalists, whose synonyms it is unnecessary to repeat; so great is the uncertainty in judging, from a mere examination of their dried or stuffed skins, of the particular tribes of birds, many of which, for several years, are constantly varying in the colours of their plumage, and, at different seasons or different ages, assuming new and very different appearances. Even the size is by no means a safe criterion, the difference in this respect between the male and female of the same species (as in the one now before us) being sometimes very considerable.

This bird arrives in Pennsylvania from the north early in October, associates with the redwings and cowpen buntings, frequents cornfields and places where grasshoppers are plenty; but Indian-corn, at that season, seems to be its principal food. It is a very silent bird, having only now and then a single note or *chuck*. We see them occasionally until about the middle of November, when they move off to the south. On the 12th of January, I overtook great numbers of these birds in the woods near Petersburgh, Virginia, and continued to see occasional parties of them almost every day as I advanced southerly, particularly in South Carolina, around the rice plantations, where they were numerous, feeding about the hogpens, and wherever Indian-corn was to be procured. They also extend to a considerable distance westward. On the 5th of March, being on the banks of the Ohio, a few miles below the mouth of the Kentucky river, in the midst of a heavy snowstorm, a flock of these birds alighted near the door of the cabin where I had taken shelter, several of which I shot, and found their stomachs, as usual, crammed with Indian-corn. Early in April they pass hastily through Pennsylvania, on their return to the north to breed.

From the accounts of persons who have resided near Hudson's Bay, it appears that these birds arrive there in the beginning of June, as soon as the ground is thawed sufficiently for them to procure their food, which is said to be worms and maggots; sing with a fine note till the time of incubation, when they have only a chucking noise, till the young take their flight, at which time they resume their song. They build their nests in trees, about eight feet from the ground, forming them with moss and grass, and lay five eggs, of a dark colour spotted with black. It is added, they gather in great flocks, and retire southerly in September.*

The male of this species, when in perfect plumage, is nine inches in length, and fourteen in extent; at a small distance appears wholly black, but, on a near examination, is of a glossy dark green; the irides of the eye are silvery, as in those of the purple grakle; the bill is black, nearly of the same form with that of the last-mentioned species; the lower mandible a little rounded, with the edges turned inward, and the upper one furnished with a sharp bony process on the inside, exactly like that of the purple species. The tongue is slender, and lacerated at the tip; legs and feet, black and strong, the hind claw the largest; the tail is slightly rounded. This is the colour of the male when of full age; but three-fourths of these birds which we meet with have the whole plumage of the breast, head, neck, and back, tinctured with brown, every feather being skirted with ferruginous; over the eye is a light line of pale brown, below that one of black passing through the eye. This brownness gradually goes off towards spring, for almost all those I shot in the southern States were but slightly marked with ferruginous. The female is nearly an inch shorter; head, neck, and breast, almost wholly brown; a light line over the eye; lores, black; belly and rump, ash; upper and under tail-coverts, skirted with brown; wings, black, edged with rust colour; tail, black, glossed with green; legs, feet, and bill, as in the male.

These birds might easily be domesticated. Several that I had winged and kept for some time, became, in a few days, quite familiar, seeming to be very easily reconciled to confinement.

* Arctic Zoology, p. 259.

PURPLE GRAKLE. (*Gracula quiscala.*)

PLATE XXI.—Fig. 4.

Linn. Syst. 165.—La Pie de la Jamaique, *Briss.* ii. 41.—*Buff.* iii. 97, *Pl. enl.* 538.—*Arct. Zool.* p. 263, No. 153.—Gracula purpurea, the Lesser Purple Jackdaw, or Crow Blackbird, *Bartram*, p. 289.—*Peale's Museum*, No. 1582.

QUISCALUS VERSICOLOR.—Vieillot.*

Quiscalus versicolor, *Vieill. Gall. des Ois.* pl. 108.—*Bonap. Synop.* p. 54.—Purple Grakle, or Common Crow Blackbird, *Aud.* pl. 7, *Orn. Biog.* i. p. 35.—Quiscalus versicolor, Common Purple Boat-tail, *North. Zool.* ii. p. 285.

This noted depredator is well known to every careful farmer of the northern and middle States. About the 20th of March the purple grakles visit Pennsylvania from the south, fly in loose flocks, frequent swamps and meadows, and follow in the furrows after the plough; their food at this season consisting of worms, grubs, and caterpillars, of which they destroy prodigious numbers, as if to recompense the husbandman beforehand for the havoc they intend to make among his crops of Indian-corn. Towards evening, they retire to the nearest cedars and pine trees to roost, making a continual chattering as they fly along. On the tallest of these trees they generally build their nests in company, about the beginning or middle of April, sometimes ten or fifteen nests being on the same tree. One of these nests, taken from a high pine tree, is now before me. It measures full five inches in diameter within, and four in depth; is composed outwardly of mud, mixed with long stalks and roots of a knotty kind of grass, and lined

* *Gracula* will be given exclusively to a form inhabiting India, of which, though one species only is described, I have every reason to believe that at least two are confounded under it. *Quiscalus* has been, on this account, taken by Vieillot for our present bird, and some others confined to America. There has been considerable confusion among the species, which has been satisfactorily cleared up by Bonaparte, and will be seen in the sequel of the work. The female is figured Plate V. of the continuation by the Prince of Musignano.—Ed.

with fine bent and horse hair. The eggs are five, of a bluish olive colour, marked with large spots and straggling streaks of black and dark brown, also with others of a fainter tinge. They rarely produce more than one brood in a season.*

The trees where these birds build are often at no great distance from the farmhouse, and overlook the plantations. From thence they issue in all directions, and with as much confidence, to make their daily depredations among the surrounding fields, as if the whole were intended for their use alone. Their chief attention, however, is directed to the Indian-corn in all its progressive stages. As soon as the infant blade of this grain begins to make its appearance above ground, the grakles hail the welcome signal with screams of peculiar satisfaction, and, without waiting for a formal invitation from the proprietor, descend on the fields, and begin to

* Audubon's account of their manner of building is at considerable variance with that given above by our author. "The lofty dead trees left standing in our newly-cultivated fields have many holes and cavities, some of which have been bored by woodpeckers, and others caused by insects or decay. These are visited and examined in succession, until, a choice being made, and a few dry weeds and feathers collected, the female deposits her eggs, which are from four to six in number, blotched and streaked with brown and black." Such is the manner of building in Louisiana; but, in the northern States, their nests are differently constructed, and, as mentioned by our author, it is a singular circumstance that a comparatively short distance should so vary this formation. "In the northern States, their nests are constructed in a more perfect manner. A pine tree, whenever it occurs in a convenient place, is selected by preference. There the grakle forms a nest, which, from the ground, might easily be mistaken for that of our robin, were it less bulky. But it is much larger, and is associated with others, often to the number of a dozen or more, on the horizontal branches of the pine, forming tier above tier, from the lowest to the highest branches. It is composed of grass, slender roots and mud, lined with hair and finer grasses." Mr Audubon has also once or twice observed them build in the fissures of rocks. "The flesh is little better than that of a crow, being dry and ill-flavoured; notwithstanding it is often used, with the addition of one or two golden-winged woodpeckers or redwings, to make what is called *pot-pie*. The eggs, on the contrary, are very delicate."—ED.

pull up and regale themselves on the seed, scattering the green blades around. While thus eagerly employed, the vengeance of the gun sometimes overtakes them ; but these disasters are soon forgotten, and those

——who live to get away,
Return to steal another day.

About the beginning of August, when the young ears are in their milky state, they are attacked with redoubled eagerness by the grakles and redwings, in formidable and combined bodies. They descend like a blackening, sweeping tempest on the corn, dig off the external covering of twelve or fifteen coats of leaves as dexterously as if done by the hand of man, and having laid bare the ear, leave little behind to the farmer but the cobs and shrivelled skins that contained their favourite fare. I have seen fields of corn of many acres where more than one-half was thus ruined. Indeed, the farmers in the immediate vicinity of the rivers Delaware and Schuylkill generally allow one-fourth of this crop to the blackbirds, among whom our grakle comes in for his full share. During these depredations, the gun is making great havoc among their numbers, which has no other effect on the survivors than to send them to another field, or to another part of the same field. This system of plunder and retaliation continues until November, when, towards the middle of that month, they begin to sheer off towards the south. The lower parts of Virginia, North and South Carolina, and Georgia, are the winter residences of these flocks. Here numerous bodies, collecting together from all quarters of the interior and northern districts, and darkening the air with their numbers, sometimes form one congregated multitude of many hundred thousands. A few miles from the banks of the Roanoke, on the 20th of January, I met with one of those prodigious armies of grakles. They rose from the surrounding fields with a noise like thunder, and, descending on the length of road before me, covered it and the fences completely with black ;

and when they again rose, and, after a few evolutions, descended on the skirts of the high-timbered woods, at that time destitute of leaves, they produced a most singular and striking effect; the whole trees for a considerable extent, from the top to the lowest branches, seeming as if hung in mourning; their notes and screaming the meanwhile resembling the distant sound of a great cataract, but in more musical cadence, swelling and dying away on the ear according to the fluctuation of the breeze. In Kentucky, and all along the Mississippi, from its juncture with the Ohio to the Balize, I found numbers of these birds, so that the purple grakle may be considered as a very general inhabitant of the territory of the United States.

Every industrious farmer complains of the mischief committed on his corn by the *crow blackbirds,* as they are usually called; though, were the same means used, as with pigeons, to take them in clap nets, multitudes of them might thus be destroyed, and the products of them in market in some measure indemnify him for their depredations. But they are most numerous and most destructive at a time when the various harvests of the husbandman demand all his attention, and all his hands to cut, cure, and take in; and so they escape with a few sweeps made among them by some of the younger boys with the gun, and by the gunners from the neighbouring towns and villages; and return from their winter quarters, sometimes early in March, to renew the like scenes over again. As some consolation, however, to the industrious cultivator, I can assure him that, were I placed in his situation, I should hesitate whether to consider these birds most as friends or enemies, as they are particularly destructive to almost all the noxious worms, grubs, and caterpillars that infest his fields, which, were they allowed to multiply unmolested, would soon consume nine-tenths of all the production of his labour, and desolate the country with the miseries of famine! Is not this another striking proof that the Deity has created nothing in vain, and that it is the duty of man, the lord of the creation, to avail himself of their usefulness, and guard against their bad effects as securely as possible, without indulging in the barbarous and even impious wish for their utter extermination?

The purple grakle is twelve inches long, and eighteen in extent; on a slight view, seems wholly black, but placed near, in a good light, the whole head, neck, and breast appear of a rich glossy steel-blue, dark violet, and silky green; the violet prevails most on the head and breast, and the green on the hind part of the neck. The back, rump, and whole lower parts, the breast excepted, reflect a strong coppery gloss; wing-coverts, secondaries, and coverts of the tail, rich light violet, in which the red prevails; the rest of the wings and rounded tail are black, glossed with steel-blue. All the above colours are extremely shining, varying as differently exposed to the light; iris of the eye, silvery; bill more than an inch long, strong, and furnished on the inside of the upper mandible with a sharp process, like the stump of a broken blade of a penknife, intended to assist the bird in macerating its food; tongue, thin, bifid at the end, and lacerated along the sides.

The female is rather less, has the upper part of the head, neck, and the back of a dark sooty brown; chin, breast, and belly, dull pale brown, lightest on the former; wings, tail, lower parts of the back and vent, black, with a few reflections of dark green; legs, feet, bill, and eyes, as in the male.

The purple grakle is easily tamed, and sings in confinement. They have also, in several instances, been taught to articulate some few words pretty distinctly.

A singular attachment frequently takes place between this bird and the fish-hawk. The nest of this latter is of very large dimensions, often from three to four feet in breadth, and from four to five feet high; composed, externally, of large sticks or fagots, among the interstices of which sometimes three or four pairs of crow blackbirds will construct their nests, while the hawk is sitting or hatching above. Here each pursues the duties of incubation and of rearing their young, living in the greatest harmony, and mutually watching and protecting each other's property from depredators.

SWAMP SPARROW. (*Fringilla palustris.*)

PLATE XXII.—Fig. 1.

Passer palustris, *Bartram*, p. 291.—*Peale's Museum*, No. 6569.

ZONOTRICHIA PALUSTRIS.—Jardine.*

Fringilla palustris, *Bonap. Synop.* p. 111.—The Swamp Sparrow, *Aud.* pl. 64, male, *Orn. Biog.* i. p. 331.

The history of this obscure and humble species is short and uninteresting. Unknown or overlooked by the naturalists of Europe, it is now for the first time introduced to the notice of the world. It is one of our summer visitants, arriving in Pennsylvania early in April, frequenting low grounds and river courses; rearing two, and sometimes three, broods in a season; and returning to the south as the cold weather commences. The immense cypress swamps and extensive grassy flats of the southern States, that border their numerous rivers, and the rich rice plantations, abounding with their favourite seeds and sustenance, appear to be the general winter resort and grand annual rendezvous of this and all the other species of sparrow that remain with us during summer. From the river Trent, in North Carolina, to that of Savannah, and still farther south, I found this species very numerous; not flying

* The four species figured on this plate will point out the form which Mr Swainson has designated as above. Of these, the present and the last will recede from the type, the one in the more slender, the other in the stronger bill, and its even cutting margins. They in every respect show a strong assimilation with the bunting, sparrow, and lark family, though they cannot properly rank with these. According to the characters now laid down, and I believe properly so, they are a most interesting form when taken in comparison with their representatives in other countries. They appear confined to America.—Ed.

Drawn from Nature by A. Wilson — Engraved by W. H. Lizars

1. Swamp Sparrow. 2. White-throated Sp. 3. Savannah Sp. 4. Fox-coloured Sp. 5. Loggerhead Shrike.

22.

in flocks, but skulking among the canes, reeds, and grass, seeming shy and timorous, and more attached to the water than any other of their tribe. In the month of April, numbers pass through Pennsylvania to the northward, which I conjecture from the circumstance of finding them at that season in particular parts of the woods, where, during the rest of the year, they are not to be seen. The few that remain frequent the swamps and reedy borders of our creeks and rivers. They form their nest in the ground, sometimes in a tussock of rank grass surrounded by water, and lay four eggs of a dirty white spotted with rufous. So late as the 15th of August, I have seen them feeding their young that were scarcely able to fly. Their principal food is grass seeds, wild oats, and insects. They have no song; are distinguished by a single *chip* or *cheep*, uttered in a rather hoarser tone than that of the song sparrow; flirt the tail as they fly; seldom or never take to the trees, but skulk from one low bush or swampy thicket to another.

The swamp sparrow is five inches and a half long, and seven inches and a half in extent; the back of the neck and front are black; crown, bright bay, bordered with black; a spot of yellowish white between the eye and nostril; sides of the neck and whole breast, dark ash; chin, white; a streak of black proceeds from the lower mandible, and another from the posterior angle of the eye; back, black, slightly skirted with bay; greater coverts also black, edged with bay; wings and tail, plain brown; belly and vent, brownish white; bill, dusky above, bluish below; eyes, hazel; legs, brown; claws, strong and sharp, for climbing the reeds. The female wants the bay on the crown, or has it indistinctly; over the eye is a line of dull white.

WHITE-THROATED SPARROW. (*Fringilla albicollis.*)

PLATE XXII.—FIG. 2.

Fringilla fusca, *Bartram*, p. 291.—*Lath.* ii. 272.—*Edwards*, 304.—*Arct. Zool.* p. 373, No. 248.—*Peale's Museum*, No. 6486.

ZONOTRICHIA PENNSYLVANICA.—SWAINSON.

Fringilla Pennsylvanica, *Lath. Ind. Orn.* i. p. 445.—*Bonap. Synop.* p. 108.—The White-Throated Sparrow, *Aud.* pl. 8, male and female, *Orn. Biog.* i. p. 42.—*North. Zool.* ii. p. 256.

THIS is the largest as well as handsomest of all our sparrows. It winters with the preceding species and several others in most of the States south of New England. From Connecticut to Savannah I found these birds numerous, particularly in the neighbourhood of the Roanoke river, and among the rice plantations. In summer they retire to the higher inland parts of the country, and also farther north, to breed. According to Pennant, they are also found at that season in Newfoundland. During their residence here in winter, they collect together in flocks, always preferring the borders of swampy thickets, creeks, and mill-ponds, skirted with alder bushes and long rank weeds, the seeds of which form their principal food. Early in spring, a little before they leave us, they have a few remarkably sweet and clear notes, generally in the morning a little after sunrise. About the 20th of April they disappear, and we see no more of them till the beginning or second week of October, when they again return, part to pass the winter with us, and part on their route farther south.

The length of the white-throated sparrow is six inches and a half; breadth, nine inches; the upper part of the back and the lesser wing-coverts are beautifully variegated with black, bay, ash, and light brown; a stripe of white passes from the base of the upper mandible to the hind head; this is bordered on each side with a stripe of black; below this again is another of white passing over each eye, and deepening into orange yellow between that and the nostril; this is again bordered by a stripe of black proceeding from the hind part of the eye; breast, ash; chin, belly, and vent, white; tail, somewhat wedged; legs, flesh coloured; bill, a bluish horn colour; eye, hazel. In the female, the white stripe on the crown is a light drab; the breast not so dark; the chin less pure; and the line of yellow before the eye scarcely half as long as in the male. All the parts that are white in the male are in the female of a light drab colour.

FOX-COLOURED SPARROW. (*Fringilla rufa.*)

PLATE XXII.—FIG. 4.

Rusty Bunting, *Arct. Zool.* p. 364, No. 231, *Ibid.* 233.—Ferruginous Finch, *Ibid.* 375, No. 251.—Fringilla rufa, *Bartram*, p. 291.—*Peale's Museum*, No. 6092.

ZONOTRICHIA ILIACA.—SWAINSON.

Fringilla iliaca, *Bonap. Synop.* p. 112.—Fringilla (zonotrichia) iliaca, *North. Zool.* ii. p. 257.

THIS plump and pretty species arrives in Pennsylvania from the north about the 20th of October; frequents low, sheltered thickets; associates in little flocks of ten or twelve; and is almost continually scraping the ground and rustling among the fallen leaves. I found this bird numerous in November among the rich cultivated flats that border the river Connecticut; and was informed that it leaves those places in spring. I also found it in the northern parts of the State of Vermont. Along the borders of the great reed and cypress swamps of Virginia and North and South Carolina, as well as around the rice plantations, I observed this bird very frequently. They also inhabit Newfoundland.* They are rather of a solitary nature, seldom feeding in the open fields, but generally under thickets, or among tall rank weeds on the edges of fields. They sometimes associate with the snow bird, but more gene-

* Pennant.

rally keep by themselves. Their manners very much resemble those of the red-eyed bunting (Plate X. fig, 4); they are silent, tame, and unsuspicious. They have generally no other note while here than a *shep, shep;* yet I suspect they have some song in the places where they breed; for I once heard a single one, a little before the time they leave us, warble out a few very sweet low notes.

The fox-coloured sparrow is six inches long, and nine and a quarter broad; the upper part of the head and neck is cinereous, edged with rust colour; back, handsomely mottled with reddish brown, and cinereous; wings and tail, bright ferruginous; the primaries, dusky within and at the tips, the first and second row of coverts, tipt with white; breast and belly, white; the former, as well as the ear-feathers, marked with large blotches of bright bay or reddish brown, and the beginning of the belly with little arrow-shaped spots of black; the tail-coverts and tail are a bright fox colour; the legs and feet, a dirty brownish white or clay colour, and very strong; the bill is strong, dusky above and yellow below; iris of the eye, hazel. The chief difference in the female is, that the wings are not of so bright a bay, inclining more to a drab; yet this is scarcely observable, unless by a comparison of the two together. They are generally very fat, live on grass seeds, eggs of insects, and gravel.

SAVANNAH SPARROW. (*Fringilla savanna.*)

PLATE XXII.—Fig. 3, Female.*

Peale's Museum, No. 6584.

ZONOTRICHIA SAVANNA.—Jardine.

Fringilla savanna, *Bonap. Synop.* p. 108.

This new species is an inhabitant of the low countries on the Atlantic coast, from Savannah, where I first discovered it,

* The male is figured Vol. II. Plate XXXIV.

to the State of New York, and is generally resident in these places, though rarely found inland, or far from the sea-shore. The drawing of this bird was in the hands of the engraver before I was aware that the male (a figure of which will appear in Vol. II.) was so much its superior in beauty of markings and in general colours. With a representation of the male will also be given particulars of their nest, eggs, and manners, which, from the season, and the few specimens I had the opportunity of procuring, I was at that time unable to collect. I have since found these birds numerous on the sea-shore in the State of New Jersey, particularly near Great Egg Harbour. A pair of these I presented to Mr Peale of this city, in whose noble collection they now occupy a place.

The female of the Savannah sparrow is five inches and a half long, and eight and a half in extent; the plumage of the back is mottled with black, bright bay, and whitish; chin, white; breast, marked with pointed spots of black, edged with bay, running in chains from each base of the lower mandible; sides, touched with long streaks of the same; temples, marked with a spot of delicate yellow; ear-feathers, slightly tinged with the same; belly, white, and a little streaked; inside of the shoulders and lining of the wing, pale yellowish; first and second rows of wing-coverts, tipt with whitish; secondaries next the body, pointed and very black, edged also with bay; tail, slightly forked, and without any white feathers; legs, pale flesh colour; hind claw, pretty long.

The very slight distinctions of colour which nature has drawn between many distinct species of this family of finches, render these minute and tedious descriptions absolutely necessary, that the particular species may be precisely discriminated.

LOGGERHEAD SHRIKE. (*Lanius Carolinensis.*)

PLATE XXII.—Fig. 5.

Peale's Museum, No. 557.

LANIUS LUDOVICIANUS.—Bonaparte.

Lanius Ludovicianus, *Bonap. Synop.* p. 72.—The Loggerhead Shrike, *Aud.* pl. 57, male and female, *Orn. Biog.* i. p. 300.

This species has a considerable resemblance to the great American shrike.* It differs, however, from that bird in size, being a full inch shorter, and in colour, being much darker on the upper parts, and in having the frontlet black. It also inhabits the warmer parts of the United States, while the great American shrike is chiefly confined to the northern regions, and seldom extends to the south of Virginia.

This species inhabits the rice plantations of Carolina and Georgia, where it is protected for its usefulness in destroying mice. It sits for hours together on the fence beside the stacks of rice, watching like a cat; and as soon as it perceives a mouse, darts on it like a hawk. It also feeds on crickets and grasshoppers. Its note in March resembles the clear creaking of a signboard in windy weather. It builds its nest, as I was informed, generally in a detached bush, much like that of the mocking bird; but as the spring was not then sufficiently advanced, I had no opportunity of seeing its eggs. It is generally known by the name of the *Loggerhead.*†

* See Plate V. fig. 1.

† In the remarks on the *Tyranninæ*, I observed that only two of the sub-families of the greater division *Laniadæ* existed in North America,—that now alluded to, and the *Lanianæ*, of which our present species, with the *L. borealis* of a former plate, and that of Europe, will form typical examples. Ornithologists have always been at variance with regard to the position of these birds, and have placed them alike with the rapacious falcons and timid thrushes. They are, however, the "falcons of the insect world;" and among the *Insessores* will be the representatives of that group.

America was seen to be the great country of the *Tyranninæ;* in like

This species is nine inches long, and thirteen in extent; the colour above is cinereous, or dark ash; scapulars and line

manner may the shrikes claim Africa for their great birthplace. They there wage incessant war on the numerous insect hosts, the larger species occasionally exercising their greater strength on some of the weaker individuals of the feathered race; and by some gamekeepers that of this country is killed as a bird of prey, being found to destroy young birds, and even to drag the weak young pheasants through the bars of the breeding coops. Small animals and reptiles also form a part of their prey. They decrease in numbers as the colder and more temperate countries are approached; and the vast extent of North America appears only to contain five species. New Holland alone is without any true *Lanius*, but is supplied by another genus, *Falcunculus*, allied in form, and now containing two species, which also unite somewhat of their habits, and feed on insects, though the mode of taking their prey shows something scansorial.

Among the tyrants, the powers of flight are developed to a great extent, as suitable to the capture of the particular prey upon which they feed. In the shrikes, the form is considerably modified; the wings become more rounded, and the tail graduated; and the general prey is the larger insects of the orders *Coleoptera* and *Hemiptera*, to capture which does not require so great an exercise of very quick or active powers, and which are often patiently watched for and pounced upon by surprise, in a similar manner to that described of the North American loggerhead.

They have all the character of being cruel and tyrannous, arising from the peculiar manner of impaling their prey upon thorns, or fastening it in the clefts of branches, often in a wanton manner, as if for the sake of murder only, thus fixing up all it can seize upon. One species is particularly remarkable for the regular exhibition of this propensity, and has become proverbial for its cruelty,—*Lanius collaris* of Southern Africa. Its habits are thus described by Le Vaillant:—"When it sees a locust, a mantis, or a small bird, it springs upon it, and immediately carries it off, in order to impale it on a thorn, which it does with great dexterity, always passing the thorn through the head of its victim. Every animal which it seizes is subjected to the same fate; and it thus continues all day long its murderous career, apparently instigated rather by the love of mischief than the desire of food. Its throne of tyranny is usually a dry and elevated branch of a tree, from which it pounces on all intruders, driving off the stronger and more troublesome, and impaling the inexperienced alive; when hungry, it visits its shambles, and helps itself to a savoury meal." The Hottentots assured Le Vaillant that it does not love fresh food, and therefore leaves its prey on the gibbet over the eye, whitish ; wings, black, with a small spot of white at the base of the primaries, and tipt with white ; a stripe of

till it becomes putrescent ; but beneath the scorching sun of Africa, the process of decomposition sometimes does not take place, from the rapid exhalation of the animal fluids in a warm and arid atmosphere, and, consequently, whatever spiny shrub may have been chosen by the butcher-bird as the place of execution is frequently found covered, not with sweet-smelling and many-coloured blossoms, but with the dried carcasses of singing-birds, and the bodies of locusts and other insects of the larger size. The species of Great Britain also exercises this propensity ; but, according to Mr Selby, it invariably kills its prey by strangulation before transfixing it. That gentleman mentions once having the gratification of witnessing this operation of the shrike upon a hedge accentor, which it had just killed. " In this instance, after killing the bird, it hovered, with its prey in its bill, for a short time over the hedge, apparently occupied in selecting a thorn fit for its purpose. Upon disturbing it, and advancing to the spot, I found the accentor firmly fixed by the tendons of the wing at the selected twig." When in confinement, this peculiarity is also displayed, in placing the food against or between the wires of the cage. They frequent woody countries, with occasional shrubs and hedges, among which they also breed ; the notes, as might be expected, are hoarse and grating, and during the season of incubation become very garrulous, particularly when alarmed ; they are very attentive to their young, and continue long to feed and attend them after they are able to shift for themselves. It may be here remarked, that the *Falconidæ*, which our present knowledge leads us to think is represented by this group, always take their prey to some eminence before commencing to devour it—a bare hillock or rock in an open country, the top of some old mound or dyke, or, if in a wood, some decayed stump ; and I have known one spot of frequent recurrence by the same individuals, thus showing some analogy to each other.

The following seem to be the species which are known to belong to North America :—

1. *L. borealis*, Vieill.—*L. excubitor*, Wils. vol. i. p. 74.—*L. borealis*, Bonap. Synop. App.*

2. *L. Ludovicianus*, Bonap.—*L. Carolinensis*, Wils. vol. iii. p. 57 ; found only in the warmer and more southern States, the Carolinas and Georgia.

3. *Lanius excubitroides*, Sw. *Nov. spec.*—American gray shrike, North. Zool. vol. ii. p. 115.

Specimens were brought to this country by the last Overland Arctic

* When writing the note at page 73 of this volume, I was not aware that Bonaparte had taken notice of the mistake mentioned there in his Appendix to the Synopsis of North American Birds.—Ed.

black passes along the front, through each eye, half way down the side of the neck; eye, dark hazel, sunk below the eyebrow; tail, cuneiform, the four middle feathers wholly black; the four exterior ones on each side tipt more and more with white to the outer one, which is nearly all white; whole lower parts white, and, in some specimens, both of males and females, marked with transverse lines of very pale brown; bill and legs, black.

The female is considerably darker both above and below, but the black does not reach so high on the front; it is also rather less in size.

Expedition. According to Dr Richardson, it is a more northern bird than *L. borealis*, and does not advance farther north in summer than the 54° of latitude, and it attains that parallel only in the meridian of the warm and sandy plains of the Saskatchewan, which enjoy an earlier spring and longer summer than the densely wooded country betwixt them and Hudson's Bay. Its manners are precisely similar to those of *L. borealis*, feeding chiefly on grasshoppers, which are exceedingly numerous. Its nest was found in a bush of willows, built of twigs of *Artemesiæ* and dried grass, and lined with feathers; the eggs, six in number, were very pale yellowish gray, with many irregular and confluent spots of oil green, interspersed with a few of smoke gray.

The merit of unravelling this species from several very closely allied to it in its native country, and from that to which it approaches nearest, the *L. excubitor* of Europe, is due to Mr Swainson. The chief distinctive characters given by that naturalist are the small proportions of the bill, the frontal feathers crossed by a narrow band of deep black, the black stripe on the side of the head encircling the upper margin of the eyelid, lateral scales of the tarsus being divided in several pieces, the shorter length of the wing when closed, and in the tail being more graduated; the total length is nine inches, six lines.

4. *Lanius elegans*, Sw.—White-crowned shrike.

Described by Mr Swainson, from a specimen in the British Museum, to which it was presented from the Fur Countries by the Hudson's Bay Company. It may at once be distinguished from the other American shrikes by the much greater quantity of white on the wings and tail, its narrower tail-feathers, longer tarsi, and less curved claws; the length is about nine inches.

5. *Lanius* (?) *natka*, Penn.—Natka shrike.

This species, the Nootka shrike of Dr Latham, from Nootka Sound, on the northwest coast of North America, seems to be of such dubious authority, that little can be said regarding it.—Ed.

BELTED KINGSFISHER. (*Alcedo alcyon.*)

PLATE XXIII.—Fig. 1, Female.

Bartram, p. 289.—*Turton*, p. 278.—*Peale's Museum*, No. 2145.

ALCEDO ALCYON.—Linnæus.*

Alcedo alcyon, *Bonap. Synop.* p. 49.—The Belted Kingsfisher, *Aud.* pl. 77, *Orn. Biog.* i. p. 394.

This is a general inhabitant of the banks and shores of all our fresh-water rivers, from Hudson's Bay to Mexico, and is the only species of its tribe found within the United States. This

* The description of Wilson, and that of Audubon, which has been added in a note from the "Ornithological Biography," give a very correct detail of the general manners of the true kingsfishers, or those resembling that of this country; there is throughout the family, however, a very considerable difference in form, and, as a matter of course, a corresponding difference in habit; this has occasioned a division of them into various groups, by almost all ornithologists; that to which our present species belongs, and of which it is the largest, contains all those of smaller size with four toes and sharp angular and lengthened bills; they feed entirely on fish and aquatic insects, and live on the banks of rivers, lakes, and creeks, and occasionally on the sea-shore. They are distributed over the world, but the warmer parts of India, Africa, and South America possess the greatest share, North America and Europe possessing only one each. The colours of the plumage, with a few exceptions, particularly the upper parts, are very bright and shining, the webs of the feathers unconnected and loose; the under parts generally white, with shades of reddish brown and orange. The division nearest to this, containing but a few species of very small size, but similar in form and colouring, has been separated on account of having three toes, and, I believe, is exclusively Indian. Another and a well-marked group is the halcyon of Mr Swainson; it differs materially in the form and manners of living, and ranges everywhere, except in North America and Europe. The birds are all above the middle size, with a stouter and more robust form; the colour sometimes very gaudy, in others of rich and pleasing shades of brown. The bill, a chief organ of distinction, is large, much dilated at the base, and, in one or two instances, very strong. They inhabit moist woods, and shady streams or creeks, where they watch on a motionless perch for the larger insects, as the common European species does for fish, and they dart upon them when passing, or when seen on

Drawn from Nature by A. Wilson *Engraved by W.H. Lizars*

1. Belted Kingsfisher. 2. Black and Yellow Warbler. 3. Blackburnian W. 4. Autumnal W. 5. Water Thrush.

23.

last circumstance, and its characteristic appearance, make it as universally known here as its elegant little brother, the common kingsfisher of Europe, is in Britain. Like the love-lorn swains, of whom poets tell us, he delights in murmuring

the ground, and return again to the same branch or rock; they also chase their prey in the manner of the Flycatchers. Notwithstanding these are their common food, fish, water insects, in a few instances crabs, are resorted to, and in all cases the vicinity of water seems requisite for their healthy support. There is an individual (*Alcedo dea*) which has been separated from this under the name of *Tanysiptera;* the only distinction now (for it has four toes) is the elongation of two tail-feathers, which exceed the length of the body considerably. It was originally discovered in the Isle of Ternate, and, according to Lesson, is abundant in New Guinea, where it is killed by the natives for ornaments; and those coming to this country, being impaled on reeds, are consequently much mutilated. Another division will comprise the very large New Holland species, under the title of *Dacelo;** this contains yet only two species, commonly known by the name of "laughing jackasses;" by the natives they are called *cuck'unda;* they are nearly as large as a common pigeon, and have all the members very powerful; the bill is much dilated, and bent at the tip; according to Lesson, their chief food is large insects, which they seize on the ground. That ornithologist extends the genus to several of the larger billed small species; we would now restrict it, as bearing better marks, to those of New Holland only, *D. gigantea* and *Leachii.* Another division has been formed among these curious birds, also by M. Lesson, of the *Alcedo rufipes* of Cuvier, under the name of *Syma,* and, as a specific appellation, that of *torotora,* by which it is known to the Papous, in its native country, New Guinea. It frequents rivers and the sea-shores, and feeds on fish; the principal distinction for which it has been separated is a serrature of the mandibles of the bill. M. Lesson, however, did not perceive anything different from its congeners to which this structure could be applied. From the above remarks it will be seen that the old genus *Alcedo* has been separated into no less than nine divisions. Four of these will, perhaps, only be necessary, and are as follows:—1. *Alcedo,* having the form of *Alcedo ispida,* feeding principally on fish; geographical distribution, the known world, except very northern latitudes; the number of species and individuals increasing from the extremes. 2. *Halcyon;* the form of *Sanctus, cinamomeus, omnicolor,*

* M. Lesson proposes a genus (*Todyrampus*) for all the smaller New Holland species, taking *A. sacra* as the type, on account, principally, of the more dilated bill. The same gentleman proposes the title *Melidora* and *Choucalcyon* to designate forms among the kingsfishers which I have not ascertained.

streams and falling waters; not, however, merely that they may soothe his ear, but for a gratification somewhat more substantial. Amidst the roar of the cataract, or over the foam of a torrent, he sits perched upon an overhanging bough, glancing his piercing eye in every direction below for his scaly prey, which, with a sudden circular plunge, he sweeps from their native element, and swallows in an instant. His voice, which is not unlike the twirling of a watchman's rattle, is naturally loud, harsh, and sudden; but is softened by the sound of the brawling streams and cascades among which he generally rambles. He courses along the windings of the brook or river, at a small height above the surface, sometimes suspending himself by the rapid action of his wings, like certain species of hawks, ready to pounce on the fry below; now and then settling on an old dead overhanging limb to reconnoitre.* Mill-dams are particularly visited by this feathered fisher; and the sound of his pipe is as well known to the miller as the rattling of his own hopper. Rapid

&c.; containing Lesson's *Todyrampus*, also perhaps his *Syma*, and the *Tanysiptera* of Vigors; the two latter groups, as species, would be at once distinguished by the peculiarities of form which are perhaps not sufficient to indicate a genus without more of like characters; geographical distribution, South America, New Holland, Africa, and India. 3. *Dacelo;* the form, *D. gigantea;* geographical distribution, New Holland. And 4. *Ceyx*, containing the three-toed kingsfisher *C. tridactyla;* geographical distribution, India.—Ed.

* Mr Audubon mentions, that this species sometimes also visits the salt-water creeks, diving after fish. When crossing from one lake to another, which it frequently does, it passes over forests in a direct line, not unfrequently by a course of twenty or thirty miles, towards the interior of the country. Its motions at this time consist of a series of slopes, about five or six in number, followed by a direct glide, without any apparent undulation.

They dig the holes for their nest with great despatch. As an instance of their working with celerity, the same gentleman mentions, that he hung a small net in front of one of their holes to entrap the bird upon the nest; but, ere morning, it had scratched its way out. On the following evening, he stopped up the hole for upwards of a foot with a stick, but the same thing again took place.—Ed.

streams, with high perpendicular banks, particularly if they be of a hard clayey or sandy nature, are also favourite places of resort for this bird ; not only because in such places the small fish are more exposed to view, but because those steep and dry banks are the chosen situations for his nest. Into these he digs with bill and claws horizontally, sometimes to the extent of four or five feet, at the distance of a foot or two from the surface. The few materials he takes in are not always placed at the extremity of the hole, that he and his mate may have room to turn with convenience. The eggs are five, pure white, and the first brood usually comes out about the beginning of June, and sometimes sooner, according to the part of the country where they reside. On the shores of Kentucky river, near the town of Frankfort, I found the female sitting early in April. They are very tenacious of their haunts, breeding for several successive years in the same hole, and do not readily forsake it, even though it be visited. An intelligent young gentleman informed me, that having found where a kingsfisher built, he took away its eggs from time to time, leaving always one behind, until he had taken no less than eighteen from the same nest. At some of these visits, the female, being within, retired to the extremity of the hole, while he withdrew the egg, and next day, when he returned, he found she had laid again as usual.

The fabulous stories related by the ancients of the nest, manner of hatching, &c., of the kingsfisher, are too trifling to be repeated here. Over the winds and the waves the humble kingsfishers of our days, at least the species now before us, have no control. Its nest is neither constructed of glue nor fish-bones; but of loose grass and a few feathers. It is not thrown on the surface of the water to float about, with its proprietor, at random, but snugly secured from the winds and the weather in the recesses of the earth; neither is its head or its feathers believed, even by the most illiterate of our clowns or seamen, to be a charm for love, a protection against witchcraft, or a security for fair weather. It is neither venerated like those of the Society Isles, nor dreaded, like those of some other countries; but is considered merely as a bird that feeds on fish; is generally fat; relished by *some* as good eating; and is now and then seen exposed for sale in our markets.

Though the kingsfisher generally remains with us, in Pennsylvania, until the commencement of cold weather, it is seldom seen here in winter; but returns to us early in April. In North and South Carolina, I observed numbers of these birds in the months of February and March. I also frequently noticed them on the shores of the Ohio in February, as high up as the mouth of the Muskingum.

I suspect this bird to be a native of the Bahama Islands, as well as of our continent. In passing between these isles and the Florida shore, in the month of July, a kingsfisher flew several times round our ship, and afterwards shot off to the south.

The length of this species is twelve inches and a half; extent, twenty; back and whole upper parts, a light bluish slate colour; round the neck is a collar of pure white, which reaches before to the chin; head, large, crested; the feathers, long and narrow, black in the centre, and generally erect; the shafts of all the feathers, except the white plumage, are black; belly and vent, white; sides under the wings, variegated with blue; round the upper part of the breast passes a band of blue, interspersed with some light brown feathers; before the eye is a small spot of white, and another immediately below it; the bill is three inches long from the point to the slit of the mouth, strong, sharp-pointed, and black, except near the base of the lower mandible, and at the tip, where it is of a horn colour; primaries and interior webs of the secondaries, black, spotted with white; the interior vanes of the tail-feathers, elegantly spotted with white on a jet black ground; lower side, light coloured; exterior vanes, blue; wing-coverts and secondaries, marked with small specks of white; legs, extremely short; when the bird perches, it generally rests on the lower side of the second joint, which is thereby thick and callous; claws, stout and black; whole leg, of a dirty yellowish colour; above the knee, bare of feathers for half an inch; the two exterior toes united together for nearly their whole length.

The female is sprinkled all over with specks of white; the band of blue around the upper part of the breast is nearly half reddish brown; and a little below this passes a band of bright reddish bay, spreading on each side under the wings. The blue and rufous feathers on the breast are strong, like scales. The head is also of a much darker blue than the back, and the white feathers on the chin and throat of an exquisite fine glossy texture, like the most beautiful satin.

BLACK AND YELLOW WARBLER. (*Sylvia magnolia.*)

PLATE XXIII.—Fig. 2.

Peale's Museum, No. 7783.

SYLVICOLA MACULOSA.—Swainson.

Sylvia maculosa, *Lath. Ind. Orn.* ii. p. 536.—*Bonap. Synop.* p. 78.—Yellow-rump Warbler, *Penn. Arct. Zool.* ii. p. 400.—The Black and Yellow Warbler (the young is figured only), *Aud.* pl. 50; *Orn. Biog.* i. p. 260.—Sylvicola maculosa, *North. Zool.* ii. p. 212.

This bird I first met with on the banks of the Little Miami, near its junction with the Ohio. I afterwards found it among the magnolias, not far from Fort Adams, on the Mississippi. These two, both of which happened to be males, are all the individuals I have ever shot of this species; from which I am justified in concluding it to be a very scarce bird in the United States. Mr Peale, however, has the merit of having been the first to discover this elegant species, which, he informs me, he found several years ago not many miles from Philadelphia. No notice has ever been taken of this bird by any European naturalist whose works I have examined. Its notes, or rather chirpings, struck me as very peculiar and characteristic; but have no claim to the title of song. It kept constantly among the higher branches, and was very active and restless.

Length, five inches; extent, seven inches and a half; front, ores, and behind the ear, black; over the eye, a fine line of white, and another small touch of the same immediately under; back, nearly all black; shoulders, thinly streaked with olive; rump, yellow; tail-coverts, jet black; inner vanes of the lateral tail-feathers, white, to within half an inch of the tip, where they are black; two middle ones, wholly black; whole lower parts, rich yellow, spotted from the throat downwards with black streaks; vent, white; tail, slightly forked; wings, black, crossed with two broad transverse bars of white; crown, fine ash; legs, brown; bill, black. Markings of the female not known.

BLACKBURNIAN WARBLER. (*Sylvia blackburniæ.*)

PLATE XXIII.—FIG. 3.

Lath. ii. p. 461, No. 67.—*Peale's Museum*, No. 7060.

SYLVICOLA BLACKBURNIÆ.—JARDINE.

Sylvia Blackburniæ, *Bonap. Synop.* p. 80.

THIS is another scarce species in Pennsylvania, making its appearance here about the beginning of May; and again in September on its return, but is seldom seen here during the middle of summer. It is an active, silent bird; inhabits also the state of New York, from whence it was first sent to Europe. Mr Latham has numbered this as a variety of the yellow-fronted warbler, a very different species. The specimen sent to Europe, and first described by Pennant, appears also to have been a female, as the breast is said to be yellow, instead of the brilliant orange with which it is ornamented. Of the nest and habits of this bird, I can give no account, as there is not more than one or two of these birds to be found here in a season, even with the most diligent search.

The blackburnian warbler is four inches and a half long, and seven in extent; crown, black, divided by a line of orange; the black again bounded on the outside by a stripe of rich orange passing over the eye; under the eye, a small touch of orange yellow; whole throat and breast, rich fiery orange, bounded by spots and streaks of black; belly, dull yellow, also streaked with black; vent, white; back, black, skirted with ash; wings, the same, marked with a large lateral spot of white; tail, slightly forked; the interior vanes of the three exterior feathers, white; cheeks, black; bill and legs, brown. The female is yellow where the male is orange; the black streaks are also more obscure and less numerous.

AUTUMNAL WARBLER. (*Sylvia autumnalis.*)

PLATE XXIII.—Fig. 4.

SYLVICOLA? AUTUMNALIS.—Jardine.

Sylvicola autumnalis, *Bonap. Synop.* p. 84.—The Autumnal Warbler, *Aud.* plate 88; *Orn.—Biog.* i. p. 447.

This plain little species regularly visits Pennsylvania from the north, in the month of October, gleaning among the willow leaves; but, what is singular, is rarely seen in spring. From the 1st to the 15th of October, they may be seen in considerable numbers, almost every day, in gardens, particularly among the branches of the weeping willow, and seem exceedingly industrious. They have some resemblance, in colour, to the pine-creeping warbler; but do not run along the trunk like that bird, neither do they give a preference to the pines. They are also less. After the first of November, they are no longer to be found, unless the season be uncommonly mild. These birds, doubtless, pass through Pennsylvania in spring, on their way to the north; but either make a very hasty journey, or frequent the tops of the tallest trees; for I have never yet met with one of them in that season, though in October I have seen more than a hundred in an afternoon's excursion.

Length, four inches and three quarters; breadth, eight inches; whole upper parts, olive green, streaked on the back with dusky stripes; tail-coverts, ash, tipt with olive; tail, black, edged with dull white; the three exterior feathers, marked near the tip with white; wings, deep dusky, edged with olive, and crossed with two bars of white; primaries, also tipt, and three secondaries next the body, edged with white; upper mandible, dusky brown; lower, as well as the chin and breast, dull yellow; belly and vent, white; legs, dusky brown; feet and claws, yellow; a pale yellow ring surrounds the eye. The males of these birds often warble out some low, but very sweet notes, while searching among the leaves in autumn.

WATER THRUSH. (*Turdus aquaticus.*)

PLATE XXIII.—FIG. 5.

Peale's Museum, No. 6896.

SEIURUS AQUATICUS.—SWAINSON.

New York Warbler, *Penn. Arct. Zool.* ii. p. 303.—Sylvia Noveboracensis, *Bonap. Synop.* p. 77.—Seiurus aquaticus, Aquatic Accentor, *North. Zool.* ii. p. 229.

THIS bird is remarkable for its partiality to brooks, rivers, shores, ponds, and streams of water; wading in the shallows in search of aquatic insects, wagging the tail almost continually, chattering as it flies; and, in short, possesses many strong traits and habits of the water wagtail. It is also exceedingly shy, darting away on the least attempt to approach it, and uttering a sharp *chip* repeatedly, as if greatly alarmed. Among the mountain streams in the state of Tennesee, I found a variety of this bird pretty numerous, with legs of a bright yellow colour; in other respects it differed not from the rest. About the beginning of May it passes through Pennsylvania to the north; is seen along the channels of our solitary streams for ten or twelve days; afterwards disappears until August. It is probable that it breeds in the higher mountainous districts even of this state, as do many other of our spring visitants that regularly pass a week or two with us in the lower parts, and then retire to the mountains and inland forests to breed.

But Pennsylvania is not the favourite resort of this species. The cane brakes, swamps, river shores, and deep watery solitudes of Louisiana, Tennesee, and the Mississippi territory, possess them in abundance; there they are eminently distinguished by the loudness, sweetness, and expressive vivacity of their notes, which begin very high and clear, falling with an almost imperceptible gradation till they are scarcely articulated. At these times the musician is perched on the middle branches of a tree over the brook or river bank, pouring out his charming melody, that may be distinctly heard for nearly half a mile. The voice of this little bird appeared to me so exquisitely sweet and expressive, that I was never tired of listening to it, while traversing the deep shaded hollows of those cane brakes where it usually resorts. I have never yet met with its nest.

The water thrush is six inches long, and nine and a half in extent; the whole upper parts are of a uniform and very dark olive, with a line of white extending over the eye, and along the sides of the neck; the lower parts are white, tinged with yellow ochre; the whole breast and sides are marked with pointed spots or streaks of black or deep brown; bill, dusky brown; legs, flesh coloured; tail, nearly even; bill, formed almost exactly like the golden-crowned thrush, above described, p. 239; and, except in frequenting the water, much resembling it in manners. Male and female nearly alike.

PAINTED BUNTING. (*Emberiza ciris.*)

PLATE XXIV.—FIG. 1.—MALE; FIG. 2.—FEMALE.

Linn. Syst. 313.—Painted Finch, *Catesb.* i. 44.—*Edw.* 130, 173.—*Arct. Zool.* p. 362, No. 226.—Le Verdier de la Louisiane, dit vulgairement le Pape, *Briss.* iii. 200. *App.* 74.—*Buff.* iv. 76. *Pl. enl.* 159.—*Lath.* ii. 206.—Linaria ciris, The Painted Finch, or Nonpareil, *Bartram*, p. 291.—*Peale's Museum*, No. 6062, and 6063.

SPIZA CIRIS.—BONAPARTE.*

Fringilla (sub-genus Spiza) ciris, *Bonap. Synop.* p. 107.—La pesserine nonpareil ou le papa, Passerina ciris, *Vieill. Gall. des Ois.* pl. 66.—The Painted Finch, *Aud.* pl. 53, male and female; *Orn. Biog.* i. 279.

THIS is one of the most numerous of the little summer birds of Lower Louisiana, where it is universally known among the French inhabitants, and called by them *Le Pape*, and by the

* From the general request of this species as a pet, it is requisite that considerable numbers should be taken, and the method used is thus described by Audubon. I may remark, in the taking of various birds alive, "call birds," or tame ones, trained for the purpose of decoy, are commonly used in all countries, and in some instances a stuffed specimen, or even a representation made of Paris plaster, is used with success.

"A male bird, in full plumage, is shot, and stuffed in a defensive attitude, and perched among some grass seed, rice, or other food, on the same platform as the trap-cage. This is taken to the fields, or near the orangeries, and placed in so open a situation, that it would be difficult for a living bird of any species to fly over it without observing it. The trap is set. A male painted finch passes, perceives it, and dives towards the stuffed bird, brings down the trap, and is made prisoner. In this manner, thousands of these birds are caught every spring; and so pertinacious are they in their attacks, that, even when the trap has closed upon them, they continue pecking at the feathers of the supposed rival."

They feed immediately, and some have been kept in confinement for ten years. They cost about sixpence in New Orleans; but, in London, three guineas are sometimes asked.

The various generic nomenclature to which this bird has been subjected, shows that ornithologists are at variance in opinion. It forms part of the first section of Bonaparte's subgenus *Spiza*, to which should also be referred the *Fringilla Cyanea* of this volume, p. 100.—ED.

1. Painted Bunting. 2. Female. 3. Prothonotary Warbler. 4. Wormeating W. 5. Yellow-winged Sparrow. 6. Blue Grosbeak.

24.

Americans *The Nonpareil.* Its gay dress and docility of manners have procured it many admirers; for these qualities are strongly attractive, and carry their own recommendations always along with them. The low countries of the southern states, in the vicinity of the sea, and along the borders of our large rivers, particularly among the rice plantations, are the favourite haunts of this elegant little bird. A few are seen in North Carolina; in South Carolina they are more numerous; and still more so in the lower parts of Georgia. To the westward, I first met them at Natchez, on the Mississippi, where they seemed rather scarce. Below Baton Rouge, along the levee, or embankment of the river, they appeared in greater numbers; and continued to become more common as I approached New Orleans, where they were warbling from almost every fence, and crossing the road before me every few minutes. Their notes very much resemble those of the indigo bird (plate vi. fig. 6), but want the strength and energy of the latter, being more feeble and more concise.

I found these birds very commonly domesticated in the houses of the French inhabitants of New Orleans; appearing to be the most common cage bird they have. The negroes often bring them to market, from the neighbouring plantations, for sale; either in cages, taken in traps, or in the nest. A wealthy French planter, who lives on the banks of the Mississippi, a few miles below Bayo Fourche, took me into his garden, which is spacious and magnificent, to show me his aviary; where, among many of our common birds, I observed several nonpareils, two of which had nests, and were then hatching.

Were the same attention bestowed on these birds as on the canary, I have no doubt but they would breed with equal facility, and become equally numerous and familiar, while the richness of their plumage might compensate for their inferiority of song. Many of them have been transported to Europe; and I think I have somewhere read, that in Holland attempts have been made to breed them, and with success. When the employments of the people of the United States become more sedentary, like those of Europe, the innocent and agreeable amusement of keeping and rearing birds in this manner will become more general than it is at present, and their manners better known. And I cannot but think, that an intercourse with these little innocent warblers is favourable to delicacy of feeling, and sentiments of humanity; for I have observed the rudest and most savage softened into benevolence while contemplating the interesting manners of these inoffensive little creatures.

Six of these birds, which I brought with me from New Orleans by sea, soon became reconciled to the cage. In good weather, the males sang with great sprightliness, though they had been caught only a few days before my departure. They were greedily fond of flies, which accompanied us in great numbers during the whole voyage; and many of the passengers amused themselves with catching these, and giving them to the Nonpareils; till, at length, the birds became so well acquainted with this amusement, that as soon as they perceived any of the people attempting to catch flies, they assembled at the front of the cage, stretching out their heads through the wires with eager expectation, evidently much interested in the issue of their efforts.

These birds arrive in Louisiana, from the south, about the middle of April, and begin to build early in May. In Savannah, according to Mr Abbot, they arrive about the 20th of April. Their nests are usually fixed in orange hedges, or on the lower branches of the orange tree; I have also found them in a common bramble or blackberry bush. They are formed exteriorly of dry grass, intermingled with the silk of caterpillars, lined with hair, and, lastly, with some extremely fine roots of plants. The eggs are four or five, white, or rather pearl coloured, marked with purplish brown specks. As some of these nests had eggs so late as the 25th of June, I think it probable that they sometimes raise two broods in the same season. The young birds of both sexes, during the first season, are of a fine green olive above, and dull yellow below. The females undergo little or no change, but that of becoming of a more brownish cast. The males, on the contrary, are long and slow in arriving at their full variety of colours. In the second season, the blue on the head begins to make its appearance, intermixed with the olive green: the next year, the yellow shews itself on the back and rump; and also the red, in detached spots, on the throat and lower parts. All these colours are completed in the fourth season, except, sometimes, that the green still continues on the tail. On the fourth and fifth season, the bird has attained his complete colours, and appears then as represented in the plate (fig. 1). No dependence, however, can be placed on the regularity of this change in birds confined in a cage, as the want of proper food, sunshine, and variety of climate, all conspire against the regular operations of nature.

The nonpareil is five inches and three quarters long, and eight inches and three quarters in extent; head, neck above, and sides of the same, a rich purplish blue; eyelid, chin, and whole lower parts, vermilion; back and scapulars, glossy yellow, stained with rich green, and in old birds with red; lesser wing-coverts, purple; larger, green; wings, dusky red, sometimes edged with green; lower part of the back, rump, and tail-coverts, deep glossy red, inclining to carmine; tail, slightly forked, purplish brown (generally green); legs and feet, leaden gray; bill, black above, pale blue below; iris of the eye, hazel.

The female (fig. 2) is five and a half inches long, and eight inches in extent; upper parts, green olive, brightest on the rump; lower parts, a dusky Naples yellow, brightest on the belly; and tinged considerably on the breast with dull green, or olive; cheeks, or ear-feathers, marked with lighter touches; bill, wholly a pale lead colour, lightest below; legs and feet, the same.

The food of these birds consists of rice, insects, and various kinds of seeds that grow luxuriantly in their native haunts.

I also observed them eating the seeds or internal grains of ripe figs. They frequent gardens, building within a few paces of the house; are particularly attached to orangeries; and chant occasionally during the whole summer. Early in October they retire to more southern climates, being extremely susceptible of cold.

PROTONOTARY WARBLER. (*Sylvia protonotarius.*)

PLATE XXIV.—Fig. 3.

Arct. Zool. p. 410.—*Buff.* v. 316.—*Lath.* ii. 494. *Pl. enl.* 704.—*Peale's Museum*, No. 7020.

VERMIVORA? PROTONOTARIUS.—Jardine.

Sylvia (sub-genus Dacnis, *Cuv.*) protonotarius, *Bonap. Synop.* p. 86.—The Protonotary Warbler, *Aud.* pl. 3, male and female; *Orn. Biog.* i. p. 22.

This is an inhabitant of the same country as the preceding species; and also a passenger from the south; with this difference, that the bird now before us seldom approaches the house or garden, but keeps among the retired, deep, and dark swampy woods, through which it flits nimbly in search of small caterpillars, uttering every now and then a few screaking notes, scarcely worthy of notice. They are abundant in the Mississippi and New Orleans territories, near the river, but are rarely found on the high ridges inland.

From the peculiar form of its bill, being roundish and remarkably pointed, this bird might, with propriety, be classed as a sub-genera, or separate family, including several others, viz., the blue-winged yellow warbler, the gold-crowned warbler, and golden-winged warbler, of Plate XV., and the worm-eating warbler of the present plate, and a few more. The bills of all these correspond nearly in form and pointedness, being generally longer, thicker at the base, and more round than those of the genus *Sylvia*, generally. The first mentioned species, in particular, greatly resembles this in its general appearance; but the bill of the protonotary is rather stouter, and the yellow much deeper, extending farther on the back; its manners, and the country it inhabits, are also different.

This species is five inches and a half long, and eight and a half in extent; the head, neck, and whole lower parts (except the vent), are of a remarkably rich and brilliant yellow, slightly inclining to orange; vent, white; back, scapulars, and lesser wing-coverts, yellow olive; wings, rump, and tail-coverts, a lead blue; interior vanes of the former, black; tail nearly even, and black, broadly edged with blue; all the feathers, except the two middle ones, are marked on their inner vanes, near the tip, with a spot of white; bill, long, stout, sharp-pointed, and wholly black; eyes, dark hazel; legs and feet, a leaden grey. The female differs in having the yellow and blue rather of a duller tint; the inferiority, however, is scarcely noticeable.

WORM-EATING WARBLER. (*Sylvia vermivora.*)

PLATE XXIV.—Fig. 4.

Arct. Zool. p. 406, No. 300.—*Edw.* 305.—*Lath.* ii. 499.—Le demi-fin mangeur de vers, *Buff.* v. 325.—*Peale's Museum*, No. 6848.

VERMIVORA PENNSYLVANICA.—Swainson.*

Ficedula Pennsylvanica, *Briss.* i. 457.—Sylvia (sub-genus Dacnis, *Cuv.*) Pennsylvanica, *Bonap. Synop.* p. 86.—The Worm-eating Warbler, *Aud.* pl. 34, male and female; *Orn. Biog.* i. p. 177.

This is one of the nimblest species of its whole family, inhabiting the same country with the preceding, but extending its migrations much farther north. It arrives in Pennsylvania about the middle of May, and leaves us in September. I have never yet met with its nest, but have seen them feeding their young about the 25th of June. This bird is remarkably fond of spiders, darting about wherever there is a probability of

* This species is the type of Mr Swainson's genus *Vermivora.* The specific title is therefore lost, and I see none better than the restoration of Brisson's old one.—Ed.

finding these insects. If there be a branch broken, and the leaves withered, it shoots among them in preference to every other part of the tree, making a great rustling, in search of its prey. I have often watched its manœuvres while thus engaged, and flying from tree to tree in search of such places. On dissection, I have uniformly found their stomachs filled with spiders or caterpillars, or both. Its note is a feeble chirp, rarely uttered.

The worm-eater is five inches and a quarter in length, and eight inches in extent; back, tail, and wings, a fine clear olive; tips and inner vanes of the wing-quills, a dusky brown; tail, slightly forked, yet the exterior feathers are somewhat shorter than the middle ones; head and whole lower parts, a dirty buff; the former marked with four streaks of black, one passing from each nostril, broadening as it descends the hind head; and one from the posterior angle of each eye; the bill is stout, straight, pretty thick at the base, roundish, and tapering to a fine point; no bristles at the side of the mouth; tongue, thin, and lacerated at the tip; the breast is most strongly tinged with the orange buff; vent, waved with dusky olive; bill, blackish above, flesh coloured below; legs and feet, a pale clay colour; eye, dark hazel. The female differs very little in colour from the male.

On this species Mr Pennant makes the following remarks: —"Does not appear in Pennsylvania till July, in its passage northward. Does not return the same way, but is supposed to go beyond the mountains which lie to the west. This seems to be the case with all the transient vernal visitants of Pennsylvania." * That a small bird should permit the whole spring, and half of the summer, to pass away before it thought of "passing to the north to breed," is a circumstance, one should think, would have excited the suspicion of so discerning a naturalist as the author of "Arctic Zoology," as to its truth. I do not know that this bird breeds to the northward of the United States. As to their returning home by "the country

* Arctic Zoology, p. 406.

beyond the mountains," this must, doubtless, be for the purpose of finishing the education of their striplings here, as is done in Europe, by making the grand tour. This, by the by, would be a much more convenient retrograde route for the ducks and geese; as, like the Kentuckians, they could take advantage of the current of the Ohio and Mississippi, to float down to the southward. Unfortunately, however, for this pretty theory, all our vernal visitants with which I am acquainted, are contented to plod home by the same regions through which they advanced, not even excepting the geese.

YELLOW-WINGED SPARROW. (*Fringilla passerina.*)

PLATE XXIV.—Fig. 5.

Peale's Museum, No. 6585.

EMBERIZA? PASSERINA.—Jardine.*

Fringilla (sub-genus Spiza) passerina, *Bonap. Synop.* p. 109.

This small species is now for the first time introduced to the notice of the public. I can, however, say little towards illustrating its history, which, like that of many individuals of the human race, would be but a dull detail of humble obscurity. It inhabits the lower parts of New York and Pennsylvania; is very numerous on Staten Island, where I first observed it; and occurs also along the sea coast of New Jersey. But, though it breeds in each of these places, it does not remain in any of them during the winter. It has a short, weak, interrupted chirrup, which it occasionally utters from the fences and tops of low bushes. Its nest is fixed on the ground

* "A few of these birds," the Prince of Musignano remarks, "can never be separated in any natural arrangement." What are now placed under the name *Emberiza*, will require a subgenus for themselves, perhaps the analogous form of that genus in the New World. In this species we have the palatial knob, and converging edges of the mandibles; and, by Bonaparte, it is placed among the finches, in the second section of his subgenus *Spiza*, as forming the passage to the buntings.—Ed.

among the grass; is formed of loose dry grass, and lined with hair and fibrous roots of plants. The eggs are five, of a greyish white sprinkled with brown. On the first of August I found the female sitting.

I cannot say what extent of range this species has, having never met with it in the southern states; though I have no doubt that it winters there, with many others of its tribe. It is the scarcest of all our summer sparrows. Its food consists principally of grass seeds, and the larvæ of insects, which it is almost continually in search of among the loose soil and on the surface; consequently it is more useful to the farmer than otherwise.

The length of this species is five inches; extent, eight inches; upper part of the head, blackish, divided by a slight line of white; hind head and neck above, marked with short lateral touches of black and white; a line of yellow extends from above the eye to the nostril; cheeks, plain brownish white; back, streaked with black, brown, and pale ash; shoulders of the wings, above and below, and lesser coverts, olive yellow; greater wing-coverts, black, edged with pale ash; primaries, light drab; tail, the same, the feathers rather pointed at the ends, the outer ones white; breast, plain yellowish white, or pale ochre, which distinguishes it from the Savannah sparrow; (plate xxii. fig. 3.); belly and vent, white; three or four slight touches of dusky at the sides of the breast; legs, flesh colour; bill, dusky above, pale bluish white below. The male and female are nearly alike in colour.

BLUE GROSBEAK. (*Loxia cœrulea.*)

PLATE XXIV.—FIG. 6.

Linn. Syst. 304.—*Lath.* iii. 116.—*Arct. Zool.* p. 351, No. 217.—*Catesb.* i. 39. *Buff.* iii. 454, *Pl. enl.* 154.—*Peale's Museum*, No. 5826.

GUIRACA CŒRULEA.—SWAINSON.*

Fringilla cœrulea, *Bonap. Synop.* p. 114.

THIS solitary and retired species inhabits the warmer parts of America, from Guiana, and probably further south,† to Virginia. Mr Bartram also saw it during a summer's residence near Lancaster, Pennsylvania. In the United States, however, it is a scarce species; and having but few notes, is more rarely observed. Their most common note is a loud *chuck;* they have also at times a few low, sweet-toned notes. They are sometimes kept in cages, in Carolina; but seldom sing in confinement. The individual represented in the plate was a very elegant specimen, in excellent order, though just arrived from Charleston, South Carolina. During its stay with me, I fed it on Indian corn, which it seemed to prefer, easily breaking with its powerful bill the hardest grains. They also feed on hemp seed, millet, and the kernels of several kinds of berries. They are timid birds, watchful, silent, and active, and generally neat in their plumage. Having never yet met with their nest, I am unable at present to describe it.

The blue grosbeak is six inches long, and ten inches in extent; lores and frontlet, black; whole upper parts, a rich purplish blue, more dull on the back, where it is streaked with dusky; greater wing-coverts, black, edged at the tip with bay; next superior row, wholly chestnut; rest of the wing, black, skirted with blue; tail, forked, black, slightly

* Loxia cœrulea is not figured in the *Pl. enl.* That bird is a *Pitylus.*
† Latham, ii. p. 116.

edged with bluish, and sometimes minutely tipt with white; legs and feet, lead colour; bill, a dusky bluish horn colour; eye, large, full, and black.

The female is of a dark drab colour, tinged with blue, and considerably lightest below. I suspect the males are subject to a change of colour during winter. The young, as usual with many other species, do not receive the blue colour until the ensuing spring, and, till then, very much resemble the female.

Latham makes two varieties of this species; the first, wholly blue, except a black spot between the bill and eye; this bird inhabits Brazil, and is figured by Brisson, "Ornithology" iii. 321, No. 6, pl. 17, fig. 2. The other is also generally of a fine deep blue, except the quills, tail, and legs, which are black; this is Edwards' "blue grosbeak, from Angola," pl. 125; which Dr Latham suspects to have been brought from some of the Brazilian settlements, and considers both as mere varieties of the first. I am sorry I cannot at present clear up this matter, but shall take some farther notice of it hereafter.

MISSISSIPPI KITE. (*Falco Mississippiensis.*)

PLATE XXV.—FIG. 1.—MALE.

Peale's Museum, No. 403.

ICTINIA PLUMBEA.—VIEILLOT.*

L'Ictinie ophiophaga, Ictinia ophiophaga, *Vieill. Gall. des Ois.* pl. 17.—Faucon ophiophaga, 2d edit. *du Nouv. Dict. d'Hist. Nat.* ii. p. 103, female (auct. *Vieill.*)—Falco plumbeus, *Bonap. Synop.* p. 30.

THIS new species I first observed in the Mississippi territory, a few miles below Natchez, on the plantation of William

* This, from every authority, appears to be the *Falco plumbeus* of Latham. Vieillot has described it in his *Gallerie des Oiseaux*, under the title of *Ictinia ophiophaga*, descriptive of its manner of feeding; but has since restored the specific name to what it should be by the right of priority entitled. The genus, however, is retained, and appears yet con-

on from Nature by A.Wilson

Engraved by W.H.Lizars

1. Mississippi Kite 2. Tennesee Warbler. 3. Kentucky W. 4. Prairie W.

25.

Dunbar, Esq., where the bird represented in the plate was obtained, after being slightly wounded; and the drawing made with great care from the living bird. To the hospitality of the gentleman above mentioned, and his amiable family, I am indebted for the opportunity afforded me of procuring this and one or two more new species. This excellent man, whose life has been devoted to science, though at that time confined to bed by a severe and dangerous indisposition, and personally unacquainted with me, no sooner heard of my arrival at the town of Natchez, than he sent a servant and horses, with an invitation and request to come and make his house my home and head-quarters, while engaged in exploring that part of the country. The few happy days I spent there I shall never forget.

In my perambulations I frequently remarked this hawk sailing about in easy circles, and at a considerable height in the air, generally in company with the turkey buzzards, whose manner of flight it so exactly imitates as to seem the same species, only in miniature, or seen at a more immense height. Why these two birds, whose food and manners, in other respects, are so different, should so frequently associate together in air, I am at a loss to comprehend. We cannot for a moment suppose them mutually deceived by the similarity of each other's flight: the keenness of their vision forbids all suspicion of this kind. They may perhaps be engaged, at such times, in mere amusement, as they are observed to soar to great heights previous to a storm; or, what is more probable, they may both be in pursuit of their respective food. One, that he may reconnoitre a vast extent of surface below, and trace the tainted atmosphere to his favourite carrion; the

fined to America, inhabiting the southern states of the northern continent, South America, and Mexico. It will be characterised by a short bill; short, slender, scutellated, and partly feathered tarsi, and with the outer toe connected by a membrane; the claws, short; wings, very long, reaching beyond the tail; the tail, even. Bonaparte thinks that it should stand intermediate between *Falco* and *Milvus*, somewhat allied to *Buteo*. —Ed.

other in search of those large beetles, or coleopterous insects, that are known often to wing the higher regions of the air; and which, in the three individuals of this species of hawk which I examined by dissection, were the only substances found in their stomachs. For several miles, as I passed near Bayo Manchak, the trees were swarming with a kind of *cicada*, or locust, that made a deafening noise; and here I observed numbers of the hawk now before us sweeping about among the trees like swallows, evidently in pursuit of these locusts; so that insects, it would appear, are the principal food of this species. Yet when we contemplate the beak and talons of this bird, both so sharp and powerful, it is difficult to believe that they were not intended by nature for some more formidable prey than beetles, locusts, or grasshoppers; and I doubt not but mice, lizards, snakes, and small birds, furnish him with an occasional repast.

This hawk, though wounded and precipitated from a vast height, exhibited, in his distress, symptoms of great strength, and an almost unconquerable spirit. I no sooner approached to pick him up than he instantly gave battle, striking rapidly with his claws, wheeling round and round as he lay partly on his rump; and defending himself with great vigilance and dexterity; while his dark red eye sparkled with rage. Notwithstanding all my caution in seizing him to carry him home, he struck his hind claw into my hand with such force as to penetrate into the bone. Anxious to preserve his life, I endeavoured gently to disengage it; but this made him only contract it the more powerfully, causing such pain that I had no other alternative but that of cutting the sinew of his heel with my penknife. The whole time he lived with me, he seemed to watch every movement I made; erecting the feathers of his hind head, and eyeing me with savage fierceness; considering me, no doubt, as the greater savage of the two. What effect education might have had on this species under the tutorship of some of the old European professors of falconry, I know not; but if extent of wing, and energy of character, and ease and rapidity of flight, would have been any recommendations to royal patronage, this species possesses all these in a very eminent degree.

The long pointed wings and forked tail point out the affinity of this bird to that family, or subdivision of the *Falco* genus, distinguished by the name of kites, which sail without flapping the wings, and eat from their talons as they glide along.

The Mississippi kite measures fourteen inches in length, and thirty-six inches, or three feet, in extent! The head, neck, and exterior webs of the secondaries, are of a hoary white; the lower parts, a whitish ash; bill, cere, lores, and narrow line round the eye, black; back, rump, scapulars, and wing-coverts, dark blackish ash; wings, very long and pointed, the third quill the longest; the primaries are black, marked down each side of the shaft with reddish sorrel; primary coverts also slightly touched with the same; all the upper plumage at the roots is white; the scapulars are also spotted with white—but this cannot be perceived unless the feathers be blown aside; tail, slightly forked, and, as well as the rump, jet black; legs, vermilion, tinged with orange, and becoming blackish towards the toes; claws, black; iris of the eye, dark red; pupil, black.

This was a male. With the female, which is expected soon from that country, I shall, in a future volume, communicate such farther information relative to their manners and incubation, as I may be able to collect.

TENNESEE WARBLER. (*Sylvia peregrina.*)

PLATE XXV.—Fig. 2.

Peale's Museum, No. 7787.

VERMIVORA PEREGRINA.—Swainson.

Sylvia peregrina, *Bonap. Synop.* p. 87.—Sylvicola (Vermivora) peregrina, *North. Zool.* ii. p. 185.

This plain little bird has hitherto remained unknown. I first found it on the banks of Cumberland river, in the state of Tennesee, and suppose it to be rare, having since met with only two individuals of the same species. It was hunting nimbly among the young leaves, and, like all the rest of the family of worm-eaters, to which, by its bill, it evidently belongs, seemed to partake a good deal of the habits of the titmouse. Its notes were few and weak; and its stomach, on dissection, contained small green caterpillars, and a few winged insects.

As this species is so very rare in the United States, it is most probably a native of a more southerly climate, where it may be equally numerous with any of the rest of its genus. The small cerulean warbler (plate xvii. fig. 5), which, in Pennsylvania, and almost all over the Atlantic States, is extremely rare, I found the most numerous of its tribe in Tennesee and West Florida; and the Carolina wren (plate xii. fig. 5), which is also scarce to the northward of Maryland, is abundant through the whole extent of country from Pittsburgh to New Orleans.

Particular species of birds, like different nations of men, have their congenial climes and favourite countries; but wanderers are common to both; some in search of better fare, some of adventures, others led by curiosity, and many driven by storms and accident.

The Tennesee warbler is four inches and three quarters long, and eight inches in extent; the back, rump, and tail-coverts are of a rich yellow olive; lesser wing-coverts, the same; wings, deep dusky, edged broadly with yellow olive; tail, forked, olive, relieved with dusky; cheeks and upper part of the head, inclining to light bluish, and tinged with olive; line from the nostrils over the eye, pale yellow, fading into white; throat and breast, pale cream colour; belly and vent, white; legs, purplish brown; bill, pointed, and thicker at the base than those of the *Sylvia* genus generally are; upper mandible, dark dusky; lower, somewhat paler; eye, hazel.

The female differs little, in the colour of her plumage, from the male; the yellow line over the eye is more obscure, and the olive not of so rich a tint.

KENTUCKY WARBLER. (*Sylvia formosa.*)

PLATE XXV.—FIG. 3.

Peale's Museum, No. 7786.

SYLVICOLA ? FORMOSA.—JARDINE.

Sylvia formosa, *Bonap. Synop.* p. 84.—The Kentucky Warbler, Aud. pl. 38, male and female; *Orn. Biog.* i. p. 196.

THIS new and beautiful species inhabits the country whose name it bears. It is also found generally in all the intermediate tracts between Nashville and New Orleans, and below that as far as the Balize, or mouths of the Mississippi; where I heard it several times twittering among the high rank grass and low bushes of those solitary and desolate-looking morasses. In Kentucky and Tennesee it is particularly numerous, frequenting low, damp woods, and builds its nest in the middle of a thick tuft of rank grass, sometimes in the fork of a low bush, and sometimes on the ground; in all of which situations I have found it. The materials are loose dry grass, mixed with the light pith of weeds, and lined with hair. The female lays four, and sometimes six eggs, pure white, sprinkled with specks of reddish. I observed her sitting early in May. This species is seldom seen among the high branches; but loves to frequent low bushes and cane swamps, and is an active, sprightly bird. Its notes are loud, and in threes, resembling *tweedle, tweedle, tweedle.* It appears in Kentucky from the south about the middle of April; and leaves the territory of New Orleans on the approach of cold weather; at least I was assured that it does not remain there during the winter. It appeared to me to be a restless, fighting species; almost always engaged in pursuing some of its fellows; though this might have been occasioned by its numbers, and the particular season of spring, when love and jealousy rage with violence in the breasts of the feathered tenants of the grove; who experience all the ardency of those passions no less than their lord and sovereign, man.

The Kentucky warbler is five inches and a half long, and eight inches in extent; the upper parts are an olive green; line over the eye, and partly under it, and whole lower parts, rich brilliant yellow; head, slightly crested, the crown, deep black, towards the hind part spotted with light ash; lores and spot curving down the neck, also black; tail, nearly *even* at the end, and of a rich olive green; interior vanes of that and the wings, dusky; legs, an almost transparent pale flesh colour.

The female wants the black under the eye, and the greater part of that on the crown, having those parts yellowish. This bird is very abundant in the moist woods along the Tennesee and Cumberland rivers.

PRAIRIE WARBLER. (*Sylvia minuta.*)

PLATE XXV.—Fig. 4.

Peale's Museum, No. 7784.

SYLVICOLA ? DISCOLOR.—Jardine.*

Sylvia discolor, *Vieill*, pl. 98. (auct. *Bonap.*)—*Bonap. Synop.* p. 82.

This pretty little species I first discovered in that singular tract of country in Kentucky, commonly called the Barrens. I shot several afterwards in the open woods of the Chactaw nation, where they were more numerous. They seem to prefer these open plains, and thinly wooded tracts; and have this singularity in their manners, that they are not easily alarmed; and search among the leaves the most leisurely of any of the tribe I have yet met with; seeming to examine every blade of grass and every leaf; uttering at short intervals a feeble *chirr*. I have observed one of these birds to sit on the lower branch of a tree for half an hour at a time, and allow me to come up nearly to the foot of the tree, without seeming to be in the least disturbed, or to discontinue the regularity of its occasional note. In activity it is the reverse of the preceding species; and is rather a scarce bird in the countries where I found it. Its food consists principally of small caterpillars and winged insects.

The prairie warbler is four inches and a half long, and six inches and a half in extent; the upper parts are olive, spotted on the back with reddish chestnut; from the nostril over and under the eye, yellow; lores, black; a broad streak of black also passes beneath the yellow under the eye; small pointed spots of black reach from a little below that along the side of the neck and under the wings; throat, breast, and belly, rich yellow; vent, cream coloured, tinged with yellow;

* Bonaparte is of opinion that this is the same with Vieillot's *Sylvia discolor*. I have not had an opportunity of examining it.—Ed.

wings, dark dusky olive; primaries and greater coverts, edged and tipt with pale yellow; second row of coverts, wholly yellow; lesser, olive; tail, deep brownish black, lighter on the edges; the three exterior feathers, broadly spotted with white.

The female is destitute of the black mark under the eye; has a few slight touches of blackish along the sides of the neck; and some faint shades of brownish red on the back.

The nest of this species is of very neat and delicate workmanship, being pensile, and generally hung on the fork of a low bush or thicket; it is formed outwardly of green moss, intermixed with rotten bits of wood and caterpillars' silk; the inside is lined with extremely fine fibres of grape-vine bark; and the whole would scarcely weigh a quarter of an ounce. The eggs are white, with a few brown spots at the great end. These birds are migratory, departing for the south in October.

CAROLINA PARROT. (*Psittacus Carolinensis.*)

PLATE XXVI.—FIG. 1.

Linn. Syst. 141.—*Catesb.* i. 11.—*Lath.* i. 227.—*Arct. Zool.* 242. No. 132. *Ibid.* 133.—*Peale's Museum*, No. 762.

CONURUS CAROLINENSIS.—KUHL.*

Conurus Carolinensis, *Kuhl. consp. psitt. Nov. act. Ceas. Leop.* tom. x. p. 4. 23.—Psittacus Carolinensis, *Bonap. Synop.* p. 41.

OF one hundred and sixty-eight kinds of parrots enumerated by European writers as inhabiting the various regions of the

* In all countries parrots have been favourites, arising from their playful and docile manners in domestication, the beauty of their plumage, and the nearly solitary example of imitating with comparative accuracy the voice and articulation of man. In ancient times, the extravagance with which these birds were sought after, either as objects of amusement and recreation, or as luxuries for the table, surpasses, if

Drawn from Nature by A. Wilson. *Engraved by W.H. Lizars.*

1 Carolina Parrot. 2 Canada Flycatcher. 3 Hooded F. 4 Green black-capt F.

26.

globe, this is the only species found native within the territory of the United States. The vast and luxuriant tracts lying within the torrid zone, seem to be the favourite residence of those noisy, numerous, and richly plumaged tribes. The Count de Buffon has, indeed, circumscribed the whole genus of parrots to a space not extending more than twenty-three degrees on each side of the equator: but later discoveries have shewn this statement to be incorrect, as these birds have been found on our continent as far south as the Straits of Magellan, and even on the remote shores of Van Diemen's Land, in Terra Australasia. The species now under consideration is also known to inhabit the interior of Louisiana, and the shores of the Mississippi and Ohio, and their tributary waters, even beyond the Illinois river, to the neighbourhood of Lake Michigan, in lat. 42 deg. north; and, contrary to the generally received opinion, is chiefly *resident* in all these places. Eastward, however, of the great range of the Alleghany, it is seldom seen farther north than the state of Maryland; though straggling parties have been occasionally observed among

possible, the many fashionable maniæ of latter days. We find frequent allusions to these birds both in the prose and poetical writers, railing against the expenses of price and maintenance, or celebrating their docility or their love and gratitude to their mistress; and at the height and splendour of the then Mistress of the World, they were brought forward to the less honourable avocation of conveying praise and flattery to the great. At the present period they are much sought after, and a "good parrot" will still bring a high price.

Intertropical countries are the natural abodes of the *Psittacidæ*, where they are gregarious, and present most conspicuous and noisy attraction, revelling in free or grotesque attitudes, among the forest and mountain glades, which, without these, and many other brilliant tenants, would present only a solitude of luxuriant vegetation. It is impossible for any one who has only seen these birds in a cage or small enclosure, to conceive what must be the gorgeous appearance of a flock, either in full flight, and performing their various evolutions, under a vertical sun, or sporting among the superb foliage of a tropical forest:

> In gaudy robes of many-colour'd patches
> The parrots swung like blossoms from the trees,
> While their harsh voices undeceived the ear.—ED.

the valleys of the Juniata; and, according to some, even twenty-five miles to the north-west of Albany, in the state of New York.* But such accidental visits furnish no certain criterion by which to judge of their usual extent of range,—those aerial voyagers, as well as others who navigate the deep, being subject to be cast away, by the violence of the elements, on distant shores and unknown countries.

From these circumstances of the northern residence of this species, we might be justified in concluding it to be a very hardy bird, more capable of sustaining cold than nine-tenths of its tribe; and so I believe it is,—having myself seen them, in the month of February, along the banks of the Ohio, in a snow storm, flying about like pigeons, and in full cry.

The preference, however, which this bird gives to the western countries, lying in the same parallel of latitude with those eastward of the Alleghany mountains, which it rarely or never visits, is worthy of remark; and has been adduced, by different writers, as a proof of the superior mildness of climate in the former to that of the latter. But there are other reasons for this partiality equally powerful, though hitherto overlooked; namely, certain peculiar features of country to which these birds are particularly and strongly attached: these are, low, rich, alluvial bottoms, along the borders of creeks, covered with a gigantic growth of sycamore trees, or button-wood; deep, and almost impenetrable swamps, where the vast and towering cypress lifts its still more majestic head; and those singular salines, or, as they are usually called, *licks*, so generally interspersed over that country, and which are regularly and eagerly visited by the paroquets. A still greater inducement is the superior abundance of their favourite fruits. That food which the paroquet prefers to all others is the seeds of the cockle bur, a plant rarely found in the lower parts of Pennsylvania or New York; but which unfortunately grows in too great abundance along the shores of the Ohio

* Barton's Fragments, &c. p. 6. Introduction.

and Mississippi; so much so as to render the wool of those sheep that pasture where it most abounds scarcely worth the cleaning, covering them with one solid mass of burs, wrought up and imbedded into the fleece, to the great annoyance of this valuable animal. The seeds of the cypress tree and hackberry, as well as beech nuts, are also great favourites with these birds; the two former of which are not commonly found in Pennsylvania, and the latter by no means so general or so productive. Here, then, are several powerful reasons, more dependent on soil than climate, for the preference given by these birds to the luxuriant regions of the west. Pennsylvania, indeed, and also Maryland, abound with excellent apple orchards, on the ripe fruit of which the paroquets occasionally feed. But I have my doubts whether their depredations in the orchard be not as much the result of wanton play and mischief, as regard for the seeds of the fruit, which they are supposed to be in pursuit of. I have known a flock of these birds alight on an apple tree, and have myself seen them twist off the fruit, one by one, strewing it in every direction around the tree, without observing that any of the depredators descended to pick them up. To a paroquet, which I wounded and kept for some considerable time, I very often offered apples, which it uniformly rejected; but burs or beech nuts, never. To another very beautiful one, which I brought from New Orleans, and which is now sitting in the room beside me, I have frequently offered this fruit, and also the seeds separately, which I never knew it to taste. Their local attachments, also, prove that food, more than climate, determines their choice of country. For even in the states of Ohio, Kentucky, and the Mississippi territory, unless in the neighbourhood of such places as have been described, it is rare to see them. The inhabitants of Lexington, as many of them assured me, scarcely ever observe them in that quarter. In passing from that place to Nashville, a distance of two hundred miles, I neither heard nor saw any, but at a place called Madison's Lick. In passing on, I next met with them on the banks and rich flats of the Tennesee river: after this I saw no more till I reached Bayo St Pierre, a distance of several hundred miles: from all which circumstances, I think we cannot, from the residence of these birds, establish with propriety any correct standard by which to judge of the comparative temperatures of different climates.

In ascending the river Ohio, by myself, in the month of February, I met with the first flock of paroquets, at the mouth of the Little Scioto. I had been informed, by an old and respectable inhabitant of Marietta, that they were sometimes, though rarely, seen there. I observed flocks of them, afterwards, at the mouth of the Great and Little Miami, and in the neighbourhood of numerous creeks that discharge themselves into the Ohio. At Big Bone Lick, thirty miles above the mouth of Kentucky river, I saw them in great numbers. They came screaming through the woods in the morning, about an hour after sunrise, to drink the salt water, of which they, as well as the pigeons, are remarkably fond. When they alighted on the ground, it appeared at a distance as if covered with a carpet of the richest green, orange, and yellow: they afterwards settled, in one body, on a neighbouring tree, which stood detached from any other, covering almost every twig of it, and the sun, shining strongly on their gay and glossy plumage, produced a very beautiful and splendid appearance. Here I had an opportunity of observing some very particular traits of their character: Having shot down a number, some of which were only wounded, the whole flock swept repeatedly around their prostrate companions, and again settled on a low tree, within twenty yards of the spot where I stood. At each successive discharge, though showers of them fell, yet the affection of the survivors seemed rather to increase; for after a few circuits around the place, they again alighted near me, looking down on their slaughtered companions with such manifest symptoms of sympathy and concern, as entirely disarmed me. I could not but take notice of the remarkable contrast between their elegant manner of flight, and their lame and crawling gait among the brancnes. They fly very much like the wild pigeon, in close compact bodies, and with great rapidity, making a loud and outrageous screaming, not unlike that of the red-headed woodpecker. Their flight is sometimes in a direct line; but most usually circuitous, making a great variety of elegant and easy serpentine meanders, as if for pleasure. They are particularly attached to the large sycamores, in the hollow of the trunks and branches of which they generally roost, thirty or forty, sometimes more, entering at the same hole. Here they cling close to the sides of the tree, holding fast by the claws and also by the bills. They appear to be fond of sleep, and often retire to their holes during the day, probably to take their regular *siesta.* They are extremely sociable, and fond of each other, often scratching each other's heads and necks, and always, at night, nestling as close as possible to each other, preferring, at that time, a perpendicular position, supported by their bill and claws. In the fall, when their favourite cockle burs are ripe, they swarm along the coast, or high grounds of the Mississippi, above New Orleans, for a great extent. At such times, they are killed and eaten by many of the inhabitants; though, I confess, I think their flesh very indifferent. I have several times dined on it from necessity, in the woods, but found it merely possible, with all the sauce of a keen appetite, to recommend it.

A very general opinion prevails, that the brains and intestines of the Carolina paroquet are a sure and fatal poison to cats. I had determined, when at Big Bone, to put this to the test of experiment; and for that purpose collected the brains and bowels of more than a dozen of them. But after close search, Mistress Puss was not to be found, being engaged, perhaps, on more agreeable business. I left the medicine with Mr Colquhoun's agent, to administer it at the first opportunity, and write me the result; but I have never yet heard from him. A respectable lady near the town of Natchez, and on whose word I can rely, assured me, that she herself had made the experiment, and that, whatever might be the cause, the cat had actually died either on that or the succeeding day. A French planter near Bayo Fourche pretended to account to me for this effect by positively asserting, that the seeds of the cockle burs on which the paroquets so eagerly feed, were deleterious to cats; and thus their death was produced by eating the intestines of the bird. These matters might easily have been ascertained on the spot, which, however, a combination of trifling circumstances prevented me from doing. I several times carried a dose of the first description in my pocket till it became insufferable, without meeting with a suitable *patient*, on whom, like other professional gentlemen, I might conveniently make a fair experiment.

I was equally unsuccessful in my endeavours to discover the time of incubation or manner of building among these birds. All agreed that they breed in hollow trees; and several affirmed to me that they had seen their nests. Some said they carried in no materials; others, that they did. Some made the eggs white; others, speckled. One man assured me that he cut down a large beech tree, which was hollow, and in which he found the broken fragments of upwards of twenty paroquets' eggs, which were of a greenish yellow colour. The nests, though destroyed in their texture by the falling of the tree, appeared, he said, to be formed of small twigs glued to each other, and to the side of the tree, in the manner of the chimney swallow. He added, that if it were the proper season, he could point out to me the weed from which they procured the gluey matter. From all these contradictory accounts nothing certain can be deduced, except that they build in companies, in hollow trees. That they commence incubation late in summer, or very early in spring, I think highly probable, from the numerous dissections I made in the months of March, April, May, and June; and the great variety which I found in the colour of the plumage of the head and neck of both sexes, during the two former of these months, convinces me, that the young birds do not receive their full colours until the early part of the succeeding summer.*

While parrots and paroquets, from foreign countries, abound in almost every street of our large cities, and become such great favourites, no attention seems to have been paid to our own, which in elegance of figure, and beauty of plumage, is certainly superior to many of them. It wants, indeed, that disposition for perpetual screaming and chattering that renders some of the former pests, not only to their keepers, but to the whole neighbourhood in which they reside. It is alike docile and sociable; soon becomes perfectly familiar; and, until equal pains be taken in its instruction, it is unfair to conclude it incapable of equal improvement in the language of man.

As so little has hitherto been known of the disposition and manners of this species, the reader will not, I hope, be displeased at my detailing some of these, in the history of a particular favourite, my sole companion in many a lonesome day's march, and of which the figure in the plate is a faithful resemblance.

Anxious to try the effects of education on one of those which I procured at Big Bone Lick, and which was but slightly wounded in the wing, I fixed up a place for it in the stern of my boat, and presented it with some cockle burs, which it freely fed on in less than an hour after being on

* Mr Audubon's information on their manner of breeding is as follows :—"Their nest, or the place in which they deposit their eggs, is simply the bottom of such cavities in trees as those to which they usually retire at night. Many females deposit their eggs together. I am of opinion that the number of eggs which each individual lays is two, although I have not been able absolutely to assure myself of this. They are nearly round, of a rich greenish white. The young are at first covered with soft down, such as is seen on young owls."

It may be remarked, that most of the parrots, whose nidification we are acquainted with, build in hollow trees, or holed banks. Few make a nest for themselves, but lay the eggs on the bare wood or earth; and when the nest is built outward, as by other birds, it is of a slight and loose structure. The eggs are always white.—Ed.

board. The intermediate time between eating and sleeping was occupied in gnawing the sticks that formed its place of confinement, in order to make a practicable breach; which it repeatedly effected. When I abandoned the river, and travelled by land, I wrapt it up closely in a silk handkerchief, tying it tightly around, and carried it in my pocket. When I stopped for refreshment, I unbound my prisoner, and gave it its allowance, which it generally despatched with great dexterity, unhusking the seeds from the bur in a twinkling; in doing which it always employed its left foot to hold the bur, as did several others that I kept for some time. I began to think that this might be peculiar to the whole tribe, and that the whole were, if I may use the expression, left-footed; but, by shooting a number afterwards while engaged in eating mulberries, I found sometimes the left, sometimes the right foot stained with the fruit; the other always clean; from which, and the constant practice of those I kept, it appears, that, like the human species in the use of their hands, they do not prefer one or the other indiscriminately, but are either left or right-footed. But to return to my prisoner: In recommitting it to "durance vile," we generally had a quarrel; during which it frequently paid me in kind for the wound I had inflicted, and for depriving it of liberty, by cutting and almost disabling several of my fingers with its sharp and powerful bill. The path through the wilderness between Nashville and Natchez is in some places bad beyond description. There are dangerous creeks to swim, miles of morass to struggle through, rendered almost as gloomy as night by a prodigious growth of timber, and an underwood of canes and evergreens; while the descent into these sluggish streams is often ten or fifteen feet perpendicular, into a bed of deep clay. In some of the worst of these places, where I had, as it were, to fight my way through, the paroquet frequently escaped from my pocket, obliging me to dismount and pursue it through the worst of the morass before I could regain it. On these occasions, I was several times tempted to abandon it; but I persisted in bringing it along. When at night I encamped in the woods I placed it on the baggage beside me, where it usually sat with great composure, dozing, and gazing at the fire till morning. In this manner I carried it upwards of a thousand miles in my pocket, where it was exposed all day to the jolting of the horse, but regularly liberated at meal times and in the evening, at which it always expressed great satisfaction. In passing through the Chickasaw and Chactaw nations, the Indians wherever I stopped to feed, collected around me, men, women, and children, laughing, and seeming wonderfully amused with the novelty of my companion. The Chickasaws called it in their language "*Kelinky;*" but when they heard me call it Poll, they soon repeated the name; and, wherever I chanced to stop among these people, we soon became familiar with each other through the medium of Poll. On arriving at Mr Dunbar's, below Natchez, I procured a cage, and placed it under the piazza, where by its call it soon attracted the passing flocks; such is the attachment they have for each other. Numerous parties frequently alighted on the trees immediately above, keeping up a constant conversation with the prisoner. One of these I wounded slightly in the wing, and the pleasure Poll expressed on meeting with this new companion was really amusing. She crept close up to it as it hung on the side of the cage; chattered to it in a low tone of voice, as if sympathising in its misfortune; scratched about its head and neck with her bill; and both at night nestled as close as possible to each other, sometimes Poll's head being thrust among the plumage of the other. On the death of this companion, she appeared restless and inconsolable for several days. On reaching New Orleans, I placed a looking glass beside the place where she usually sat, and the instant she perceived her image, all her former fondness seemed to return, so that she could scarcely absent herself from it a moment. It was evident that she was completely deceived. Always when evening drew on, and often during the day, she laid her head close to that of the image in the glass, and began to doze with great composure and satisfaction. In this short space she had learnt to know her name; to answer, and come when called on; to climb up my clothes, sit on my shoulder, and eat from my mouth. I took her with me to sea, determined to persevere in her education; but, destined to another fate, poor Poll, having one morning, about daybreak, wrought her way through the cage while I was asleep, instantly flew overboard and perished in the Gulf of Mexico.

The Carolina, or Illinois parrot (for it has been described under both these appellations), is thirteen inches long, and twenty-one in extent; forehead and cheeks, orange red; beyond this, for an inch and a half, down and round the neck, a rich and pure yellow; shoulder and bend of the wing, also edged with rich orange red. The general colour of the rest of the plumage is a bright yellowish silky green, with light blue reflections, lightest and most diluted with yellow below; greater wing-coverts and roots of the primaries yellow, slightly tinged with green; interior webs of the primaries, deep dusky purple, almost black, exterior ones bluish green; tail, long, cuneiform, consisting of twelve feathers, the exterior one only half the length, the others increasing to the middle ones which are streaked along the middle with light blue; shafts of all the larger feathers, and of most part of the green plumage, black; knees and vent, orange yellow; feet, a pale whitish flesh colour; claws, black; bill, white, or slightly tinged with pale cream; iris of the eye hazel; round the eye is a small space without feathers, covered with a whitish skin; nostrils placed in an elevated membrane at the base of the bill, and covered with feathers; chin, wholly bare of feathers, but concealed by those descending on each side; from each side of the palate hangs a lobe or skin of a blackish colour; tongue, thick and fleshy; inside of the upper mandible near the point, grooved exactly like a file, that it may hold with more security.

The female differs very little in her colours and markings

from the male. After examining numerous specimens, the following appear to be the principal differences: The yellow on the neck of the female does not descend quite so far; the interior vanes of the primaries are brownish, instead of black, and the orange red on the bend and edges of the wing, is considerably narrower; in other respects the colours and markings are nearly the same.

The young birds of the preceding year, of both sexes, are generally destitute of the yellow on the head and neck, until about the beginning or middle of March, having those parts wholly green, except the front and cheeks, which are orange red in them as in the full-grown birds. Towards the middle of March the yellow begins to appear, in detached feathers, interspersed among the green, varying in different individuals. In some which I killed about the last of that month, only a few green feathers remained among the yellow; and these were fast assuming the yellow tint: for the colour changes without change of plumage. A number of these birds, in all their grades of progressive change from green to yellow, have been deposited in Mr Peale's museum.

What is called by Europeans the Illinois parrot (*Psittacus pertinax*), is evidently the young bird in its imperfect colours. Whether the present species be found as far south as Brazil, as these writers pretend, I am unable to say; but from the great extent of country in which I have myself killed and examined these birds, I am satisfied that the present species, now described, is the only one inhabiting the United States.

Since the foregoing was written, I have had an opportunity, by the death of a tame Carolina paroquet, to ascertain the fact of the poisonous effects of their head and intestines on cats. Having shut up a cat and her two kittens (the latter only a few days old), in a room with the head, neck, and whole intestines of the paroquet, I found, on the next morning, the whole eaten except a small part of the bill. The cat exhibited no symptom of sickness; and, at this moment, three days after the experiment has been made, she and her kittens are in their usual health. Still, however, the effect might have been different, had the daily food of the bird been cockle burs, instead of Indian corn.

CANADA FLYCATCHER. (*Muscicapa Canadensis.*)

PLATE XXVI.—Fig. 2.

Linn. Syst. 324.—*Arct. Zool.* p. 338, No. 273.—*Lath.* ii. 354.
Peale's Museum, No. 6969.

SETOPHAGA CANADENSIS.—Swainson.*

Sylvia pardalina, *Bonap. Synop.* p. 79.

This is a solitary, and, in the lower parts of Pennsylvania, rather a rare species; being more numerous in the interior, particularly near the mountains, where the only two I ever met with were shot. They are silent birds, as far as I could observe; and were busily darting among the branches after insects. From the specific name given them, it is probable that they are more plenty in Canada than in the United

* Mr Swainson in a note to the "Northern Zoology," has hinted his suspicion that this bird and *Muscicapa Bonapartii* of Audubon are the same: as far as we can judge from the two plates, there does not seem any resemblance. Mr Swainson adds, "As regards the generic *name* (of *Setophaga Bonapartii*), we consider the whole structure of the bird as obviously intermediate between the *Sylvicolæ* and the typical *Setophagæ*, but more closely allied to the latter than the former." For the present, we shall place the two following species in *Setophaga*, but suspect that this intermediate form will hereafter rank in the value of a sub-genus.[1] To this also may be referred the *Muscicapa Selbii* of Audubon, which seems to approach nearer *Setophaga* in the more flattened representation of the bill, and stronger bristles. Mr Audubon has only met with it three times in Louisiana. The upper parts are of a dark olive colour; the whole under parts, with a streak over each eye, rich yellow. The length is about five inches and a half; it was very active in pursuit of flies, and the snapping of the bill, when seizing them, was distinctly heard at some distance.—Ed.

[1] **They are all furnished with rictorial bristles; but the bill is not so much depressed. The habits are those of *Setophaga*.**

States; where it is doubtful whether they be not mere passengers in spring and autumn.

This species is four inches and a half long, and eight in extent; front, black; crown, dappled with small streaks of grey and spots of black; line from the nostril to and around the eye, yellow; below the eye, a streak or spot of black, descending along the sides of the throat, which, as well as the breast and belly, is brilliant yellow, the breast being marked with a broad rounding band of black, composed of large irregular streaks; back, wings, and tail, cinereous brown; vent, white; upper mandible, dusky; lower, flesh coloured; legs and feet, the same; eye, hazel.

Never having met with the female of this bird, I am unable, at present, to say in what its colours differ from those of the male.

HOODED FLYCATCHER. (*Muscicapa cucullata.*)

PLATE XXVI.—FIG. 3.

Le gobe-mouche citrin, *Buff.* iv. 538, *Pl. enl.* 666.—Hooded Warbler, *Arct. Zool.* p. 400, No. 287.—*Lath.* ii. 462.—*Catesb.* i. 60.—Mitred Warbler, *Turton*, i. 601.—Hooded Warbler, *Ibid.*—*Peale's Museum*, No. 7062.

SETOPHAGA MITRATA.—SWAINSON.

Sylvia mitrata, *Bonap. Synop.* p. 79.

WHY those two judicious naturalists, Pennant and Latham, should have arranged this bird with the warblers, is to me unaccountable, as few of the *muscicapæ* are more distinctly marked than the species now before us. The bill is broad at the base, where it is beset with bristles; the upper mandible, notched, and slightly overhanging at the tip; and the manners of the bird, in every respect, those of a flycatcher. This species is seldom seen in Pennsylvania and the northern states; but through the whole extent of country south of Maryland, from the Atlantic to the Mississippi, is very abundant. It is, however, most partial to low situations, where there is plenty of thick underwood; abounds among the canes in the state of Tennessee, and in the Mississippi territory; and seems perpetually in pursuit of winged insects; now and then uttering three loud, not unmusical, and very lively notes, resembling *twee, twee, twitchie,* while engaged in the chase. Like almost all its tribe, it is full of spirit, and exceedingly active. It builds a very neat and compact nest, generally in the fork of a small bush, forms it outwardly of moss and flax, or broken hemp, and lines it with hair, and sometimes feathers; the eggs are five, of a greyish white, with red spots towards the great end. In all parts of the United States, where it inhabits, it is a bird of passage. At Savannah I met with it about the 20th of March; so that it probably retires to the West India islands, and perhaps Mexico, during winter. I also heard this bird among the rank reeds and rushes within a few miles of the mouth of the Mississippi. It has been sometimes seen in the neighbourhood of Philadelphia, but rarely; and, on such occasions, has all the mute timidity of a stranger at a distance from home.

This species is five inches and a half long, and eight in extent; forehead, cheeks, and chin, yellow, surrounded with a hood of black, that covers the crown, hind head, and part of the neck, and descends, rounding over the breast; all the rest of the lower parts are rich yellow; upper parts of the wings, the tail, and back, yellow olive; interior vanes, and tips of the wing and tail, dusky; bill, black; legs, flesh coloured; inner webs of the three exterior tail feathers, white for half their length from the tips; the next, slightly touched with white; the tail slightly forked, and exteriorly edged with rich yellow olive.

The female has the throat and breast yellow, slightly tinged with blackish; the black does not reach so far down the upper part of the neck, and is not of so deep a tint. In the other parts of her plumage she exactly resembles the male. I have found some females that had little or no black on the head or neck above; but these I took to be young birds, not yet arrived at their full tints.

GREEN BLACK-CAPT FLYCATCHER. (*Muscicapa pusilla.*)

PLATE XXVI.—Fig. 4.

Peale's Museum, No. 7785.

SETOPHAGA? WILSONII.—Jardine.*

Sylvia Wilsonii, *Bonap. Synop.* p. 86.—*Nomenclature*, No. 127.

This neat and active little species I have never met with in the works of any European naturalist. It is an inhabitant of the swamps of the southern states, and has been several times seen in the lower parts of the states of New Jersey and Delaware. Amidst almost unapproachable thickets of deep morasses it commonly spends its time during summer, and has a sharp squeaking note, nowise musical. It leaves the southern states early in October.

This species is four inches and a half long, and six and a half in extent; front line over the eye, and whole lower parts, yellow, brightest over the eye, and dullest on the cheeks, belly, and vent, where it is tinged with olive; upper parts, olive green; wings and tail, dusky brown, the former very short; legs and bill, flesh coloured; crown, covered with a patch of deep black; iris of the eye, hazel.

The female is without the black crown, having that part of a dull yellow olive, and is frequently mistaken for a distinct species. From her great resemblance, however, in other respects, to the male, now first figured, she cannot hereafter be mistaken.

* The Prince of Musignano has never seen this species, but was of opinion that it would prove a *Sylvia;* and the specific name being preoccupied, he chose that of its discoverer. I have retained his specific name, though the reason of the change will not now be available. The services of Wilson, however, can scarcely be overpaid, and the reputation of no one is here implicated.—Ed.

PINNATED GROUSE. (*Tetrao cupido.*)

PLATE XXVII.—FIG. 1.

Linn. Syst. i. p. 274-5.—*Lath.* ii. p. 740.—*Arct. Zool.*—La Gelinote huppée d'Amérique, *Briss. Orn.* i. p. 212, 10.—Urogalus minor, fuscus cervice, plumis alas imitantibus donatà, *Catesb. Car. App.* pl. 1.—Tetrao lagogus, the Mountain Cock, or Grouse, *Bartram*, p. 290.—Heath Hen, Prairie Hen, Barren Hen.—*Peale's Museum*, No. 4700, male; 4701, female.

TETRAO CUPIDO.—LINNÆUS.

Attagan Americana, *Brisson*, i. p. 59.—Pinnated Heathcock, Bonasa cupido, *Steph. Sh. Cont.* xi. p. 299.—Tetrao cupido, *Bonap. Synop.* p. 126.

BEFORE I enter on a detail of the observations which I have myself personally made on this singular species, I shall lay before the reader a comprehensive and very circumstantial memoir on the subject, communicated to me by the writer, Dr Samuel L. Mitchill, of New York, whose exertions, both in his public and private capacity, in behalf of science, and in elucidating the natural history of his country, are well known, and highly honourable to his distinguished situation and abilities. That peculiar tract, generally known by the name of the Brushy Plains of Long Island, having been, for time immemorial, the resort of the bird now before us, some account of this particular range of country seemed necessarily connected with the subject, and has, accordingly, been obligingly attended to by the learned professor.

"NEW YORK, *Sept.* 19, 1810.

"DEAR SIR,—It gives me much pleasure to reply to your letter of the 12th instant, asking of me information concerning the grouse of Long Island.

"The birds which are known there emphatically by the name of grouse, inhabit chiefly the forest range. This district of the island may be estimated as being between forty and fifty miles in length, extending from Bethphage, in Queen's County, to the neighbourhood of the court-house, in Suffolk. Its

Drawn from Nature by A. Wilson

Engraved by W. H. Lizars

1. Pinnated Grous. 2. Blue green Warbler. 3. Nashville W.

27

breadth is not more than six or seven. For, although the island is bounded by the Sound, separating it from Connecticut on the north, and by the Atlantic Ocean on the south, there is a margin of several miles, on each side, in the actual possession of human beings.

"The region in which these birds reside, lies mostly within the towns of Oysterbay, Huntington, Islip, Smithtown, and Brookhaven; though it would be incorrect to say, that they were not to be met with sometimes in Riverhead and Southampton. Their territory has been defined by some sportsmen, as situated between Hampstead Plain on the west, and Shinnecock Plain on the east.

"The more popular name for them is heath-hens. By this they are designated in the act of our legislature for the preservation of them and of other game. I well remember the passing of this law. The bill was introduced by Cornelius J. Bogert, Esq., a member of the Assembly from the city of New York. It was in the month of February 1791, the year when, as a representative from my native county of Queens, I sat for the first time, in a legislature.

"The statute declares, among other things, that the person who shall kill any heath-hen within the counties of Suffolk or Queens, between the 1st day of April and the 5th day of October, shall, for every such offence, forfeit and pay the sum of two dollars and a half, to be recovered, with costs of suit, by any person who shall prosecute for the same, before any justice of the peace, in either of the said counties; the one half to be paid to the plaintiff, and the other half to the overseers of the poor; and if any heath-hen, so killed, shall be found in the possession of any person, he shall be deemed guilty of the offence, and suffer the penalty. But it is provided, that no defendant shall be convicted, unless the action shall be brought within three months after the violation of the law.*

* The doctor has probably forgotten a circumstance of rather a ludicrous kind, that occurred at the passing of this law, and which was,

"The country selected by these exquisite birds, requires a more particular description. You already understand it to be the midland and interior district of the island. The soil of this island is, generally speaking, a sandy or gravelly loam. In the parts less adapted to tillage, it is more of an unmixed sand. This is so much the case, that the shore of the beaches beaten by the ocean affords a material from which glass has been prepared. Silicious grains and particles predominate in the region chosen by the heath-hens or grouse. Here there are no rocks, and very few stones of any kind. This sandy tract appears to be a dereliction of the ocean, but is, nevertheless, not doomed to total sterility. Many thousand acres have been reclaimed from the wild state, and rendered very productive to man ; and within the towns frequented by these birds, there are numerous inhabitants, and among them, some of our most wealthy farmers.

"But within the same limits, there are also tracts of great extent where men have no settlements, and others where the population is spare and scanty. These are, however, by no means, naked deserts: they are, on the contrary, covered with trees, shrubs, and smaller plants. The trees are mostly pitch-pines of inferior size, and white oaks of a small growth. They are of a quality very fit for burning. Thousands of cords of both sorts of firewood are annually exported from these barrens. Vast quantities are occasionally destroyed by the fires which, through carelessness or accident, spread far and wide through the woods. The city of New York will probably, for ages, derive fuel from the grouse grounds. The land, after having been cleared, yields to the cultivator poor crops. Unless, therefore, he can help it by manure, the best

not long ago, related to me by my friend Mr Gardiner, of Gardiner's Island, Long Island. The bill was entitled, "An Act for the preservation of Heath-hen and other game." The honest chairman of the Assembly—no sportsman, I suppose—read the title, "An Act for the preservation of *Heathen*, and other game !" which seemed to astonish the northern members, who could not see the propriety of preserving *Indians*, or any other heathen.

disposition is to let it grow up to forest again. Experience has proved, that, in a term of forty or fifty years, the new growth of timber will be fit for the axe. Hence it may be perceived, that the reproduction of trees, and the protection they afford to heath-hens, would be perpetual, or, in other words, not circumscribed by any calculable time, provided the persecutors of the latter would be quiet.

"Beneath these trees grow more dwarfish oaks, overspreading the surface, sometimes with here and there a shrub, and sometimes a thicket. These latter are from about two to ten feet in height. Where they are the principal product, they are called, in common conversation, *brush,* as the flats on which they grow are termed *brushy plains.* Among this hardy shrubbery may frequently be seen the creeping vegetable named the partridgeberry, covering the sand with its lasting verdure. In many spots, the plant which produces hurtleberries sprouts up among the other natives of the soil. These are the more important; though I ought to inform you, that the hills reaching from east to west, and forming the spine of the island, support kalmias, hickories, and many other species; that I have seen azalias and andromedas, as I passed through the wilderness; and that, where there is water, craneberries, alders, beeches, maples, and other lovers of moisture, take their stations.

"This region, situated thus between the more thickly inhabited strips, or belts, on the north and south sides of the island, is much travelled by waggons, and intersected, accordingly, by a great number of paths.

"As to the birds themselves, the information I possess scarcely amounts to an entire history. You, who know the difficulty of collecting facts, will be the most ready to excuse my deficiencies. The information I give you is such as I rely on. For the purpose of gathering the materials, I have repeatedly visited their haunts. I have likewise conversed with several men who were brought up at the precincts of the grouse ground, who had been witnesses of their habits and manners, who were accustomed to shoot them for the market, and who have acted as guides to gentlemen who go there for sport.

"*Bulk.*—An adult grouse, when fat, weighs as much as a barn-door fowl of moderate size, or about three pounds avoirdupois. But the eagerness of the sportsmen is so great, that a large proportion of those they kill are but a few months old, and have not attained their complete growth. Notwithstanding the protection of the law, it is very common to disregard it. The retired nature of the situation favours this. It is well understood that an arrangement can be made which will blind and silence informers, and that the gun is fired with impunity for weeks before the time prescribed in the act. To prevent this unfair and unlawful practice, an association was formed a few years ago, under the title of the *Brush Club*, with the express and avowed intention of enforcing the game law. Little benefit, however, has resulted from its laudable exertions; and under a conviction that it was impossible to keep the poachers away, the society declined. At present the statute may be considered as operating very little towards their preservation. Grouse, especially full-grown ones, are becoming less frequent. Their numbers are gradually diminishing; and, assailed as they are on all sides, almost without cessation, their scarcity may be viewed as foreboding their eventual extermination.

"*Price.*—Twenty years ago, a brace of grouse could be bought for a dollar. They now cost from three to five dollars. A handsome pair seldom sells in the New York market now-a-days for less than thirty shillings [three dollars, seventy-five cents], nor for more than forty [five dollars]. These prices indicate, indeed, the depreciation of money and the luxury of eating. They prove, at the same time, that grouse are become rare; and this fact is admitted by every man who seeks them, whether for pleasure or for profit.

"*Amours.*—The season for pairing is in March, and the breeding time is continued through April and May. Then the male grouse distinguishes himself by a peculiar sound.

When he utters it, the parts about the throat are sensibly inflated and swelled. It may be heard on a still morning for three or more miles; some say they have perceived it as far as five or six. This noise is a sort of ventriloquism. It does not strike the ear of a bystander with much force, but impresses him with the idea, though produced within a few rods of him, of a voice a mile or two distant. This note is highly characteristic. Though very peculiar, it is termed *tooting*, from its resemblance to the blowing of a conch or horn from a remote quarter. The female makes the nest on the ground, in recesses very rarely discovered by men. She usually lays from ten to twelve eggs. Their colour is of a brownish, much resembling those of a guinea hen. When hatched, the brood is protected by her alone. Surrounded by her young, the mother bird exceedingly resembles a domestic hen and chickens. She frequently leads them to feed in the roads crossing the woods, on the remains of maize and oats contained in the dung dropped by the travelling horses. In that employment they are often surprised by the passengers. On such occasions the dam utters a cry of alarm. The little ones immediately scamper to the brush; and while they are skulking into places of safety, their anxious parent beguiles the spectator by drooping and fluttering her wings, limping along the path, rolling over in the dirt, and other pretences of inability to walk or fly.

"*Food.*—A favourite article of their diet is the *heath-hen plum*, or partridgeberry before mentioned. They are fond of hurtleberries and craneberries. Worms and insects of several kinds are occasionally found in their crops. But, in the winter, they subsist chiefly on acorns, and the buds of trees which have shed their leaves. In their stomachs have been sometimes observed the leaves of a plant supposed to be a winter green; and it is said, when they are much pinched, they betake themselves to the buds of the pine. In convenient places, they have been known to enter cleared fields, and regale themselves on the leaves of clover; and old gunners have reported, that they have been known to trespass upon patches of buckwheat, and pick up the grains.

"*Migration.*—They are stationary, and never known to quit their abode. There are no facts shewing in them any disposition to migration. On frosty mornings, and during snows, they perch on the upper branches of pine trees. They avoid wet and swampy places, and are remarkably attached to dry ground. The low and open brush is preferred to high shrubbery and thickets. Into these latter places they fly for refuge when closely pressed by the hunters; and here, under a stiff and impenetrable cover, they escape the pursuit of dogs and men. Water is so seldom met with on the true grouse ground, that it is necessary to carry it along for the pointers to drink. The flights of grouse are short, but sudden, rapid, and whirring. I have not heard of any success in taming them. They seem to resist all attempts at domestication. In this, as well as in many other respects, they resemble the quail of New York, or the partridge of Pennslyvania.

"*Manners.*—During the period of mating, and while the females are occupied in incubation, the males have a practice of assembling, principally by themselves. To some select and central spot, where there is very little underwood, they repair from the adjoining district. From the exercises performed there, this is called a *scratching place.* The time of meeting is the break of day. As soon as the light appears, the company assembles from every side, sometimes to the number of forty or fifty. When the dawn is past, the ceremony begins by a low tooting from one of the cocks. This is answered by another. They then come forth one by one from the bushes, and strut about with all the pride and ostentation they can display. Their necks are incurvated; the feathers on them are erected into a sort of ruff; the plumes of the tails are expanded like fans; they strut about in a style resembling, as nearly as small may be illustrated by great, the pomp of the turkey cock. They seem to vie with each other in stateliness; and, as they pass each other, frequently cast looks of insult, and utter notes of defiance. These are the signals for battles. They engage with wonderful spirit and fierceness. During these contests, they leap a foot or two from the ground, and utter a cackling, screaming, and discordant cry.

"They have been found in these places of resort even earlier than the appearance of light in the east. This fact has led to the belief that a part of them assemble over night. The rest join them in the morning. This leads to the farther belief that they roost on the ground. And the opinion is confirmed by the discovery of little rings of dung, apparently deposited by a flock which had passed the night together. After the appearance of the sun they disperse.

"These places of exhibition have been often discovered by the hunters; and a fatal discovery it has been for the poor grouse. Their destroyers construct for themselves lurking holes made of pine branches, called *bough houses,* within a few yards of the parade. Hither they repair with their fowling-pieces, in the latter part of the night, and wait the appearance of the birds. Watching the moment when two are proudly eyeing each other, or engaged in battle, or when a greater number can be seen in a range, they pour on them a destructive charge of shot. This annoyance has been given in so many places, and to such extent, that the grouse, after having been repeatedly disturbed, are afraid to assemble. On approaching the spot to which their instinct prompts them, they perch on the neighbouring trees, instead of alighting at the scratching place. And it remains to be observed, how far the restless and tormenting spirit of the marksmen may alter the native habits of the grouse, and oblige them to betake themselves to new ways of life.

"They commonly keep together in coveys, or packs, as the phrase is, until the pairing season. A full pack consists, of course, of ten or a dozen. Two packs have been known to associate. I lately heard of one whose number amounted to twenty-two. They are so unapt to be startled, that a hunter assisted by a dog, has been able to shoot almost a whole pack, without making any of them take wing. In like manner, the men lying in concealment near the scratching places have been known to discharge several guns before either the report of the explosion, or the sight of their wounded and dead fellows, would rouse them to flight. It has farther been remarked, that when a company of sportsmen have surrounded a pack of grouse, the birds seldom or never rise upon their pinions while they are encircled; but each runs along until it passes the person that is nearest, and then flutters off with the utmost expedition.

"As you have made no inquiry of me concerning the ornithological character of these birds, I have not mentioned it, presuming that you are already perfectly acquainted with their classification and description. In a short memoir written in 1803, and printed in the eighth volume of the *Medical Repository*, I ventured an opinion as to the genus and species. Whether I was correct is a technical matter, which I leave you to adjust. I am well aware that European accounts of our productions are often erroneous, and require revision and amendment. This you must perform. For me it remains to repeat my joy at the opportunity your invitation has afforded me to contribute somewhat to your elegant work, and at the same time to assure you of my earnest hope that you may be favoured with ample means to complete it.

"SAMUEL L. MITCHILL."

Duly sensible of the honour of the foregoing communication, and grateful for the good wishes with which it is concluded, I shall now, in farther elucidation of the subject, subjoin a few particulars properly belonging to my own department.

It is somewhat extraordinary that the European naturalists, in their various accounts of our different species of grouse, should have said little or nothing of the one now before us, which, in its voice, manners, and peculiarity of plumage, is the most singular, and, in its flesh, the most excellent, of all those of its tribe that inhabit the territory of the United States.

It seems to have escaped Catesby during his residence and different tours through this country, and it was not till more than twenty years after his return to England, viz., in 1743, that he first saw some of these birds, as he informs us, at Cheswick, the seat of the Earl of Wilmington. His lordship said they came from America; but from what particular part, could not tell.* Buffon has confounded it with the ruffed grouse, the common partridge of New England, or pheasant of Pennsylvania (*Tetrao umbellus*); Edwards and Pennant have, however, discovered that it is a different species; but have said little of its note, of its flesh, or peculiarities; for, alas! there was neither voice, nor action, nor delicacy of flavour in the shrunk and decayed skin from which the former took his figure, and the latter his description; and to this circumstance must be attributed the barrenness and defects of both.

That the curious may have an opportunity of examining to more advantage this singular bird, a figure of the male is here given, as large as life, drawn with great care from the most perfect of several elegant specimens shot in the Barrens of Kentucky. He is represented in the act of *strutting*, as it is called, while with inflated throat he produces that extraordinary sound so familiar to every one who resides in his vicinity, and which has been described in the foregoing account. So very novel and characteristic did the action of these birds appear to me at first sight, that, instead of shooting them down, I sketched their attitude hastily on the spot, while concealed among a brush heap, with seven or eight of them within a short distance. Three of these I afterwards carried home with me.

This rare bird, though an inhabitant of different and very distant districts of North America, is extremely particular in selecting his place of residence; pitching only upon those tracts whose features and productions correspond with his modes of life, and avoiding immense intermediate regions that he never visits. Open dry plains, thinly interspersed with trees, or partially overgrown with shrub oak, are his favourite haunts. Accordingly we find these birds on the

* Catesby, Car. p. 101 App.

grouse plains of New Jersey, in Burlington county, as well as on the brushy plains of Long Island; among the pines and shrub oaks of Pocano, in Northampton county, Pennsylvania; over the whole extent of the Barrens of Kentucky; on the luxuriant plains and prairies of the Indiana Territory, and Upper Louisiana; and, according to the information of the late Governor Lewis, on the vast and remote plains of the Columbia River: in all these places preserving the same singular habits.

Their predilection for such situations will be best accounted for by considering the following facts and circumstances:—First, their mode of flight is generally direct, and laborious, and ill calculated for the labyrinth of a high and thick forest, crowded and intersected with trunks and arms of trees, that require continual angular evolution of wing, or sudden turnings, to which they are by no means accustomed. I have always observed them to avoid the high timbered groves that occur here and there in the Barrens. Connected with this fact, is a circumstance related to me by a very respectable inhabitant of that country, viz., that one forenoon a cock grouse struck the stone chimney of his house with such force, as instantly to fall dead to the ground.

Secondly, their known dislike of ponds, marshes, or watery places, which they avoid on all occasions, drinking but seldom, and, it is believed, never from such places. Even in confinement this peculiarity has been taken notice of. While I was in the state of Tennesee, a person living within a few miles of Nashville had caught an old hen grouse in a trap; and, being obliged to keep her in a large cage, as she struck and abused the rest of the poultry, he remarked that she never drank, and that she even avoided that quarter of the cage where the cup containing the water was placed. Happening, one day, to let some water fall on the cage, it trickled down in drops along the bars, which the bird no sooner observed, than she eagerly picked them off, drop by drop, with a dexterity that showed she had been habituated to this mode of quenching her thirst; and probably, to this mode only, in those dry and barren tracts, where, except the drops of dew, and drops of rain, water is very rarely to be met with. For the space of a week he watched her closely, to discover whether she still refused to drink; but, though she was constantly fed on Indian-corn, the cup and water still remained untouched and untasted. Yet no sooner did he again sprinkle water on the bars of the cage, than she eagerly and rapidly picked them off as before.

The last, and, probably, the strongest inducement to their preferring these plains, is the small acorn of the shrub oak; the strawberries, huckleberries, and partridge-berries, with which they abound, and which constitute the principal part of the food of these birds. These brushy thickets also afford them excellent shelter, being almost impenetrable to dogs or birds of prey.

In all these places where they inhabit, they are, in the strictest sense of the word, resident; having their particular haunts, and places of rendezvous (as described in the preceding account), to which they are strongly attached. Yet they have been known to abandon an entire tract of such country, when, from whatever cause it might proceed, it became again covered with forest. A few miles south of the town of York, in Pennsylvania, commences an extent of country, formerly of the character described, now chiefly covered with wood, but still retaining the name of Barrens. In the recollection of an old man born in that part of the country, this tract abounded with grouse. The timber growing up, in progress of years, these birds totally disappeared; and, for a long period of time, he had seen none of them, until, migrating with his family to Kentucky, on entering the Barrens, he one morning recognised the well-known music of his old acquaintance, the grouse; which, he assures me, are the very same with those he had known in Pennsylvania.

But what appears to me the most remarkable circumstance relative to this bird, is, that not one of all those writers who have attempted its history, have taken the least notice of those two extraordinary bags of yellow skin which mark the neck of the male, and which constitute so striking a peculiarity. These appear to be formed by an expansion of the gullet, as well as of the exterior skin of the neck, which, when the bird is at rest, hangs in loose, pendulous, wrinkled folds, along the side of the neck, the supplemental wings, at the same time, as well as when the bird is flying, lying along the neck, in the manner represented in one of the distant figures on the plate. But when these bags are inflated with air, in breeding time, they are equal in size, and very much resemble in colour, a middle sized fully ripe orange. By means of this curious apparatus, which is very observable several hundred yards off, he is enabled to produce the extraordinary sound mentioned above, which, though it may easily be imitated, is yet difficult to describe by words. It consists of three notes, of the same tone, resembling those produced by the night hawks in their rapid descent; each strongly accented, the last being twice as long as the others. When several are thus engaged, the ear is unable to distinguish the regularity of these triple notes, there being, at such times, one continued bumming, which is disagreeable and perplexing, from the impossibility of ascertaining from what distance or even quarter it proceeds. While uttering this, the bird exhibits all the ostentatious gesticulations of a turkey cock; erecting and fluttering his neck wings, wheeling and passing before the female, and close before his fellows, as in defiance. Now and then are heard some rapid crackling notes, not unlike that of a person tickled to excessive laughter; and, in short, one can scarcely listen to them without feeling disposed to laugh from sympathy. These are uttered by the males while engaged in fight, on which occasion they leap up against each other, exactly in the manner of turkeys, seemingly with more malice than effect. This bumming continues from a little before daybreak to eight or nine o'clock in the morning, when the parties separate to seek for food.

Fresh ploughed fields, in the vicinity of their resorts, are sure to be visited by these birds every morning, and frequently also in the evening. On one of these I counted, at one time, seventeen males, most of whom were in the attitude represented in the plate; making such a continued sound, as, I am persuaded, might have been heard for more than a mile off.

The people of the Barrens informed me, that, when the weather becomes severe with snow, they approach the barn and farmhouse, are sometimes seen sitting on the fences in dozens, mix with the poultry, and glean up the scattered grains of Indian-corn, seeming almost half domesticated. At such times, great numbers are taken in traps. No pains, however, or regular plan, has ever been persisted in, as far as I was informed, to domesticate these delicious birds. A Mr Reed, who lives between the Pilot Knobs and Bairdstown, told me, that, a few years ago, one of his sons found a grouse's nest, with fifteen eggs, which he brought home, and immediately placed below a hen then sitting, taking away her own. The nest of the grouse was on the ground, under a tussock of long grass, formed with very little art, and few materials; the eggs were brownish white, and about the size of a pullet's. In three or four days the whole were hatched. Instead of following the hen, they compelled her to run after them, distracting her with the extent and diversity of their wanderings; and it was a day or two before they seemed to understand her language, or consent to be guided by her. They were let out to the fields, where they paid little regard to their nurse; and, in a few days, only three of them remained. These became extremely tame and familiar; were most expert flycatchers; but, soon after, they also disappeared.

The pinnated grouse is nineteen inches long, twenty-seven inches in extent, and, when in good order, weighs about three pounds and a half; the neck is furnished with supplemental wings, each composed of eighteen feathers, five of which are black, and about three inches long; the rest shorter, also black, streaked laterally with brown, and of unequal lengths; the head is slightly crested; over the eye is an elegant semicircular comb of rich orange, which the bird has the power of raising or relaxing; under the neck wings are two loose, pendulous, and wrinkled skins, extending along the side of the neck for two-thirds of its length; each of which, when inflated with air, resembles in bulk, colour, and surface, a middle-sized orange; chin, cream coloured; under the eye runs a dark streak of brown ; whole upper parts, mottled transversely with black, reddish brown, and white ; tail short, very much rounded, and of a plain brownish soot colour ; throat, elegantly marked with touches of reddish brown, white, and black ; lower part of the breast and belly, pale brown, marked transversely with white ; legs, covered to the toes with hairy down of a dirty drab colour ; feet, dull yellow ; toes, pectinated ; vent, whitish ; bill, brownish horn colour ; eye, reddish hazel. The female is considerably less ; of a lighter colour, destitute of the neck wings, the naked yellow skin on the neck, and the semicircular comb of yellow over the eye.

On dissecting these birds, the gizzard was found extremely muscular, having almost the hardness of a stone ; the heart remarkably large ; the crop was filled with brier knots, containing the larvæ of some insect, quantities of a species of green lichen, small hard seeds, and some grains of Indian corn.

BLUE-GREEN WARBLER. (*Sylvia rara.*)

PLATE XXVII. –Fig. 2.

Peale's Museum, No. 7788.

VERMIVORA RARA.—Jardine.*

Sylvia rara, *Bonap. Synop.* p. 82.—*Aud.* pl. 49, male ; *Orn. Biog.* i. p. 258.

This new species, the only one of its sort I have yet met with, was shot on the banks of Cumberland River, about the be-

* This species was discovered by Wilson, and does not seem to have been again met with by any ornithologist except Mr Audubon, who has figured it, and added somewhat to our knowledge of its manners.

"It is rare in the middle districts, and is only found in the dark recesses of the Pine Swamp. On its passage through the States, it appears in Louisiana in April. They are met with in Kentucky, in Ohio, upon the Missouri, and along Lake Erie." Mr Audubon has never seen the nest. In spring the song is soft and mellow, and not heard beyond the distance of a few paces ; it is performed at intervals, between the times at which the bird secures an insect, which it does with great expertness, either on the wing, or among the leaves of the trees and bushes. While catching it on the wing, it produces a slight clicking sound with its bill, like *Vireo.* It also, like them, eats small berries,

ginning of April, and the drawing made with care immediately after. Whether male or female, I am uncertain. It is one of those birds that usually glean among the high branches of the tallest trees, which renders it difficult to be procured. It was darting about with great nimbleness among the leaves, and appeared to have many of the habits of the flycatcher. After several ineffectual excursions in search of another of the same kind, with which I might compare the present, I am obliged to introduce it with this brief account.

The specimen has been deposited in Mr Peale's museum.

The blue green warbler is four inches and a half long, and seven and a half in extent; the upper parts are verditer, tinged with pale green, brightest on the front and forehead; lores, line over the eye, throat, and whole lower parts, very pale cream; cheeks, slightly tinged with greenish; bill and legs, bright light blue, except the upper mandible, which is dusky; tail, forked, and, as well as the wings, brownish black; the former marked on the three exterior vanes with white, and edged with greenish; the latter having the first and second row of coverts tipt with white. Note, a feeble chirp.

NASHVILLE WARBLER. (*Sylvia ruficapilla.*)

PLATE XXVII.—FIG. 3.

Peale's Museum, No. 7789.

VERMIVORA RUBRICAPILLA.—SWAINSON.*

Sylvia rubricapilla, *Wils. Catal.*—*Bonap. Synop.* p. 87.—Sylvicola (Vermivora) rubricapilla, *North. Zool.* ii. p. 220.—The Nashville Warbler, *Aud.* pl. 89; *Orn. Biog.* i. p. 450.

THE very uncommon notes of this little bird were familiar to me for several days before I succeeded in obtaining it. These

particularly towards autumn, when insects begin to fail. There seems little difference between the sexes. Such is the most important information given by Mr Audubon.—ED.

* Wilson discovered this species, and afterwards, in his "Catalogue of Birds in the United States," changed the specific name as above. Like

notes very much resembled the breaking of small dry twigs, or the striking of small pebbles of different sizes smartly against each other for six or seven times, and loud enough to be heard at the distance of thirty or forty yards. It was some time before I could ascertain whether the sound proceeded from a bird or an insect. At length I discovered the bird, and was not a little gratified at finding it an entire new and hitherto undescribed species. I was also fortunate enough to meet afterwards with two others exactly corresponding with the first, all of them being males. These were shot in the State of Tennessee, not far from Nashville. It had all the agility and active habits of its family, the worm-eaters.

The length of this species is four inches and a half, breadth, seven inches; the upper parts of the head and neck, light ash, a little inclining to olive; crown, spotted with deep chestnut in small touches; a pale yellowish ring round the eye; whole lower parts, vivid yellow, except the middle of the belly, which is white; back, yellow olive, slightly skirted with ash; rump and tail-coverts, rich yellow olive; wings, nearly black, broadly edged with olive; tail, slightly forked, and very dark olive; legs, ash; feet, dirty yellow; bill, tapering to a fine point, and dusky ash; no white on wings or tail; eye, hazel.

the last, it seems very rare; Wilson saw only three; Audubon, three or four; and a single individual was shot by the Overland Arctic Expedition. "The latter was killed hopping about the branches of a tree, and emitting a creaking noise something like the whetting of a saw." The nest does not yet seem to be known.—ED.

END OF VOLUME I.

PRINTED BY BALLANTYNE AND COMPANY
EDINBURGH AND LONDON